Flexible Micromanipulators and Micromanipulation

Flexible Micromanipulators and Micromanipulation

Editor

Alessandro Cammarata

MDPI • Basel • Beijing • Wuhan • Barcelona • Belgrade • Manchester • Tokyo • Cluj • Tianjin

Editor
Alessandro Cammarata
Dipartimento di Ingegneria
Civile e Architettura
Università degli studi di
Catania
Catania
Italy

Editorial Office
MDPI
St. Alban-Anlage 66
4052 Basel, Switzerland

This is a reprint of articles from the Special Issue published online in the open access journal *Micromachines* (ISSN 2072-666X) (available at: www.mdpi.com/journal/micromachines/special_issues/flexible_micromanipulators).

For citation purposes, cite each article independently as indicated on the article page online and as indicated below:

LastName, A.A.; LastName, B.B.; LastName, C.C. Article Title. *Journal Name* **Year**, *Volume Number*, Page Range.

ISBN 978-3-0365-7069-3 (Hbk)
ISBN 978-3-0365-7068-6 (PDF)

Contents

About the Editor

Alessandro Cammarata

Alessandro Cammarata graduated in Mechanical Engineering in 2004. In 2008, he obtained his Ph.D. in Structural Mechanics. Since March 2020, he has been an associate professor in Applied Mechanics and belongs to the Department of Civil Engineering and Architecture (DICAR) of the University of Catania. His research interests range from flexible multibody systems to compliant mechanisms and robotics. He is the author and co-author of over 60 publications in national and international journals and conference proceedings.

Preface to "Flexible Micromanipulators and Micromanipulation"

With the increasing demand for more precise and accurate control in the fabrication of micro- and nanoscale structures, micromanipulators have become an indispensable tool that has revolutionized a wide range of applications in various fields.

One of the critical areas where micromanipulators have made significant contributions is in the development of flexible micromanipulators. These micromanipulators are designed to provide precise and flexible control over the manipulation of micro- and nanoscale objects, such as cells, microorganisms, nanoparticles, and even single atoms. Furthermore, flexible manipulators can exploit compliant structures and fabrication methods that are widespread in the MEMS technology industry.

This reprint, "Flexible Micromanipulators and Micromanipulation,"provides 12 papers and is intended for researchers and engineers interested in research areas including biology, medicine, physics, and engineering, and their applications.

Alessandro Cammarata
Editor

 *micromachines*

Editorial

Editorial for the Special Issue on Flexible Micromanipulators and Micromanipulation

Alessandro Cammarata 

Department of Civil Engineering and Architecture (DICAR), University of Catania, 95125 Catania, Italy; alessandro.cammarata@unict.it

The field of micromanipulation is rapidly growing and evolving thanks to advancements in microfabrication technologies and the increased demand for precise and accurate manipulation of microscale objects. The Special Issue on "Flexible Micromanipulators and Micromanipulation" in the *Micromachines* journal provides a platform for researchers to share their latest developments, ideas, and results in this field. This issue includes 12 papers that address various challenges and opportunities in the design, fabrication, and control of flexible micromanipulators. The papers in this Special Issue highlight the latest developments in the field and demonstrate the potential of flexible micromanipulators in addressing some of the most challenging problems in science and engineering.

Kazemzadeh Heris et al. [1] present the design and fabrication of a magnetic actuator for torque and force control using an artificial neural network (ANN)/simulated annealing (SA) algorithm. Kumar et al. [2] propose an electromagnetic micromanipulator levitation system for metal additive manufacturing applications. These two papers demonstrate the potential of magnetic and electromagnetic actuation for micromanipulation and their application in various fields, including additive manufacturing. Xie et al. [3] introduce a novel triaxial parallel compliant manipulator inspired by the tripteron mechanism. This design enables high precision and a large workspace for micromanipulation tasks, which can benefit various fields, including microassembly, microsurgery, and microscale material characterization. Kilikevicius et al. [4] present omnidirectional manipulation of microparticles on a platform subjected to circular motion using dynamic dry friction control. This work provides a promising approach for manipulating microparticles with high precision and speed, which is important in various applications, including biomedical, microelectronics, and environmental fields. Zhang et al. [5] present a micromanipulation and automatic data analysis method to determine the mechanical strength of microparticles. This work provides a fast and accurate method for characterizing the mechanical properties of microscale objects, which can benefit various fields, including material science, biology, and chemistry. Cammarata et al. [6] present a dynamic model of a conjugate-surface flexure hinge, which considers the impacts between cylinders. This work provides an improved understanding of the dynamics of flexure hinges, which is important for the design and control of micromanipulators. Ito et al. [7] propose a vision feedback control for the automation of the pick-and-place of a capillary force gripper. This work provides a promising approach for automating micromanipulation tasks, which is important for reducing human error and increasing efficiency in various fields, including microassembly and biotechnology. Ren et al. [8] propose an optimal design for a 3-PSS flexible parallel micromanipulator based on kinematic and dynamic characteristics. This work provides a systematic approach for designing micromanipulators with high precision and a large workspace, which is important for various fields, including microsurgery and microscale material characterization. Wu et al. [9] propose a condensed substructure approach for kinetostatic modeling of compliant mechanisms with complex topology. This work provides an efficient and accurate method for modeling the behavior of compliant mechanisms, which is important for the design and control of micromanipulators. Cammarata et al. [10] present a direct

Citation: Cammarata, A. Editorial for the Special Issue on Flexible Micromanipulators and Micromanipulation. *Micromachines* **2023**, *14*, 597. https://doi.org/10.3390/mi14030597

Received: 27 February 2023
Accepted: 1 March 2023
Published: 4 March 2023

kinetostatic analysis of a gripper with curved flexures. This work provides a detailed understanding of the behavior of curved flexures, which is important for the design and control of micromanipulators. Baiocco et al. [11] developed a method to measure the Young's moduli of microcapsules' shell materials based on diametric compression between two parallel surfaces and numerical modeling. They found a linear relationship between the moduli of the whole microcapsule and the shell material. Tanabe et al. [12] designed a holonomic inchworm robot that can be precisely controlled using four optical encoders. The robot moves in any direction, making it suitable for tasks such as inspection, assembly, and maintenance in tight spaces.

One of the key themes that emerges from the 12 papers is the importance of precise and accurate control in micromanipulation. Several papers present novel control algorithms that enable real-time feedback and adjustment of the micromanipulator's position, enabling highly precise manipulation of objects at the microscale. Developing new control algorithms and feedback systems is critical to the continued advancement of flexible micromanipulators and their applications.

Another important theme that emerges from the papers in this Special Issue is the diversity of applications for flexible micromanipulators. While some papers focus on biological and medical applications, others explore using flexible micromanipulators in manufacturing, microengineering, and materials science. The ability to manipulate objects at the microscale has enormous potential for a wide range of applications, and the papers in this Special Issue demonstrate the versatility of flexible micromanipulators.

Finally, the papers in this Special Issue also highlight the importance of collaboration between different fields of science and engineering. The development of flexible micromanipulators requires expertise in a range of areas, including materials science, control theory, and robotics. The papers demonstrate the value of collaboration between researchers in these different fields and the potential for interdisciplinary research to drive advances in micromanipulation.

In conclusion, the papers in this Special Issue demonstrate the exciting developments in flexible micromanipulation and the potential of this technology to revolutionize a range of fields. The continued development of flexible micromanipulators and the exploration of new applications will require collaboration between researchers in various disciplines, and this Special Issue highlights the importance of this collaborative approach. We hope that this Special Issue will inspire further research and development in the field of flexible micromanipulation, and we look forward to seeing the future advances that will emerge from this exciting area of research.

Funding: This research received no external funding.

Data Availability Statement: Some data are available from the authors upon request.

Conflicts of Interest: The author declares no conflict of interest.

References

1. Kazemzadeh Heris, P.; Khamesee, M. Design and Fabrication of a Magnetic Actuator for Torque and Force Control Estimated by the ANN/SA Algorithm. *Micromachines* **2022**, *13*, 327. [CrossRef] [PubMed]
2. Kumar, P.; Malik, S.; Toyserkani, E.; Khamesee, M. Development of an Electromagnetic Micromanipulator Levitation System for Metal Additive Manufacturing Applications. *Micromachines* **2022**, *13*, 585. [CrossRef] [PubMed]
3. Xie, Y.; Li, Y.; Cheung, C. Design and Modeling of a Novel Tripteron-Inspired Triaxial Parallel Compliant Manipulator with Compact Structure. *Micromachines* **2022**, *13*, 678. [CrossRef]
4. Kilikevičius, S.; Liutkauskienė, K.; Uldinskas, E.; El Banna, R.; Fedaravičius, A. Omnidirectional Manipulation of Microparticles on a Platform Subjected to Circular Motion Applying Dynamic Dry Friction Control. *Micromachines* **2022**, *13*, 711. [CrossRef]
5. Zhang, Z.; He, Y.; Zhang, Z. Micromanipulation and Automatic Data Analysis to Determine the Mechanical Strength of Microparticles. *Micromachines* **2022**, *13*, 751. [CrossRef]
6. Cammarata, A.; Maddìo, P.; Sinatra, R.; Rossi, A.; Belfiore, N. Dynamic Model of a Conjugate-Surface Flexure Hinge Considering Impacts between Cylinders. *Micromachines* **2022**, *13*, 957. [CrossRef] [PubMed]
7. Ito, T.; Fukuchi, E.; Tanaka, K.; Nishiyama, Y.; Watanabe, N.; Fuchiwaki, O. Vision Feedback Control for the Automation of the Pick-and-Place of a Capillary Force Gripper. *Micromachines* **2022**, *13*, 1270. [CrossRef] [PubMed]

8. Ren, J.; Li, Q.; Wu, H.; Cao, Q. Optimal Design for 3-PSS Flexible Parallel Micromanipulator Based on Kinematic and Dynamic Characteristics. *Micromachines* **2022**, *13*, 1457. [CrossRef] [PubMed]
9. Wu, S.; Shao, Z.; Fu, H. A Substructure Condensed Approach for Kinetostatic Modeling of Compliant Mechanisms with Complex Topology. *Micromachines* **2022**, *13*, 1734. [CrossRef] [PubMed]
10. Cammarata, A.; Maddio, P.; Sinatra, R.; Belfiore, N. Direct Kinetostatic Analysis of a Gripper with Curved Flexures. *Micromachines* **2022**, *13*, 2172. [CrossRef] [PubMed]
11. Baiocco, D.; Zhang, Z.; He, Y.; Zhang, Z. Relationship between the Young's Moduli of Whole Microcapsules and Their Shell Material Established by Micromanipulation Measurements Based on Diametric Compression between Two Parallel Surfaces and Numerical Modelling. *Micromachines* **2023**, *14*, 123. [CrossRef] [PubMed]
12. Tanabe, K.; Shiota, M.; Kusui, E.; Iida, Y.; Kusama, H.; Kinoshita, R.; Tsukui, Y.; Minegishi, R.; Sunohara, Y.; Fuchiwaki, O. Precise Position Control of Holonomic Inchworm Robot Using Four Optical Encoders. *Micromachines* **2023**, *14*, 375. [CrossRef] [PubMed]

Article

Precise Position Control of Holonomic Inchworm Robot Using Four Optical Encoders

Kengo Tanabe †, Masato Shiota †, Eiji Kusui, Yohei Iida, Hazumu Kusama, Ryosuke Kinoshita, Yohei Tsukui, Rintaro Minegishi, Yuta Sunohara and Ohmi Fuchiwaki *

Department of Mechanical Engineering, Yokohama National University, 79-5 Tokiwadai, Hodogaya-ku, Yokohama 2408051, Kanagawa, Japan
* Correspondence: ohmif@ynu.ac.jp; Tel.: +81-45-339-3693
† These authors contributed equally to this work.

Abstract: In this study, an XYθ position sensor is designed/proposed to realize the precise control of the XYθ position of a holonomic inchworm robot in the centimeter to submicrometer range using four optical encoders. The sensor was designed to be sufficiently compact for mounting on a centimeter-sized robot for closed-loop control. To simultaneously measure the XYθ displacements, we designed an integrated two-degrees-of-freedom scale for the four encoders. We also derived a calibration equation to decrease the crosstalk errors among the XYθ axes. To investigate the feasibility of this approach, we placed the scale as a measurement target for a holonomic robot. We demonstrated closed-loop sequence control of a star-shaped trajectory for multiple-step motion in the centimeter to micrometer range. We also demonstrated simultaneous three-axis proportional–integral–derivative control for one-step motion in the micrometer to sub-micrometer range. The close-up trajectories were examined to determine the detailed behavior with sub-micrometer and sub-millidegree resolutions in the MHz measurement cycle. This study is an important step toward wide-range flexible control of precise holonomic robots for various applications in which multiple tools work precisely within the limited space of instruments and microscopes.

Keywords: XYθ position control; holonomic inchworm robot; optical encoder; closed-loop control; calibration; crosstalk error

Citation: Tanabe, K.; Shiota, M.; Kusui, E.; Iida, Y.; Kusama, H.; Kinoshita, R.; Tsukui, Y.; Minegishi, R.; Sunohara, Y.; Fuchiwaki, O. Precise Position Control of Holonomic Inchworm Robot Using Four Optical Encoders. *Micromachines* **2023**, *14*, 375. https://doi.org/10.3390/mi14020375

Academic Editor: Fabio Di Pietrantonio

Received: 12 January 2023
Revised: 30 January 2023
Accepted: 30 January 2023
Published: 2 February 2023

1. Introduction

In recent years, electronic devices and their components have been miniaturized in microelectromechanical systems (MEMS) and mobile computers [1–3]. In biology, micromanipulation is necessary for delicate and fragile objects, such as biological cells, microfossils, and microorganisms [4–7].

Micromanipulation requires tools to accurately manipulate objects and multi-axis positioners to move the tool to an appropriate position and orientation with a sub-micrometer resolution. Various tools have been developed, such as vacuum nozzles [8], microtweezers with force sensors [4,9], and grippers that use capillary force [10,11]. Regarding the multi-axis positioner, the combination of a single-axis stage driven by linear synchronous motors with linear guides and stepping motors with ball-screw mechanisms are widely used in the industry [12,13]. These machines satisfy the requirements for micromanipulation in terms of accuracy, payload, and durability. However, they are considerably larger and heavier than the tiny parts. To adapt to a wider range of applications, multiaxis positioning stages should be sufficiently compact to use multiple tools simultaneously within the limited space of various instruments and microscopes.

Micromanipulation using small self-propelled robots has been studied to circumvent these problems [14–18]. This type of robot has three-degree-of-freedom (3-DoF) movements and is capable of holonomic movements. Therefore, it is expected to be used for monotonous XY-axis movement and complicated tasks, such as revolution around a tooltip within a narrow microscopic image.

Various types of self-propelled robots have employed different principles and structures. Typical types of motion principles include stick–slip [19–22], centrifugal force [23], USM [24,25], and inchworm [26]. Figure 1 shows a comparison of the representative micromanipulation robots and the XY stage [22,24,27]. These robots have distinct advantages in terms of their specific performance, such as motion range, velocity, carrying capacity, and motion resolution. The performance of the inchworm robot was above average. Therefore, this robot can perform various micromanipulation tasks.

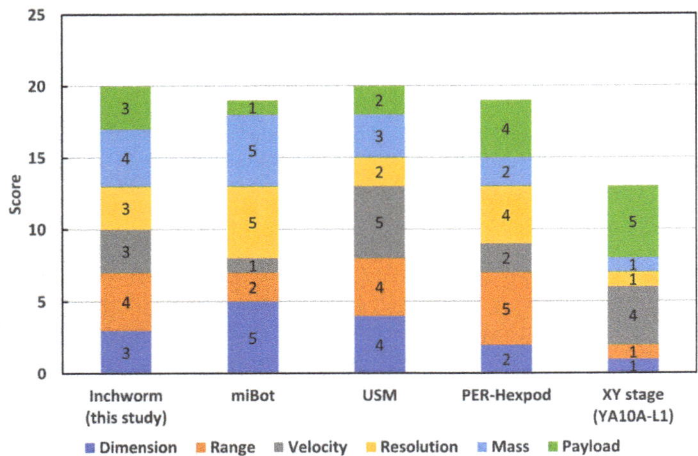

Figure 1. Comparison of typical micromanipulation robots and XY stage [22,24,27]. Scores are evaluated based on the reference [27] (refer to Table S1 of the Supplementary Materials for the quantitative comparison of the performances).

Previous research has also confirmed that the holonomic inchworm robots that we developed can be driven omnidirectionally under open-loop control [28]. However, self-propelled robots are susceptible to external disturbances, such as unevenness in the work-table and tension from the feeding wires. Therefore, precise and fast measurement of XYθ axis displacements with the sub-micrometer resolution is important for determining the exact position and orientation of robots.

Various methods have been developed to measure displacement precisely [29]. One of the most commonly used methods is pattern matching using camera images; however, this method has practical problems concerning a tradeoff among the measuring resolution, range, and cycle [30,31]. Strain gauges [32], lasers [33], and encoders [34,35] have also been used.

Optical encoders are feasible as multiaxis and fast position sensors because of their short measurement cycle, long range, and high resolution. In recent years, encoders have become significantly miniaturized, allowing them to be mounted on small robots. We previously reported an outline of an XYθ displacement sensor using four optical encoders, although we omitted the evaluation of the measurement performance [28,36]. In this study, we describe the evaluation and demonstration of the precise position control of a holonomic inchworm robot in the centimeter to the sub-micrometer range.

The remainder of this paper is organized as follows. In Section 2, we describe the holonomic inchworm robot, and we evaluate the XYθ displacement sensor in Section 3. The closed-loop sequence control for a multiple-step motion in the centimeter to micrometer range and simultaneous three-axis proportional–integral–derivative (PID) control for a one-step motion in the micrometer to sub-micrometer range are presented in Sections 4 and 5, respectively. Finally, conclusions and future prospects are presented in Section 6.

2. Holonomic Inchworm Robot

2.1. Structure

Figure 2a shows the structure of the holonomic inchworm robot. *EM-1* and *EM-2* are pairs of Y-shaped electromagnets (EMs) that are separated to avoid mutual attraction. These EMs form a closed loop via a ferromagnetic surface to obtain a sufficient magnetic force to fix the floor. *PA-F*, *PA-B*, *PA-L1*, *PA-L2*, *PA-R1*, and *PA-R2* are piezoelectric actuators (PAs) with a mechanical displacement amplification mechanism. We arranged *EM-1* and *EM-2* to cross each other and connect them to the six PAs to move precisely in all directions according to the inchworm principle.

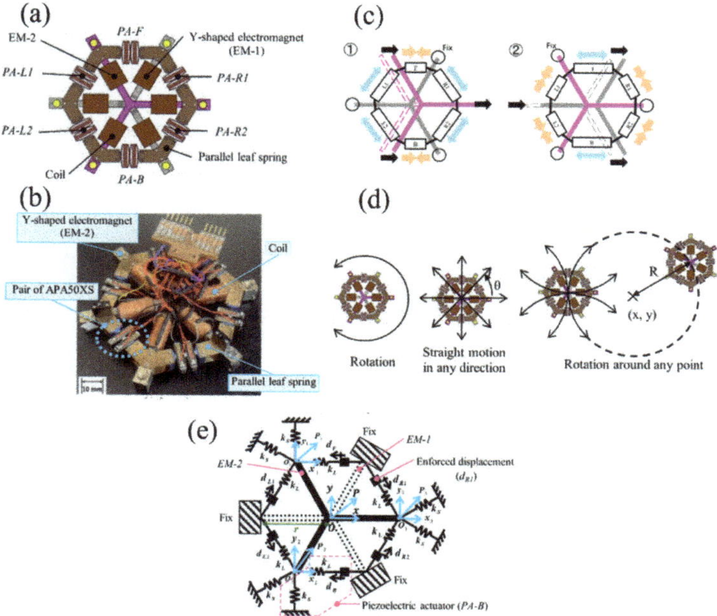

Figure 2. Holonomic inchworm robot: (**a**) structure (top view); (**b**) overview; (**c**) principle of motion (rightward movement); (**d**) motion patterns; (**e**) dynamic model.

Figure 2b shows a photograph of the holonomic inchworm robot. The robot weighed 100 g and measured 86 mm × 86 mm × 11 mm. It had parallel leaf springs for the simultaneous smooth contact of all legs on the surface. Tables 1 and 2 list the specifications of the robot and the PAs, respectively. We used an APA50XS "Moonie" PA (Cedrat Inc., Meylan, France) connected in series.

Table 1. Performance of the inchworm mobile robot.

Characteristic Value	Quantity
Step length (120 V)	~65 μm
Resolution (15–25 °C, less than 50% rH)	Less than 10 nm
DoF	X, Y, θ
Natural Frequency (blocked free)	X: 413, Y: 418, θ: 476 Hz
Maximum Velocity [frequency]	~6.5 mm/s [100 Hz]
Repeatability (CV; ratio of SD of final points to a path length with 10 mm path) [frequency]	~3% [100 Hz]
Maximum payload	<1000 g
Dimension	86 × 86 × 15 mm
Weight	100 g

Table 2. Performance of the piezoelectric actuator.

Characteristic Value	Quantity
Displacement (100 V)	95.5 ± 5 μm
Generative Force (100 V)	18.0 N
Spring constant	115,000 N/m
Capacitance	1.04 μF
Resolution (15–25 °C, less than 50% rH)	1.52 nm
Natural Frequency (blocked free)	1.45 kHz
Dimension	12.9 × 6.4 × 9.2 mm
Weight	4 g

2.2. Principle

Figure 2c shows the motion sequence of the rightward orthogonal motions. It moves as an inchworm while retaining the synchronism between the rectangular forces of the two EMs and the vibrations of the six PAs. When the amplitudes of the vibrations are precisely controlled, it moves in any 3-DoF direction with a resolution of less than 10 nm.

As shown in Figure 2d, the robot can move linearly in any direction and rotate around any point with precision on well-polished ferromagnetic surfaces. If one EM is fixed to the floor, the other can be precisely positioned using six PAs.

2.3. Dynamic Model

As shown in Figure 2e, we defined a 3-DoF dynamic model of the robot when *EM-2* was free, and *EM-1* was fixed. Here, k_L and k_S are the spring constants of the six PAs in the compression and shear deformations, respectively. The d_F is the enforced displacement of *PZT-F*, and d_B, d_{R1}, d_{R2}, d_{L1}, and d_{L2} are similar.

EM-1 and *EM-2* were connected by hinged joints at P_1, P_2, and P_3. O_1, O_2, and O_3 are the initial positions of P_1, P_2, and P_3, respectively. Furthermore, x_1 and y_1 are the coordinates of P_1; (x_2, y_2) and (x_3, y_3) similarly define P_2 and P_3, respectively. P is the center of gravity of *EM-1*. The position of P is represented by the orthogonal coordinate system used for x and y. O is the origin of P. We assume that EMs are rigid bodies.

The six PAs and 3-DoF motions were determined based on the positions of P_1, P_2, and P_3. Therefore, the PAs move to the free leg; m is the mass of the EMs, I is the moment of inertia of the gravity centers of the EMs, and r is the distance between the center and end of the EMs. From Figure 1, Newton's equations of motion for *EM-1* and the related parameters are represented by Equations (1)–(7).

$$M\ddot{y} + K\,y = Ku \tag{1}$$

$$M = \begin{bmatrix} m & 0 & 0 \\ 0 & m & 0 \\ 0 & 0 & I/r^2 \end{bmatrix} \tag{2}$$

$$y = \begin{bmatrix} x \\ y \\ r\theta \end{bmatrix} \tag{3}$$

$$u \equiv \begin{bmatrix} u_x \\ u_y \\ ru_\theta \end{bmatrix} = K^{-1}K_d\,d \tag{4}$$

$$K = \begin{bmatrix} 3(k_L + k_S) & 0 & 0 \\ 0 & 3(k_L + k_S) & 0 \\ 0 & 0 & 3(3k_L + k_S)/2 \end{bmatrix} \tag{5}$$

$$K_d = k_L \begin{bmatrix} -1 & -1 & 1/2 & 1/2 & 1/2 & 1/2 \\ 0 & 0 & \sqrt{3}/2 & -\sqrt{3}/2 & -\sqrt{3}/2 & \sqrt{3}/2 \\ \sqrt{3}/2 & -\sqrt{3}/2 & -\sqrt{3}/2 & \sqrt{3}/2 & -\sqrt{3}/2 & \sqrt{3}/2 \end{bmatrix} \quad (6)$$

$$d = \begin{bmatrix} d_F & d_B & d_{L1} & d_{L2} & d_{R1} & d_{R2} \end{bmatrix}^t. \quad (7)$$

$$V = C \cdot \sin \omega t \begin{bmatrix} -W\cos\phi/3K_1 + r\Theta/\sqrt{3}K_2 \\ -W\cos\phi/3K_1 - r\Theta/\sqrt{3}K_2 \\ W\left(\cos\phi + \sqrt{3}\sin\phi\right)/2K_1 - r\Theta/\sqrt{3}K_2 \\ W\left(\cos\phi - \sqrt{3}\sin\phi\right)/2K_1 + r\Theta/\sqrt{3}K_2 \\ W\left(\cos\phi - \sqrt{3}\sin\phi\right)/2K_1 - r\Theta/\sqrt{3}K_2 \\ W\left(\cos\phi + \sqrt{3}\sin\phi\right)/2K_1 + r\Theta/\sqrt{3}K_2 \end{bmatrix} + V_0 \begin{bmatrix} 1 \\ 1 \\ 1 \\ 1 \\ 1 \\ 1 \end{bmatrix}, \quad (8)$$

$$V = \begin{bmatrix} V_F & V_B & V_{L1} & V_{L2} & V_{R1} & V_{R2} \end{bmatrix}^t. \quad (9)$$

$$\begin{bmatrix} K_1 \\ K_2 \end{bmatrix} = \begin{bmatrix} k_L/\{3(k_S + k_L) - m\omega^2\} \\ k_L/\{3(3k_L + k_S)/2 - I\omega^2/r^2\} \end{bmatrix} \quad (10)$$

2.4. Input Voltages

Simplified voltages to the PAs for the translational motion were reported in [26]. In Equation (9), V_F–V_{R2} are the input voltages to the PAs. V_F–V_{R2} was obtained by solving Equation (1) using the approximation of harmonic oscillations with no damping.

Here, the moving direction is defined as ϕ, stride length during a half-step motion is W, half-orientation change of one step is Θ, angular frequency of the inchworm motion is ω, and the approximate proportional coefficient between the enforced displacements and voltage is C. A sinusoidal voltage was applied to the PAs up to 120 $V_{p\text{-}p}$. The offset voltage was determined to be V_0 = 60 V, which is half the maximum voltage of 120 V. K_1 and K_2 are given by Equation (10).

3. XYθ Position Sensor

3.1. Structure

Figure 3 shows the proposed XYθ position sensor. Figure 3a shows a magnified view of the measurement area, and Figure 3b shows the assembly of the measurement components. As discussed previously, the robot moved on a ferromagnetic plate.

Four linear optical encoders (TA-200, Technohands, Yokohama Kanagawa, Japan), each with a resolution of 0.1 μm and a maximum measurement speed of 800 mm/s, were installed on the encoder installation plate, as shown in Figure 3c. We fixed the encoder installation plate above the robot to measure XYθ displacements in a stationary coordinate system.

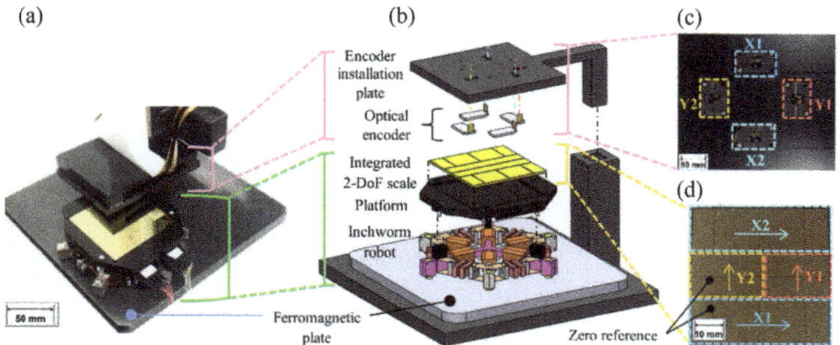

Figure 3. XYθ position sensor: (**a**) magnified view of the measurement area; (**b**) assembly drawing; (**c**) encoder installation plate; (**d**) integrated 2-DoF scale.

The integrated 2-DoF scale was placed as a measurement target on top of one of the two legs, as shown in Figure 3d. The encoder specifications are listed in Table 3. The measurement range along the X–Y axis was 16 mm × 16 mm.

Table 3. Performance of encoder (TA-200, Technohands).

Characteristic Value	Quantity
Resolution [μm/count]	0.1
Maximum measurement speed [mm/s]	800
Dimension [mm]	15 × 10 × 1.5

3.2. Signal Processing

As shown in Figure 4, we used a field-programmable gate array (FPGA) (C-RIO-9049, NI, TX, USA) to generate the control voltages of the robot and convert the transistor-transistor logic (TTL) signals from the encoders to XYθ displacements. The control voltages were magnified by 30 times using an amplifier circuit. The magnified signals are the input voltages to the PAs (*V*), as shown in Equation (9). The minimum measurement cycle of the FPGA was 0.35 μs, and the maximum calculated speed was 2800 mm/s. The FPGA measured the displacement of a sinusoidal vibration up to 1273 Hz with a displacement amplitude of 30 μm.

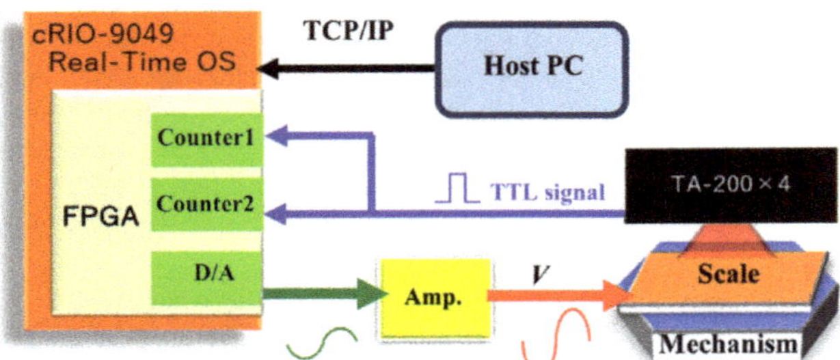

Figure 4. Signal processing.

We assumed that the measurement cycle was sufficient because we moved the robot with an inchworm frequency of 100 Hz using sinusoidal displacement with an amplitude of 30 μm and a mechanical resonance frequency of less than 500 Hz.

Table 4 lists the performances of the position sensors. The measurement resolution was 0.1 μm with an uncertainty of ±0.2 μm for the X and Y axes in the static state. The measurement resolution was 0.3 millidegrees with an uncertainty of ±0.6 millidegrees for the θ-axis.

Table 4. Specifications of the XYθ position sensor.

Characteristic Value	Quantity
Measurement range X × Y [mm], θ [°]	16 × 16, ±25
Measurement resolution in X (Y) [μm], θ [millidegrees]	0.1, 0.3
Uncertainty in static state in X (Y) [μm], θ [millidegrees]	±0.2, ±0.6
Measurement frequency [MHz]	2.86
Maximum measurable speed [mm/s]	800
Principle of measurement	Incremental
Measurement accuracy in X and Y (−8~8mm) [%]	0.08–0.18
Measurement accuracy in θ (−25~25°) [%]	0.06–0.19

3.3. Measurement Principle

In this section, the conversion of the four measured distances into a 3-DoF motion of the scale is explained, and the major equations are presented. As shown in Figure 5, we performed vector analysis when the encoder installation plate was fixed to the coordinate system at rest and the scale was moved by the PAs, as shown in Figure 3.

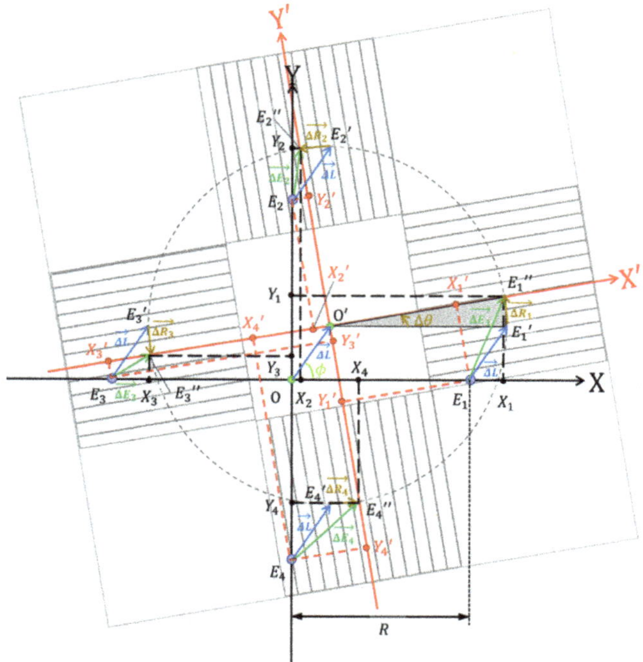

Figure 5. Vector diagram when the scale moves.

We defined the XY and $X'Y'$ coordinate systems as the coordinate systems at rest and on the scale, respectively. The coordinate transformation between XY and $X'Y'$ is given as follows:

$$\begin{bmatrix} X \\ Y \end{bmatrix} = \begin{bmatrix} \cos\Delta\theta & -\sin\Delta\theta \\ \sin\Delta\theta & \cos\Delta\theta \end{bmatrix} \begin{bmatrix} X' \\ Y' \end{bmatrix} + W \begin{bmatrix} \cos\phi \\ \sin\phi \end{bmatrix}. \tag{11}$$

where $E_1{}'$ is the position after moving $\Delta\vec{L}$ from its initial position. $\Delta\vec{L}$ is defined as a vector component of the translational movement as follows:

$$\Delta\vec{L} \equiv W \begin{bmatrix} \cos\phi \\ \sin\phi \end{bmatrix} \equiv \begin{bmatrix} \Delta X \\ \Delta Y \end{bmatrix} \tag{12}$$

where $\Delta\theta$ is the posture angle displacement of the scale, W is the translational movement, ϕ is the translational direction, ΔX is the translational movement along the X-axis, and ΔY is the translational movement along the Y-axis.

The 3-DoF motion is divided into translational and rotational motions. E_1 is the initial position of encoder-1, which is defined as (X_{10}, Y_{10}) in the XY coordinate system. E_2, E_3, and E_4 are defined similarly. R is the geometrical center of E_1, E_2, E_3, and E_4.

$$\begin{bmatrix} X_{10} & X_{20} & X_{30} & X_{40} \\ Y_{10} & Y_{20} & Y_{30} & Y_{40} \end{bmatrix} = \begin{bmatrix} R & 0 & -R & 0 \\ 0 & R & 0 & -R \end{bmatrix} \tag{13}$$

where E_1'' is the position of encoder-1 after moving $\overrightarrow{\Delta R_1}$ from E_1'. $\overrightarrow{\Delta R_1}$ is defined as the vector component of rotational movement. We define (X_1, Y_1) as the XY coordinates of E_1'', and (X_1', Y_1') as the $X'Y'$ coordinates of E_1. The other parameters are defined similarly. We obtain the following coordinate transformation of the initial positions of E_k from Equations (11) and (12) as follows:

$$\begin{bmatrix} X_{k0} \\ Y_{k0} \end{bmatrix} = \begin{bmatrix} \cos\Delta\theta & -\sin\Delta\theta \\ \sin\Delta\theta & \cos\Delta\theta \end{bmatrix} \begin{bmatrix} X_k' \\ Y_k' \end{bmatrix} + W \begin{bmatrix} \cos\phi \\ \sin\phi \end{bmatrix}. \tag{14}$$

The measurable parameters are Y_1', X_2', Y_3', and X_4' using the four encoders, which are represented by Equation (14) as follows:

$$\begin{cases} Y_1' = -R\sin\Delta\theta - W\sin(\phi - \Delta\theta) \\ X_2' = R\sin\Delta\theta - W\cos(\phi - \Delta\theta) \\ Y_3' = R\sin\Delta\theta - W\sin(\phi - \Delta\theta) \\ X_4' = -R\sin\Delta\theta - W\cos(\phi - \Delta\theta) \end{cases} \tag{15}$$

From Equation (15), the displacements are represented as follows:

$$\begin{bmatrix} \Delta X \\ \Delta Y \end{bmatrix} = -\begin{bmatrix} \cos\Delta\theta & -\sin\Delta\theta \\ \sin\Delta\theta & \cos\Delta\theta \end{bmatrix} \begin{bmatrix} \frac{X_2'+X_4'}{2} \\ \frac{Y_1'+Y_3'}{2} \end{bmatrix} \tag{16}$$

$$\widetilde{\Delta\theta} = \frac{-1}{2}\left\{\sin^{-1}\left(\frac{X_4' - X_2'}{2R}\right) + \sin^{-1}\left(\frac{Y_1' - Y_3'}{2R}\right)\right\} \tag{17}$$

where $\widetilde{\Delta\theta}$ is the best estimator of $\Delta\theta$ and is the average of two expressions. When $\widetilde{\Delta\theta} \ll 1$, $(\Delta X, \Delta Y, \widetilde{\Delta\theta})$ are approximated as follows:

$$\begin{bmatrix} \Delta X \\ \Delta Y \\ \widetilde{\Delta\theta} \end{bmatrix} \cong -\begin{bmatrix} 1 & -\widetilde{\Delta\theta} & 0 \\ \widetilde{\Delta\theta} & 1 & 0 \\ 0 & 0 & 1 \end{bmatrix} \begin{bmatrix} 0 & \frac{1}{2} & 0 & \frac{1}{2} \\ \frac{1}{2} & 0 & \frac{1}{2} & 0 \\ \frac{1}{4R} & -\frac{1}{4R} & -\frac{1}{4R} & \frac{1}{4R} \end{bmatrix} \begin{bmatrix} Y_1' \\ X_2' \\ Y_3' \\ X_4' \end{bmatrix}. \tag{18}$$

3.4. Experimental Results

The measurement performance of the sensor was evaluated by comparing it with the machine coordinates of a conventional precise XYθ stage. The XYθ stage was composed of a ball-screw type precise XY stage (KOHZU, Kawasaki Kanagawa, Japan, YA10A-L1) and θ stage (KOHZU, Japan, RA04A-W01) with the repeatability of 0.5 μm and 0.002°, respectively.

Figure 6a shows the experimental setup. We replaced the robot with an XYθ stage, as shown in Figure 3. We fixed the encoder installation plate above the XYθ stage to measure the XYθ displacements from a stationary coordinate system. We placed the 2-DoF scale as the measurement target at the top of the θ stage.

Figure 6b shows the measurement errors of the XYθ-axes when the stage moves along the X-axis. Figure 6c,d show the errors in the XYθ-axes when the stage moves along the Y- and θ-axes, respectively. Nonlinear errors were caused by the XYθ-position sensor in the X- and Y-directions. The maximum error of the X-axis was approximately ±1.2 μm during the X-directional motion, and the maximum error of the Y-axis was approximately ±1.0 μm during the Y-directional motion. When the stage was moved along the θ-axis, the errors were closer to linear. The maximum error is approximately ±0.04° in the θ direction. The variations in the errors of the Y- and θ-axes are shown in Figure 6b; in other words, the crosstalk errors of the XYθ-position sensor. The maximum crosstalk errors were approximately ±40 μm and ±0.002° along the Y- and θ-axes. To reduce this crosstalk error, we calibrated the XYθ position sensors. We define the crosstalk error components as C_x, C_y, and θ_0, where C_x and C_y are the eccentricities between the rotation axis of the θ stage and the geometrical center of the 2-DoF scale, θ_0, is the posture angle of the misalignment between the axes of the sensor head and those of the 2-DoF scale. The displacements after calibration are expressed as follows:

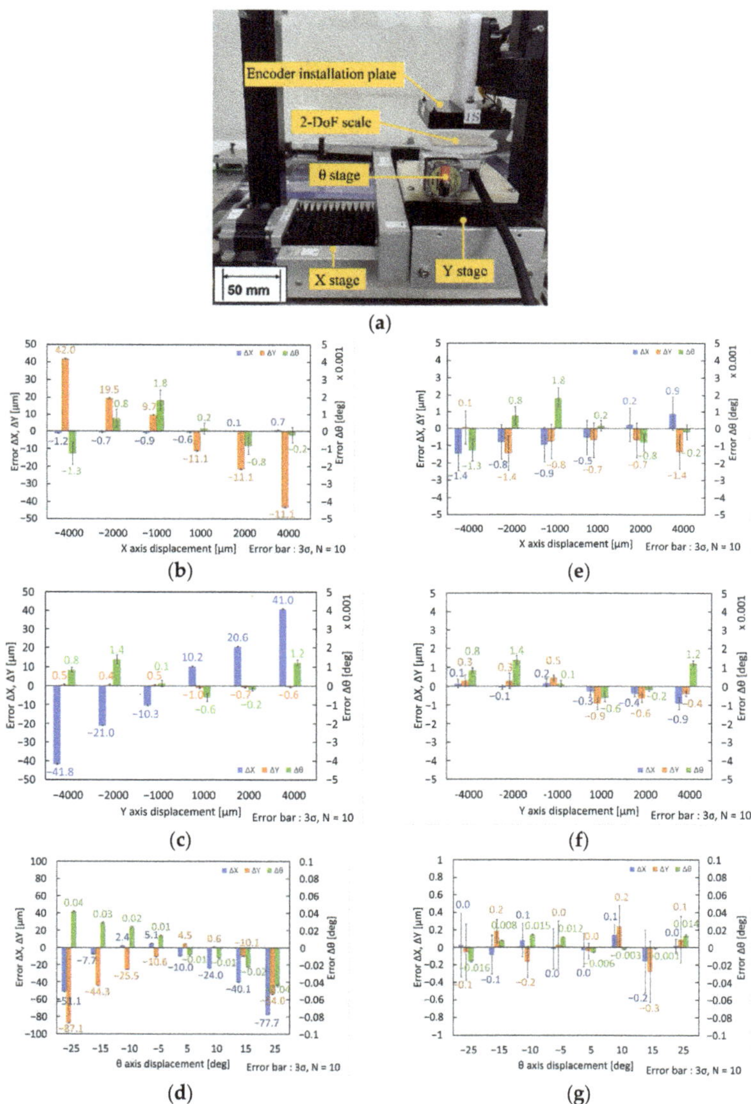

Figure 6. Measuring errors of XYθ-position sensor and variations in the ΔX, ΔY, and Δθ outputs after calibration: (**a**) Experimental setup for evaluation of the measuring performance of XYθ-position sensor; (**b**) Plots of XYθ-axes errors vs. X-axis displacement before calibration; (**c**) Plots of XYθ-axes errors vs. Y-axis displacement before calibration; (**d**) Plots of XYθ-axes errors vs. θ-axis displacement before calibration; (**e**) Plots of XYθ-axes errors vs. X-axis displacement after calibration; (**f**) Plots of XYθ-axes errors vs. Y-axis displacement after calibration; (**g**) Plots of XYθ-axes errors vs. θ-axis displacement after calibration.

$$\begin{bmatrix} \Delta X_c \\ \Delta Y_c \\ \Delta \theta_c \end{bmatrix} = \begin{bmatrix} \cos\theta_0 & \sin\theta_0 & 0 \\ -\sin\theta_0 & \cos\theta_0 & 0 \\ 0 & 0 & \frac{\sin\Delta\theta}{\sin(\Delta\theta+\theta_0)-\sin\theta_0} \end{bmatrix} \begin{bmatrix} \Delta X \\ \Delta Y \\ \Delta\theta \end{bmatrix} + \begin{bmatrix} -C_x(1-\cos\Delta\theta) - C_y\sin\Delta\theta \\ C_x\sin\Delta\theta - C_y(1-\cos\Delta\theta) \\ 0 \end{bmatrix} \quad (19)$$

Figure 6e shows the measurement errors of the XYθ-axes after calibration when the stage was moved along the X-axis. Figure 6f,g shows those along the Y- and θ-axes, respectively. As shown in the figure, we performed a calibration using Equation (19), and the crosstalk errors significantly decreased in the X- and Y-directions. However, the crosstalk errors of the θ-axis in the X- and Y-directional motions did not improve significantly. The maximum residual crosstalk error of the X-directional movement was approximately 1.3 μm in the Y-direction, and that of the Y-directional movement was approximately 0.9 μm in the X-direction. In θ-directional motion, the maximum residual crosstalk errors were approximately 0.15 μm and 0.25 μm in the X- and Y-direction.

(See also Figure S1 for evaluating XYθ-axes errors for XYθ-axes simultaneous displacement for (X, Y, θ) = (4000 μm, 4000 μm, 5°)).

4. Sequence Control of Multiple Step Motions

4.1. Control Sequence

We demonstrated closed-loop control of star-shaped translational motion in the centimeter range. Figure 7 shows the control sequence that switches the combination of coarse and fine motions. Here, P_T is the target position, and e_{T1} is the threshold of the distance error e for coarse motion; e_{T2} is the fine motion, $\delta\theta_T$ is the threshold of the orientation error $\delta\theta$, and W_1 and W_2 are the step lengths of the coarse and fine motions, respectively. We determined e_{T1} = 80 μm, e_{T2} = 15 μm, $\delta\theta_T$ = 0.06°, W_1 = 60.0 μm, and W_2 = 5.0 μm. The robot moved to an inchworm frequency of 100 Hz.

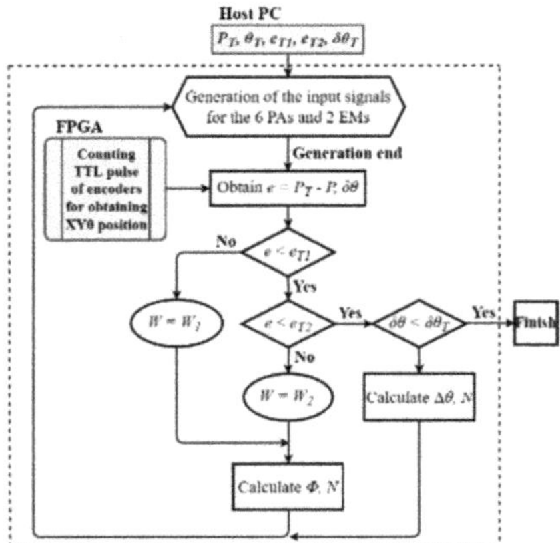

Figure 7. Sequence of navigation.

4.2. Experimental Results

Based on the aforementioned conditions, the robot was controlled to draw a star shape with five corners as target points. The experiment was performed twice, and a map of the trajectories of the center of each robot is shown in Figure 8. As shown in Figure 8a, the first and second trajectories for each experiment almost overlapped; therefore, this mobile robot had high repeatability. In both trajectories, the average velocity was approximately 4.3 mm/s, whereas the velocity of the coarse motion was approximately 6.5 mm/s. In addition, close-up views around each corner of the first trajectory are shown. Regarding the trajectory, blue, green, and red indicate coarse, fine, and rotational motions, respectively. Considering corner four as an example, the robot first entered the range of e_{T1} with coarse movement, (1) Coarse1. It switched to fine movement and moved near the center of the range of e_{T2}, (2) Fine1. If the center was within the range of e_{T2}, the orientation angle θ was measured. If it was out of the range of $\delta\theta_T$, as shown in this case, the orientation was corrected by rotation, (3) Rotation1. If it moved outside the range of e_{T2}, it approached the corner again using fine movements, (4) Fine2. If both the center and orientation were within the corresponding ranges of e_{T2} and $\delta\theta_T$, respectively, the robot changed to a coarse movement toward the next corner, (5) Coarse2.

As represented by the areas highlighted in yellow in Figure 8, cyclic vibrations occurred along the trajectory. We confirmed that these also appeared in the fine and rotational movements during every step by observing an enlarged view of the trajectories shown in Figure 8. Therefore, it is assumed that slippages of the electromagnetic legs occur during switching between the fixing and moving legs.

The plots of X, Y, and θ vs. time in Figure 8b,c show that the settling times around the corners were 0.2–0.8 s. We assume that the settling time can be reduced by improving the navigation sequence and motion compensation.

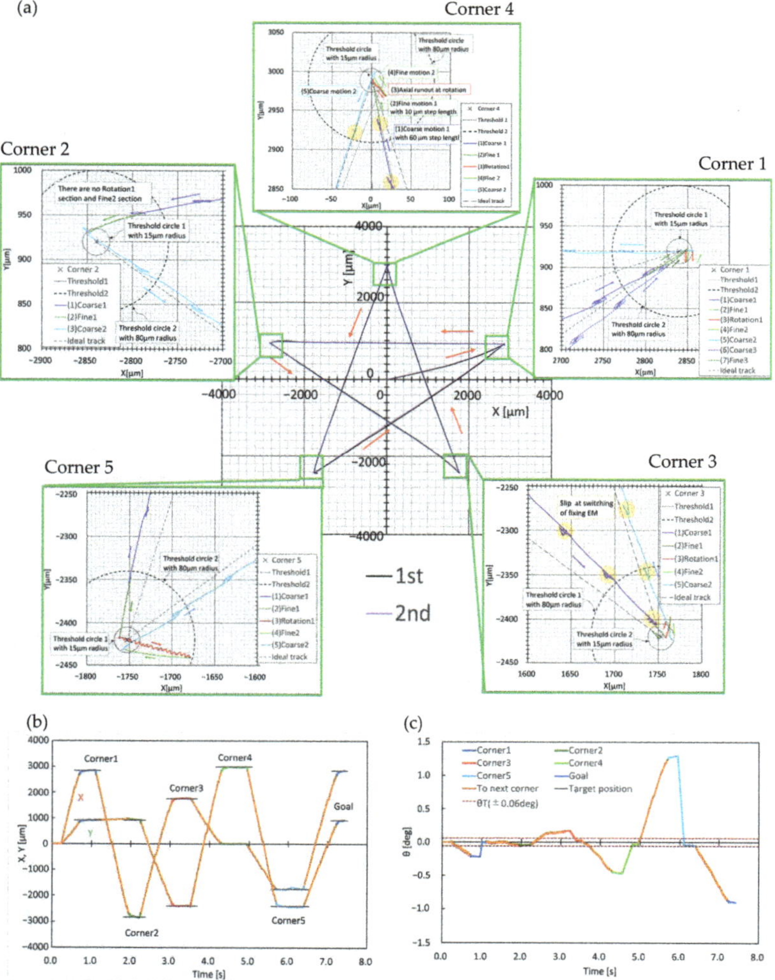

Figure 8. Star-shaped trajectories with five target points: (**a**) XY trajectory; (**b**) X, Y vs. time; (**c**) θ vs. time.

5. 3-Axis PID Control of One-Step Motion

5.1. Transfer Function

For precise positioning in the micrometer to sub-micrometer range, we demonstrated three-axis PID control of the one-step motion of *EM-2* when *EM-1* was fixed on the floor, as shown in Figure 1e. Considering the Laplace transform of Equation (1), we obtain the transfer function matrix $\boldsymbol{P}(\boldsymbol{s})$ as a second-order system:

$$Y(s) \equiv \left[s^2 M + K\right]^{-1} K U(s) \tag{20}$$

$$P(s) = \left[s^2 M + K\right]^{-1} K \tag{21}$$

$$U(s) = K^{-1} K_d \, D(s) \tag{22}$$

Here, we define the Laplace transform of $\boldsymbol{y}(t)$ as $\boldsymbol{Y}(s)$. The Laplace transforms of the other parameters are defined similarly.

Figure 9 shows a block diagram of the PID control. We define $\boldsymbol{d_m}(t)$ and $\boldsymbol{w}(t)$ as vectors composed of the modeling error and uncertainties of the measured displacement, respectively. Here, $\boldsymbol{k_I}$, $\boldsymbol{k_P}$, and $\boldsymbol{k_D}$ are vectors composed of integral, proportional, and derivative gains along the x-, y-, and θ-axes, respectively; $\boldsymbol{r}(t)$ is the target position, and $\boldsymbol{e}(t)$ is the deviation between $\boldsymbol{r}(t)$ and $\boldsymbol{y}(t)$. We applied a first-order Butterworth filter with a cutoff filter of 50 Hz to $\boldsymbol{E}(s)$ before derivative (D) control to minimize the time delay of the primary experiments.

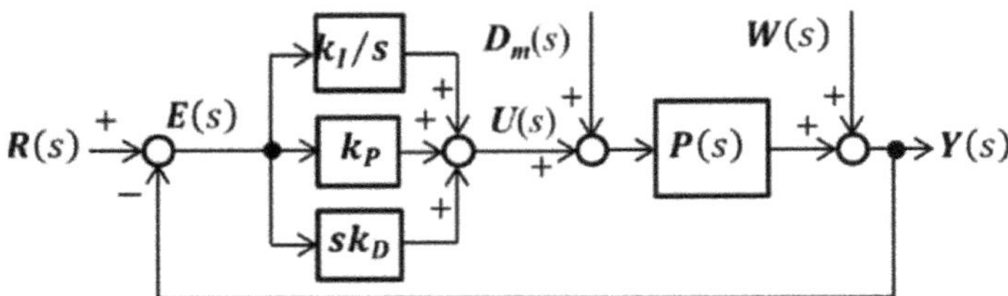

Figure 9. Block diagram of three-axe PID control.

5.2. Experimental Results

In the experiments, we adjusted $\boldsymbol{k_I}$, $\boldsymbol{k_P}$, and $\boldsymbol{k_D}$ for each experimental condition by using a heuristic method to obtain a no-overshoot trajectory. We conducted experiments five times for each condition, and similar results were obtained. We determined the target travel lengths r_i as 1, 5, and 10 µm and the target moving directions φ_i as 0°, 30°, 60°, and 90° for translational movements. We also determined the target rotational displacement of θ_i as 0 and 14.8 millidegrees in the θ-axis. In addition, we determined the thresholds of the X-, Y-, and θ-axes as $X_T = \pm 0.14$ µm, $Y_T = \pm 0.14$ µm, and $\theta_T = \pm 0.4$ millidegrees around their corresponding targets.

Figure 10 shows the experimental results for PID control of a single motion. The step-shaped targets of the X-axis were determined as X_S, Y_S, and θ_S were similarly defined. As shown in Figure 10a for the XY trajectories, we succeeded in precisely controlling the position of the free leg, although non-negligible oscillations in the X- and Y-axes were generated for the directions of 60° and 90°. Figure 10b,c compares the plots of X, X_S, Y, Y_S, θ, and θ_S vs. time for the 0 °and 60 °directions, and their settling times were approximately 65 and 78 ms, respectively. We assumed that the oscillation was attributed to electrical noise from a commercial 50 Hz AC power supply, as observed in Figure 10b,c,e,f.

Figure 10d–f shows the experimental results for the parabolic-shaped references with a rise time of 100 ms. Similarly, we define the parabolic-shaped target of the X-axis as X_P, Y_P, and θ_P. Figure 10d shows the XY trajectories. Figure 10e,f shows plots of X, X_P, Y, Y_P, θ, and θ_P vs. time for the 0° and 60° directions, and their settling times were approximately

122 and 130 ms, respectively. The oscillations decreased to ±0.5 μm in the X and Y axes, as shown in Figure 10.

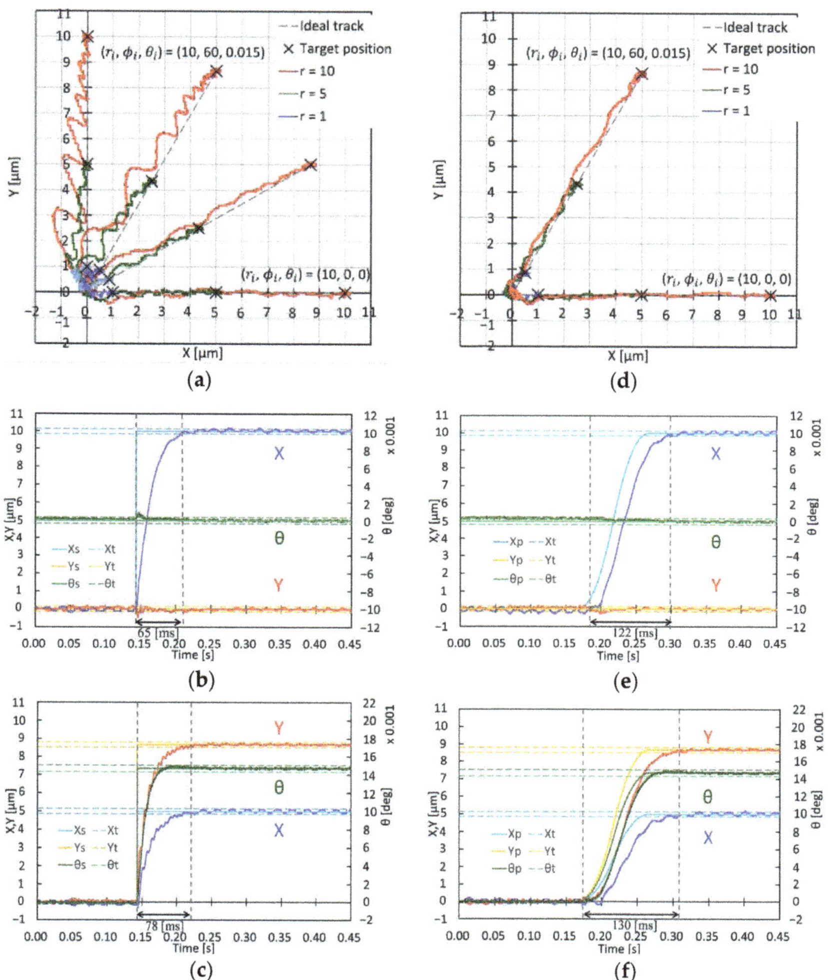

Figure 10. PID control of one step motion (left: step reference, right: parabolic reference with rise time of 100 ms): (**a**) plots of XY trajectories along ϕ_i = 0, 30, 60, and 90°; (**b**) plots of XYθ vs. time for (r_i, ϕ_i, θ_i) = (10 μm, 0°, 0 m°); (**c**) plots of XYθ vs. time for (r_i, ϕ_i, θ_i) = (10 μm, 60°, 14.8 m°); (**d**) plots of XY trajectories along ϕ_i = 0 and 60°; (**e**) plots of XYθ vs. time for (r_i, ϕ_i, θ_i) = (10 μm, 0°, 0 m°); (**f**) plots of XYθ vs. time for (r_i, ϕ_i, θ_i) = (10 μm, 60°, 14.8 m°).

6. Conclusions and Future Prospects

The design and experiments described in this study proved that the realization of a centimeter-sized XYθ position sensor is possible, which is sufficiently compact to attach a centimeter-sized holonomic inchworm robot and can simultaneously and precisely measure XYθ displacements, with a measuring cycle of 0.1 μs, resolution of 0.1 μm and 0.4 millidegrees, measurement accuracy of 0.06–0.19%, and range of 16 × 16 mm^2 and −25–25°. We demonstrated the sequential positioning control of the multiple-step motion from the centimeter to the micrometer range. We have also demonstrated PID control of the one-step motion from the micrometer to the sub-micrometer range.

In future research, we plan to decrease the noise, obtain the frequency response, improve the pseudo-differential filter, and automatically tune the PID gains. For multiple-step motion, we plan to decrease the threshold around the target position to the sub-micrometer range to achieve more precise positioning control. We also plan to shorten the time required for precise positioning by applying motion compensation. Eliminating cyclic vibrations during the switching of the supporting leg is also an important approach for more accurate control.

For precise one-step motion, we plan to develop a systematic method for obtaining PID gains and investigate more efficient control sequences, such as optimal, model-following, and acceleration feedback control.

Finally, the improvement of the calibration of the XYθ displacement sensor, reduction of the friction between the legs and floor, and reduction of the hysteretic nonlinearity of the PAs are required for more precise and faster positioning.

To expand the applicability of the holonomic inchworm robot, we developed manipulators to be mounted on the robot, which can precisely manipulate minute parts, such as electronic chip components, MEMS, biological cells, microorganisms, and microfossils. We also developed a precise control method for the revolution around the manipulator tip within a narrow microscopic image.

The final goal of this research is to realize the automation of a mobile robotic factory organized by multiple centimeter-sized robots equipped with various tools for the multi-axial processing of biological samples and the assembly of millimeter-sized micro-robots and mechanisms.

Supplementary Materials: The following supporting information can be downloaded at: https://www.mdpi.com/article/10.3390/mi14020375/s1; Table S1—Specification and score of the representative micromanipulation robots.; Figure S1—Plots of XYθ-axes errors vs. XYθ-axes simultaneous displacement.; Video S1 3DOF motion of holonomic inchworm robot.

Author Contributions: Conceptualization, K.T., M.S. and O.F.; methodology, K.T., M.S. and O.F.; software, K.T. and M. S.; validation, K.T., M.S., E.K. and O.F.; formal analysis, K.T., M.S. and O.F.; investigation, K.T., M.S. and O.F.; resources, K.T., M.S. and O.F.; data curation, K.T., M.S. and O.F.; writing—original draft preparation, K.T., M.S. and O. F.; writing—review and editing, E.K., Y.I., H.K., R.K., Y.T., R.M., Y.S. and O.F.; visualization, K.T. and M.S.; supervision, O.F.; project administration, O.F.; funding acquisition, O.F. All authors have read and agreed to the published version of the manuscript.

Funding: This work was supported by the Grant-in-Aid for Scientific Research (C) [grant number 16K06178], Japan Keirin Autorace Foundation [grant number 27JKA-Mech-232], Chubu Electric Power Foundation [grant number R-29229], Takano Foundation, MAZAK Foundation, and Tsugawa Foundation.

Data Availability Statement: The data that support the findings of this study are available from the corresponding author, O.F., upon reasonable request.

Acknowledgments: We thank Hitoshi Funatsu, Kenta Fukumoto, Yuuhei Hara, Ryuuga Ito, and Arato Ogawa for developing the measurement and control device with optical encoders to observe and control the elusive sub-micrometer-scale XYθ displacements.

Conflicts of Interest: The authors declare no conflict of interest. The funders had no role in the design of this study, in the collection, analyses, interpretation of data, the writing of the manuscript, or in the decision to publish the results.

References

1. Shin, G.; Gomez, A.M.; Al-Hasani, R.; Jeong, Y.R.; Kim, J.; Xie, Z.; Banks, A.; Lee, S.M.; Han, S.Y.; Yoo, C.J.; et al. Miniaturized flexible electronic systems with wireless power and near-field communication capabilities. *Adv. Funct. Mater.* **2015**, *25*, 4761–4767.
2. Templier, F. GaN-based emissive microdisplays: A very promising technology for compact, ultra-high brightness display systems. *J. Soc. Inf. Disp.* **2016**, *24*, 669–675. [CrossRef]

3. Henley, F.J. 52-3: Invited Paper: Combining Engineered EPI Growth Substrate Materials with Novel Test and Mass-Transfer Equipment to Enable MicroLED Mass-Production. In *SID Symposium Digest of Technical Papers*; John Wiley & Sons, Inc.: New York, NY, USA, 2018.
4. Boudaoud, M.; Regnier, S. An overview on gripping force measurement at the micro and nano-scales using two-fingered microrobotic systems. *Int. J. Adv. Robot. Syst.* **2014**, *11*, 45. [CrossRef]
5. Itaki, T.; Taira, Y.; Kuwamori, N.; Maebayashi, T.; Takeshima, S.; Toya, K. Automated collection of single species of microfossils using a deep learning–micromanipulator system. *Prog. Earth Planet Sci.* **2020**, *7*, 19. [CrossRef]
6. Ishimoto, K. Bacterial spinning top. *J. Fluid Mech.* **2019**, *880*, 620–652. [CrossRef]
7. Yamada, T.; Iima, M. Hydrodynamic turning mechanism of microorganism by solitary loop propagation on a single flagellum. *J. Phys. Soc. Jpn.* **2018**, *88*, 114401. [CrossRef]
8. Rong, W.; Fan, Z.; Wang, L.; Xie, H.; Sun, L. A vacuum microgripping tool with integrated vibration releasing capability. *Rev. Sci. Instrum.* **2014**, *85*, 085002. [CrossRef]
9. Tiwari, B.; Billot, M.; Clévy, C.; Agnus, J.; Piat, E.; Lutz, P. A two-axis piezoresistive force sensing tool for microgripping. *Sensors* **2021**, *21*, 6059. [CrossRef]
10. Tanaka, K.; Ito, T.; Nishiyama, Y.; Fukuchi, R.; Fuchiwaki, O. Double-nozzle capillary force gripper for cubical, triangular prismatic, and helical 1-mm-sized-objects. *IEEE Robot Autom. Lett.* **2022**, *7*, 1324–1331. [CrossRef]
11. Overton, J.K.; Kinnell, P.K.; Lawes, S.D.A.; Ratchev, S. High precision self-alignment using liquid surface tension for additively manufactured micro components. *Prec. Eng.* **2015**, *40*, 230–240. [CrossRef]
12. Maeda, G.; Sato, K. Practical control method for ultra-precision positioning using a ballscrew mechanism. *Prec. Eng.* **2008**, *32*, 309–318. [CrossRef]
13. Chen, S.; Chiang, H. Precision motion control of permanent magnet linear synchronous motors using adaptive fuzzy fractional-order sliding-mode control. *IEEE/ASME Trans. Mechatron.* **2019**, *24*, 741–752. [CrossRef]
14. Fuchiwaki, O.; Tanaka, Y.; Notsu, H.; Hyakutake, T. Multi-axial non-contact in situ micromanipulation by steady streaming around two oscillating cylinders on holonomic miniature robots. *Microfluid. Nanofluidics* **2018**, *22*, 80. [CrossRef]
15. Tamadazte, B.; Piat, N.L.F.; Dembélé, S. Robotic micromanipulation and microassembly using monoview and multiscale visual servoing. *IEEE/ASME Trans. Mechatron.* **2011**, *16*, 277–287. [CrossRef]
16. Fuchiwaki, O.; Ito, A.; Misaki, D.; Aoyama, H. Multi-axial micromanipulation organized by versatile micro robots and micro tweezers. In Proceedings of the 2008 IEEE International Conference on Robotics and Automation, Pasadena, CA, USA, 19–23 May 2008; pp. 893–898.
17. Fatikow, S.; Wich, T.; Hulsen, H.; Sievers, T.; Jahnisch, M. Microrobot system for automatic nanohandling inside a scanning electron microscope. *IEEE/ASME Trans. Mechatron.* **2007**, *12*, 244–252. [CrossRef]
18. Brufau, J.; Puig-Vidal, M.; Lopez-Sanchez, J.; Samitier, J.; Snis, N.; Simu, U.; Johansson, S.; Driesen, W.; Breguet, J.M.; Gao, J.; et al. MICRON: Small autonomous robot for cell manipulation applications. In Proceedings of the IEEE International Conference On Robotics and Automation (ICRA 2015), Barcelona, Spain, 18–22 April 2005; pp. 844–849.
19. Saadabad, N.A.; Moradi, H.; Vossoughi, G. Dynamic modeling, optimized design, and fabrication of a 2DOF piezo-actuated stick-slip mobile microrobot. *Mech. Mach. Theory* **2019**, *133*, 514–530. [CrossRef]
20. Ji, X.; Liu, X.; Cacucciolo, V.; Imboden, M.; Civet, Y.; El Haitami, A.; Cantin, S.; Perriard, Y.; Shea, H. An autonomous untethered fast soft robotic insect driven by low-voltage dielectric elastomer actuators. *Sci. Robot.* **2019**, *4*, eaaz6451. [CrossRef]
21. Zhong, B.; Zhu, J.; Jin, Z.; He, H.; Wang, Z.; Sun, L. A large thrust trans-scale precision positioning stage based on the inertial stick–slip driving. *Microsyst. Technol.* **2019**, *25*, 3713–3721. [CrossRef]
22. Imina. Available online: https://imina.ch/en/products/nano-robotics-solutions-electron-microscopes (accessed on 21 December 2022).
23. Vartholomeos, P.; Vlachos, K.; Papadopoulos, E. Analysis and motion control of a centrifugal-force microrobotic platform. *IEEE Trans. Autom. Sci. Eng.* **2013**, *10*, 545–553. [CrossRef]
24. Zhou, J.; Suzuki, M.; Takahashi, R.; Tanabe, K.; Nishiyama, Y.; Sugiuchi, H.; Maeda, Y.; Fuchiwaki, O. Development of a Δ-type mobile robot driven by thee standing-wave-type piezoelectric ultrasonic motors. *IEEE Robot. Autom. Lett.* **2020**, *5*, 6717–6723. [CrossRef]
25. Jia, B.; Wang, L.; Wang, R.; Jin, J.; Zhao, Z.; Wu, D. A novel traveling wave piezoelectric actuated wheeled robot: Design, theoretical analysis, and experimental investigation. *Smart Mater. Struct.* **2021**, *30*, 035016. [CrossRef]
26. Fuchiwaki, O.; Yatsurugi, M.; Sato, T. The basic performance of a miniature omnidirectional 6-legged inchworm robot from cm-to μm-scale precise positioning. *Trans. Mater. Res. Soc. Jpn.* **2014**, *39*, 211–215. [CrossRef]
27. Yu, H.; Liu, Y.; Deng, J.; Li, J.; Zhang, S.; Chen, W.; Zhao, J. Bioinspired Multilegged piezoelectric robot: The design philosophy aiming at high-performance micromanipulation. *Adv. Intell. Syst.* **2021**, *4*, 2100142. [CrossRef]
28. Tanabe, K.; Masato, S.; Eiji, K.; Fuchiwaki, O. Precise trajectory measurement of self-propelled mechanism by XYθ 3-axis displacement sensor consisting of multiple optical encoders. In Proceedings of the 2021 60th Annual Conference of the Society of Instrument and Control Engineers of Japan (SICE), Tokyo, Japan, 8–10 September 2021; pp. 998–1003.
29. Gao, W.; Kim, S.W.; Bosse, H.; Haitjema, H.; Chen, Y.L.; Lu, X.D.; Knapp, W.; Weckenmann, A.; Estler, W.T.; Kunzmann, H. Measurement technologies for precision positioning. *CIRP Ann.—Manuf. Technol.* **2015**, *64*, 773–796. [CrossRef]

30. Estana, R.; Woern, H. Moire-based positioning system for microrobots. In *Optical Measurement Systems for Industrial Inspection III*; SPIE: Bellingham, WA, USA, 2003. Available online: https://ui.adsabs.harvard.edu/link_gateway/2003SPIE.5144 (accessed on 21 December 2022).
31. Fuchiwaki, O.; Yamagiwa, T.; Omura, S.; Hara, Y. In-situ repetitive calibration of microscopic probes maneuvered by holonomic inchworm robot for flexible microscopic operations. In Proceedings of the IEEE International Conference on Intelligent Robots and Systems (IROS 2015), Hamburg, Germany, 28 September–2 October 2015; pp. 1467–1472. [CrossRef]
32. Peng, Y.; Ito, S.; Shimizu, Y.; Azuma, T.; Gao, W.; Niwa, E. A Cr-N thin film displacement sensor for precision positioning of a micro-stage. *Sens. Actuator A Phys.* **2014**, *211*, 89–97. [CrossRef]
33. Jang, Y.; Kim, S.W. Compensation of the refractive index of air in laser interferometer for distance measurement: A review. *Int. J. Precis. Eng. Manuf.* **2017**, *18*, 1881–1890. [CrossRef]
34. Li, X. A six-degree-of-freedom surface encoder for precision positioning of a planar motion stage. *Precis. Eng.* **2013**, *37*, 771–781. [CrossRef]
35. Dejima, S.; Gao, W.; Shimizu, H.; Kiyono, S.; Tomita, Y. Precision positioning of a five degree-of-freedom planar motion stage. *Mechatronics* **2005**, *15*, 969–987. [CrossRef]
36. Fuchiwaki, O.; Yatsurugi, M.; Ogawa, A. Design of an integrated 3DoF inner position sensor and 2DoF feedforward control for a 3DoF precision inchworm mechanism. In Proceedings of the IEEE International Conference Robotics and Automation (ICRA 2013), Karlsruhe, Germany, 6–10 May 2013; pp. 5475–5481. [CrossRef]

 micromachines

Article

Relationship between the Young's Moduli of Whole Microcapsules and Their Shell Material Established by Micromanipulation Measurements Based on Diametric Compression between Two Parallel Surfaces and Numerical Modelling

Daniele Baiocco [1], Zhihua Zhang [1,2], Yanping He [3] and Zhibing Zhang [1,*]

1 School of Chemical Engineering, University of Birmingham, Edgbaston, Birmingham B15 2TT, UK
2 Changzhou Institute of Advanced Manufacturing Technology, Changzhou 213164, China
3 School of Chemical Engineering, Kunming University of Science and Technology, Chenggong Campus, Kunming 650504, China
* Correspondence: z.zhang@bham.ac.uk

Abstract: Micromanipulation is a powerful technique to measure the mechanical properties of microparticles including microcapsules. For microparticles with a homogenous structure, their apparent Young's modulus can be determined from the force versus displacement data fitted by the classical Hertz model. Microcapsules can consist of a liquid core surrounded by a solid shell. Two Young's modulus values can be defined, i.e., the one is that determined using the Hertz model and another is the intrinsic Young's modulus of the shell material, which can be calculated from finite element analysis (FEA). In this study, the two Young's modulus values of microplastic-free plant-based microcapsules with a core of perfume oil (hexyl salicylate) were calculated using the aforementioned approaches. The apparent Young's modulus value of the whole microcapsules determined by the classical Hertz model was found to be $E_A = 0.095 \pm 0.014$ GPa by treating each individual microcapsule as a homogeneous solid spherical particle. The previously obtained simulation results from FEA were utilised to fit the micromanipulation data of individual core–shell microcapsules, enabling to determine their unique shell thickness to radius ratio $(h/r)_{FEA} = 0.132 \pm 0.009$ and the intrinsic Young's modulus of their shell ($E_{FEA} = 1.02 \pm 0.13$ GPa). Moreover, a novel theoretical relationship between the two Young's modulus values has been derived. It is found that the ratio of the two Young's module values (E_A/E_{FEA}) is only a function on the ratio of the shell thickness to radius (*h/r*) of the individual microcapsule, which can be fitted by a third-degree polynomial function of *h/r*. Such relationship has proven applicable to a broad spectrum of microcapsules (i.e., non-synthetic, synthetic, and double coated shells) regardless of their shell chemistry.

Keywords: plant-based; non-synthetic; microcapsules; micromanipulation; intrinsic mechanical properties; apparent elastic modulus; mathematical modelling; finite element analysis

Citation: Baiocco, D.; Zhang, Z.; He, Y.; Zhang, Z. Relationship between the Young's Moduli of Whole Microcapsules and Their Shell Material Established by Micromanipulation Measurements Based on Diametric Compression between Two Parallel Surfaces and Numerical Modelling. *Micromachines* **2023**, *14*, 123. https://doi.org/10.3390/mi14010123

Academic Editor: Alessandro Cammarata

Received: 29 November 2022
Revised: 28 December 2022
Accepted: 30 December 2022
Published: 1 January 2023

1. Introduction

Nowadays, microcapsules are being incorporated in many industrial fast-moving consumer goods (FMCG), with particular emphasis on laundry formulations and functionalised textiles [1]. Interestingly, microcapsules have proven effective at minimising the amount and then the cost of fragrance actives required in formulation [2], since they act as protective carriers of the active [3], thereby boosting its physico-chemical stability and shelf-life [4]. Perfume microcapsules (PMCs) for laundry applications are aimed to selectively deliver the fragrance active onto fabrics following their deposition during a washing-drying cycle [5,6]. When a mechanical action is applied by rubbing or caressing dry garments, PMCs can release their perfuming load for an enhanced costumer experience [3]. Over the last decades, melamine formaldehyde (MF) has been the leading shell material for the fabrication of PMCs [7], due to its desirable properties (e.g., water resistance and great acid-base tolerance) and relative inexpensiveness at an industrial scale [3]. Although a broad variety of PMCs can be efficiently prepared with MF [8], outstanding environmental, health & safety (EHS) concerns have arisen around microplastic-based (melamine) and presumably health-threatening (formaldehyde) materials [9]. Increased self-awareness against the potential carcinogenic effects and poor indoor air quality due to formaldehyde has been raised [10]. Moreover, adverse environmental impact of released formaldehyde has been reported, although its residue in products is still within the legal limit [11]. As a first step towards complying with the regulations enforced against carcinogenic materials, novel coating materials have been attempted for laundry and textile applications, such as polysulfone [4], polyurethanes-urea and polyesters [12]. Although the above mentioned materials can provide several desirable features (e.g., thermal stability and mechanical resistance) [13], they are costly (raw materials and processing) [14], poorly/non-biodegradable, and contain respiratory toxic and asthma-inducing isocyanates [15], which highly restrict their potential applications. Therefore, intense efforts towards encapsulating fragrance oils using natural polymers have been undertaken. Bruyninckx and Dusselier [16] have reviewed several sustainable and high performance alternatives for encapsulation of volatile organic compounds (VOCs). Hazard-free gelatine (Gl), chitosan (Ch), and gum Arabic (GA) can form the shell around VOCs by complex coacervation yielding high payloads [17,18]. However, the animal origin of Gl and Ch may still affect their global acceptance in consumer products following the pathogenesis of novel diseases, such as human prion protein misfolding [19]. As a further step towards overcoming MF-related concerns and the biodegradability issues of synthetic polymers, as well as complying with personal and religious beliefs, microplastic-free plant-based microcapsules from safe biopolymers are currently being developed [2]. Therefore, in-depth understanding of the intrinsic mechanical properties of the shell materials of such microcapsules including the Young's modulus is crucial to assess their performance during fabrication and at potential end-use applications [20].

As reported in the literature, the application of a load onto individual microcapsules enables to generate compression force versus displacement profiles [21]. Interestingly, these can be further analysed to determine the mechanical properties, including the Young's modulus, which is essential to predict the deformational behaviour of a material when facing an external force.

Several system-specific techniques for the mechanical characterisation of objects have arisen over the years, including strip-ring extensiometry for hydrogels [22], micropipette aspiration for cellular elasticity [23], microinjection of biological cells [24], optical tweezers for human red blood cells and lipid membranes [25,26], microelectromechanical systems (MEMS) for investigating the biophysical properties of human breast cancer cells [27], and microfluidic channels for liquid-loaded capsules [28]. Notwithstanding, atomic force microscopy (AFM) and micromanipulation via compression between two parallel surfaces are still the preferred techniques owing to their adaptability, and broad range of applicable loads [20,21]. AFM has proven suitable for investigating the mechanical behaviour and Young's modulus of many biological cells, and active-laden microcapsules. Pinpoint forces can be exerted with superb precision by AFM, which detects deformations in the range of a few nanometres [29]. Specifically, the deformational behaviour of soft particles in food science [30], microbial cells with a core–shell configuration [31], and biofilm build-ups [32] have been investigated by AFM, as well as the biomechanical properties of cellular membranes during mitosis and ligand-interacting membrane proteins [33], and the nanomechanical response of trapped bacteria or viruses [34]. The apparent Young's modulus of microcapsules (E_A) entrapping phase change materials (PCMs) within acrylate shells was also quantified by AFM. This was fulfilled by employing a SiO_2 probe with a Young's modulus value E_{SiO2}~75 GPa (baseline). Considering the baseline and the response generated by the pinpoint probe onto a specific area of the microcapsule shell, the apparent

Young's modulus of microcapsules was evaluated by the Hertz model, which yielded values with a significant variability (~0.15–1.5 GPa) [35]. However, the Young's modulus values of microcapsules determined using the AFM measurement combined with the Hertz model should be interpreted with caution since they can depend on the penetration depth of the tip attached to the cantilever (nanoindentation). The greater the penetration depth, the more impact the liquid core can generate on the measurement results. Moreover, since AFM tends to measure local mechanical behaviours of materials, the results may be greatly affected by inhomogeneous and non-smooth shells, especially when the shell thickness and the liquid reservoir depth of a microcapsule are unknown. Another general disadvantage of AFM is mainly due to its incapability in covering more extended surface areas at a time, regardless of the tip geometry (e.g., spherical and conical shape) [21]. In the literature, there are various reports on the Young's modulus of microcapsules. For example, the instantaneous Young's modulus from Poly(d,l-lactide-co-glycolide) (PLGA) based microcapsules was measured by Sarrazin et al. [36] at minute indentation depths (0–12 nm) with significant viscoplastic effects being observed from an indentation of ~5 nm. The Young's modulus values of poly(styrene sulfonate) –poly(allylamine) microcapsules and poly(urea–formaldehyde) composite microcapsules [37] were found to be 2.5–4.0 GPa. Moreover, it is difficult to maintain the alignment between the AFM colloid probe and spherical microcapsules which can then slip away easily [20].

In contrast, micromanipulation has proven more effective for the compression of single microparticles including microcapsules with a core–shell configuration between two parallel surfaces, since a typical force range (100 nN–1.0 N) greater than that of AFM (from pN to μN) can be applied [21,38]. Moreover, it allows displacements greater than the object of study (e.g., size of microcapsule) to be generated [39]. Furthermore, micromanipulation can yield the rupture of individual particles under compression, which is difficult to achieve using other techniques [40].

Interestingly, the resulting force-displacement data can be fitted to specific models to estimate both apparent and intrinsic Young's modulus of particles, which is conditional upon their structural configurations (e.g., solid microbeads and core–shell microcapsule) [5,41]. Accordingly, the apparent Young's modulus of relatively porous calcium-shellac microbeads was estimated by the Hertz model (0.54 ± 0.09 GPa) [42,43], whereas the intrinsic Young's modulus poly(methyl-methacrylate) microcapsules with a core–shell configuration was quantified via a finite element model (0.75 ± 0.3 GPa) [6]. Over the years, simplistic mathematical solutions have been proposed to determine the Young's modulus of microparticles/microcapsules. Smith et al. [44] first pioneered the application of a model based on finite element analysis (FEA) to compression experiments in order to determine the mechanical properties of fully elastic cellular walls at different fractional deformations (ε). However, when dealing with core–shell microcapsules, a fully elastic behaviour of their shell cannot be assumed, especially at high fractional deformations [45]. Core-shell MF microcapsules have proven plastic deformation at $\varepsilon \geq 0.2$, which may not rupture until $\varepsilon \geq 0.5$ [46]. Indeed, any stretching, bending, and wrinkling effect of the shell under compression should be taken into account. Interestingly, Mercade-Prieto et al. [3] have developed a powerful FEA model for core–shell microcapsules, which is capable of evaluating the intrinsic Young's modulus of the shell materials (E_{FEA}), as well as calculating the unique shell thickness to radius ratio (h/r) of individual elastoplastic microcapsules using the force versus displacement data corresponding to relatively low fractional deformation ($\varepsilon \leq 0.1$ within the elastic region). This model relying on ad-hoc h/r-dependent polynomial functions (Table 1) has proven highly accurate with its output h/r values being validated against a large number (186) of cross-sectional images of individual microcapsules via Transmission Electron Microscopy (TEM) [5]. The establishment of a specific core–shell model has represented a breakthrough over the classical interpretation of the apparent Hertzian Young's modulus (E_A) which can approximately address microcapsules with a liquid core as whole solid-like particles. With that being said, for the same microcapsules, two different Young's modulus values can be obtained using the Hertz model and FEA based on the same force versus displacement data obtained by micromanipulation, which can differ by an order of magnitude [47]. To the authors' best knowledge, their interrelationship has not yet been established. The present study therefore aims to develop and establish a theoretical relationship between the apparent Young's modulus (E_A) of single whole microcapsules calculated by using the Hertz model, and their intrinsic Young's modulus (E_{FEA}) of the shell material determined using FEA, which is intimately related to the *h/r* value of microcapsules. The experimental force versus displacement data were generated using an advanced micromanipulation technique based on diametrical compression of individual microcapsules between two parallel surfaces. Ad hoc simulations were run to validate the results. As presented in our previous papers [2,48], microplastic-free biopolymer-based microcapsules with a core of perfume oil (hexyl salicylate) have been employed herein, and their micromanipulation dataset was used for direct validation. In addition, different microcapsules with a synthetic and composite shell were also used to investigate the broad spectrum applicability of the novel model regardless of their microencapsulation process and/or shell chemistry. This results can be used to interpret various Young's modulus data of microcapsules reported in the literature and the approach taken represents a unique and unambiguous methodology to characterise the elastic properties of microcapsules with a core–shell structure.

Table 1. Complex polynomial functions of *h/r* by Mercadé-Prieto et al. [5].

Coefficient	Polynomial Function
f_1	$95071.891\,(h/r)^5 - 28426.030\,(h/r)^4 + 2411.056\,(h/r)^3 - 7.476\,(h/r)^2 - 10.829\,(h/r) + 1.52882$
f_2	$-318.702\,(h/r)^4 + 120.784\,(h/r)^3 - 11.380\,(h/r)^2 + 2.518\,(h/r) - 0.05792$
f_3	$-0.004242\,(h/r) + 0.00107$

2. Materials and Methods

2.1. Materials

Gum Arabic and fungal chitosan (fCh; deacetylation degree 79%, molecular weight ~150 kDa) were purchased from Nexira Food (Rouen Cedex, France, EU) and Kitozyme (Herstal, Belgium, EU), respectively. Analytical grade chemicals being hexylsalicylate (HS; 1.04 g·mL^{-1}), sorbitan triolate (Span85), triethanolamine (TEA), aqueous glutaraldehyde (GLT; 50% w/w), fuming hydrochloric acid (HCl; 36% w/v), LR white acrylic resin were bought from Sigma-Aldrich (Gillingham, Dorset, UK), stored according to the safety data sheet (SDS) instructions, and used without any additional purification. All admixtures were prepared with demineralised water (18.2 MΩ·cm at 25 °C).

2.2. Preparation of Microcapsules

The microcapsules were fabricated via one-step complex coacervation according to Baiocco et al. [2]. Briefly, HS (40 g) dyed with fluorescence sensing Nile Red (~5 mg) was added to an aqueous admixture (730 mL) at pH 1.95 (acidified by HCl_{aq}) containing GA (2.0% w/v) and fCh (0.5% w/v), which led to two phases. Sorbitan triolate (0.8 g) as the emulsifier was added. Homogenisation (1000 rpm; IKA Eurostar 20, Germany, Staufen, EU) was carried out to achieve oil-in-water (o/w) droplets with a target mean size of ~30 µm measured by laser diffraction. Complex coacervation between GA and fCh was induced by increasing the pH to 3.4 via gradual addition (160 mL) of TEA until a shell encircling the oil droplet was evident under a bright-field microscope (Leica DM500, Buffalo Grove, IL, USA). GLT (0.3 g/g-biopolymer) was added to trigger the crosslinking with amines along the microcapsule shells, hence their reticulation. The suspension of microcapsules was left crosslinking under stirring (300 rpm; ~15 h).

2.3. Mechanical Properties

A micromanipulation technique based on parallel compression of individual microcapsules was utilised to determine the mechanical properties of single microcapsules [21,42,46].

A flat glass slide (~2.5 cm^2; thickness ~ 1.5 mm) was covered with two drops of suspended microcapsules, which were then left to dry out at ambient temperature (23.5 ± 1.5 °C). Microcapsules were observed by a side-view camera (high performance charge-coupled device camera, Model 4912-5010/000, Cohu, Poway, CA, USA). A flat-end glass probe was attached to a force transducer (Model 403A, force scale 5 mN; Aurora Scientific Inc., Aurora, ON, Canada) with a measurement resolution of ±0.1 μN. The probe was enabled to move down/upwards by a fine micromanipulator with a displacement resolution (± 0.1 μm). A descending speed of 2.0 $\mu m \cdot s^{-1}$ was selected to compress each particle. Thirty randomly chosen microcapsules were compressed in order to generate statistically representative results [46]. Since particles can exhibit elastic, viscoelastic, and elastoplastic shells, their behaviour should be determined through compression-holding-unloading experiments and different speeds. Previous literature has demonstrated that polymeric excipients [49] and thin-shell MF-based microcapsules [50] with different cores exhibited mainly elastoplastic deformations [46]. Moreover, a negligible viscous character was reported for GA, thereby it mainly exhibited elastic deformations [51]. Similarly, Ch has proven elastic properties [52], especially when combined with xanthan gum [53]. Accordingly, only one compression velocity (2.0 $\mu m \cdot s^{-1}$) was used to compress the microcapsules.

2.4. Morphology

Bright-field optical (PL-Fluotar 5×/0.12 and 10×/0.30 lens) and fluorescence (CoolLED pE-300 blue light beam, wavelength λ = 460 nm) microscopy, as well as accelerated voltage (15–30 kV) scanning electron microscopy (SEM; JEOL 6060, Peabody, MA, USA) were employed to assay the microcapsules for their morphology, surface topography, and structural properties.

2.5. Particle Sizing

Laser diffraction technique was used to quantify the mean moist diameter (i.e., volume-to-surface Sauter diameter $D_{[3,2]}$) and size distribution of the microcapsules (Mastersizer2000, Malvern Instruments, Malvern, UK). The microcapsule suspension (~3 g) was added into the sample dispersion unit under stirring (2500 rpm) coupled with the instrument. The tests were performed at ambient temperature employing a He–Ne laser (measurement range of 50 nm–0.9 mm). The number-based diameter ($D_{n\text{-}b}$) of dry microcapsules was measured via on-screen image analysis following calibration of the micromanipulation side view camera with a calibrating eyepiece graticule slide (10 μm microcalibrating ruler, Graticules Ltd., Tonbridge, Kent, UK) [40].

3. Results and Discussion

3.1. Morphology

Figure 1A displays HS-laden microcapsules fabricated by CC. Most of microcapsules encircled single oil droplets (yolk-white like structures) within their fungal chitosan-gum Arabic (fCh-GA) shells. The microcapsules appeared to be individual, hence neither clustering nor agglutination phenomena among the microcapsules were observed. Interestingly, a morphological analysis revealed the presence of relatively spherical microcapsules. Specifically, slightly elongated shells with an eye-shape were observed, which has been similarly reported by Leclercq et al. [17] for gelatine based microcapsules encapsulating limonene. This eye-shaped configuration could be ascribable to fast stirring while inducing coacervation, thereby triggering an alignment of the forming shells (mobile shells) of microcapsules with the flow pattern within the agitated vessel. Alternatively, Baiocco et al. [2] have elucidated that some excess polymeric material could be deformed around the oil droplets by the agitation during the development of the shell, hence generating the eye shape. Surface topography was investigated by SEM observation, which highlighted the presence of HS-microcapsules with a relatively smooth shell [48]. However, several rough and indented areas were also visible at the shell surface (Figure 1B,C). Specifically, multiple surface vacuoles were observed. This might be a result of the high vacuum inside the SEM chamber, which affected the structural chassis of microcapsules, according to Farshchi et al. [54]. Thus, such surface vacuoles might act as inside-out bridges for the core oil (HS) to suddenly diffuse out through the shell in the chamber of the scanning electron microscope (Figure 1C).

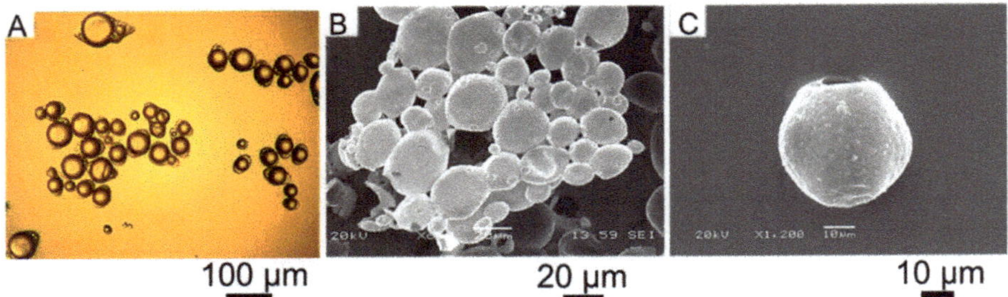

Figure 1. Suspended HS-microcapsules in water by bright-field microscopy (**A**); SEM micrographs of (**B**) HS-microcapsules (overview), (**C**) an incomplete HS microcapsule with surface vacuoles.

3.2. Particle Sizing

The microcapsules were assayed for their Sauter diameter and size distribution (Supplementary Figure S1). They were determined to range between 14 µm and 100 µm ($D_{[3,2]}$ = 36.5 ± 1.2 µm) with a SPAN value of 1.1, which is consistent with our previous works [2,48,55]. Statistical analysis suggested that the size distribution could be fitted to a lognormal distribution curve with 95% confidence. In addition, the size of microcapsules detected by laser diffraction was found to be in good agreement with their SEM micrographs (Figure 1B,C).

Figure 2 presents the optical images of the wet microcapsules which were left air-drying (25 ± 1.5 °C) whilst being monitored via real-time optical microscopy imaging before (0 h; Figure 2(A1–C1)) and after full drying (~1 h; Figure 2(A2–C2)). In moist conditions the spherical microcapsule (A1) exhibited a moist diameter of 22.0 ± 0.7 µm whilst it was found to be 17.6 ± 0.5 µm (~20% smaller) under dry conditions. When dealing with B1 and C1, the size measurements were conducted along the cross-sectional diameters of each microcapsule. As can be seen, the shell material was formed abundantly around the oil droplets, which might therefore trap a great number of minute water molecules within [56]. The moist diameters of Figure 2(B1,C1) were determined to be 14.5 ± 0.3 µm and 15.3 ± 0.2 µm, respectively. As anticipated, their dry diameters were 20–25% lower, being 10.6 ± 0.3 µm and 12.1 ± 0.4 µm, respectively. Moreover, Figure 2(A3–C3) display the fluorescence sensing response of the three microcapsules in dry conditions. The photomicrographs clearly elucidate the effective retention of the perfume oil emitting a green visible spectrum when being excited by the fluorescent light source. This phenomenon is attributable to the solvatochromatic response of the dye (i.e., Nile Red) within a relatively polar environment (i.e., hexyl salicylate), as also discussed by Baiocco et al. [55] and Zhang et al. [43]. The corresponding diameters obtained from the fluorescent micrographs were 16.9 ± 0.7 µm, 10.2 ± 0.2 µm and 10.5 ± 0.3 µm for Figure 2(A3,B3,C3), respectively, which appeared not to be statistically different from those attained under optical light in dry conditions for the same microcapsules (Figure 2(A2,B2,C2), with 95% confidence. Clearly, it was not possible to estimate the shell thickness via direct comparison between the bright filed and fluorescence photomicrographs. However, this result seems to suggest that the perfume oil might possess some natural affinity for the coacervate network, possibly migrating into the shell via partial solubilisation, thereby enabling the fluorescent signal to be detected within the shell as well. Overall, it could be inferred that the shell material retained some water leading to a partial moisturisation/swelling of the shell (moist diameter), which shrank to a certain extent upon drying.

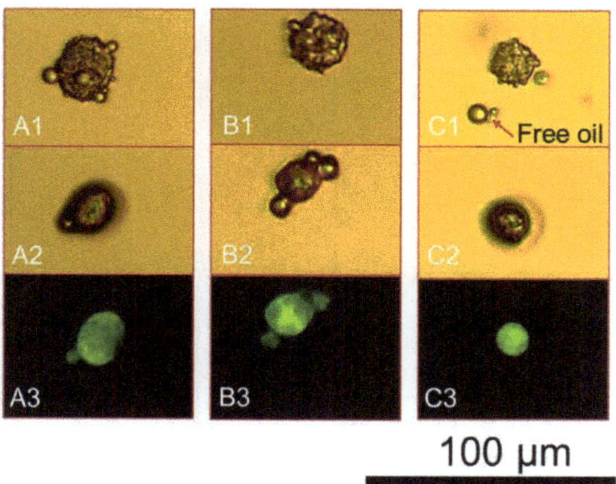

Figure 2. Bright-field optical microphotographs of microcapsules with a spherical morphology suspended in water (**A1–C1**); after drying (**A2–C2**) and under fluorescence (**A3–C3**).

3.3. *Mechanical Properties*

3.3.1. Apparent Young's Modulus of Whole Microcapsules Determined by the Hertz Model

The classic Hertz model has proven effective at describing the relationship between the compression force (F) and the axial displacement (δ) of a spherical particle at minute deformations ($\leq$10%) [38]. The model assumes (i) frictionless and smooth contact surfaces, (ii) homogeneous, isotropic and linearly elastic (Hooke's law) contacting materials, and (iii) negligible geometric non-linearities due to larger strains [47]. While being aware of the aforementioned assumptions, the Hertz model has been applied to determine the apparent Young's modulus (E_A) of core–shell microcapsules including their liquid core, which are treated as solid spheres:

$$F = \frac{\psi}{1-\nu^2} E_A r^2 \left(\frac{\delta}{2r}\right)^{3/2} \tag{1}$$

where ψ is the spherical shape factor equal to $4/3$, ν is the Poisson ratio that is assumed to be $1/2$ for non-compressible polymeric microcapsules, whereas the group $\delta/(2r)$ represents the fractional deformation (ε). An example of the Hertz model fitting to the force versus displacement data obtained from micromanipulation for a single microcapsule is illustrated in Figure 3A, whereas the mean apparent E_A value along with the corresponding mean coefficient of determination (R^2) of the model performance for the 30 tested microcapsules can be found in Table 2. The mean R^2 value from all the microcapsules was determined to be $\geq$0.93, which indicates that the compression force versus displacement data of single microcapsules can be fitted by the Hertz model reliably [57]. Figure 4A-(i) displays the apparent (E_H) Young's modulus of microcapsules as a function of their diameter. Interestingly, the apparent Young's modulus of the microcapsules did not seem to vary with the diameter significantly on average with 95% confidence (mean value E_A = 0.095 $\pm$ 0.014 GPa). However, E_A cannot fully represent the Young's modulus of the pure shell material since the microcapsule contains a liquid core which in theory has no elasticity but can also contribute the force response under compression. In light of the above, it has become increasingly imperative to develop a methodology to overcome the shortcomings of the basic Hertzian theory, therefore predicting the intrinsic elastic modulus value of the shell material [5].

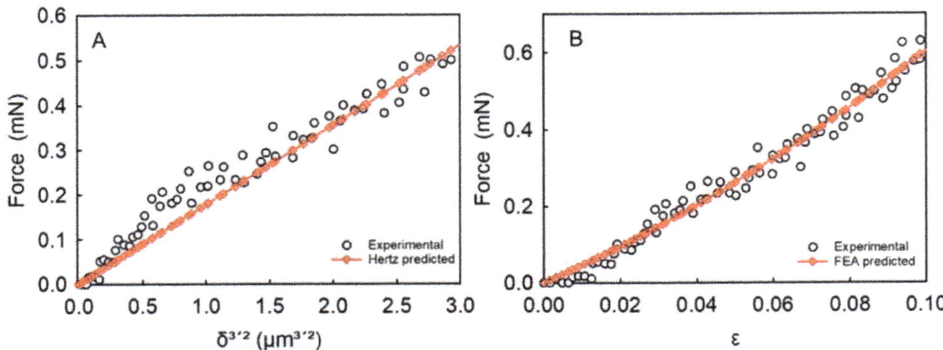

Figure 3. Typical fitting of a (**A**) force-displacement curve by Hertz model (R^2 = 0.95) and (**B**) force-fractional deformation ($\varepsilon \leq 0.1$) curve by FEA simulation results (R^2 = 0.98) for a single microcapsule (d = 23.7 μm).

Table 2. Intrinsic mechanical property parameters of HS laden microcapsules (Mean ± St. Error). The symbol * represents the predicted intrinsic Young's modulus via modelling.

Title 1	HS Laden Microcapsules	
Number based diameter (μm)	27.6 ± 1.7	
E_{FEA} (GPa)	1.02 ± 0.13	R^2 = 0.98
E_A (GPa)	0.095 ± 0.014	R^2 = 0.95
E_A/E_{FEA} (GPa)	$0.085 \pm 2 \times 10^{-3}$	
E_{FEA*} (GPa)	1.09 ± 0.14	
$(h/r)_{FEA}$	$0.132 \pm 9 \times 10^{-3}$	

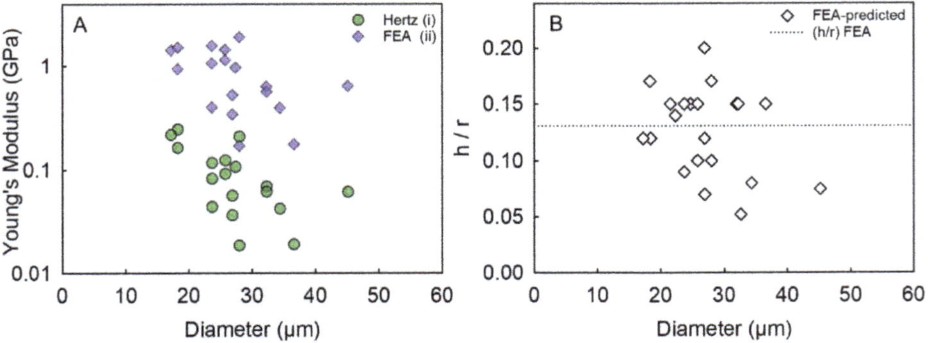

Figure 4. (**A**) Hertzian and FEA-derived Young's modulus and (**B**) FEA-predicted h/r of single microcapsules versus diameter; the dashed line represents the average FEA-predicted $(h/r)_{FEA}$ = 0.132 ± 0.009.

3.3.2. Determination of the Shell Young's Modulus of Microcapsules by FEA

In order to determine the intrinsic Young's modulus value of the shell material of microcapsules, the FEA model developed by Mercadé-Prieto et al. [5] includes both bending and stretching of the microcapsule shell under compression. Mathematically, it is presented as follows:

$$F = E_{FEA}\frac{h}{r}\left[f_1\left(\frac{\delta}{2}\right)^2 + f_2\left(\frac{\delta}{2}\right)r + f_3\,r^2\right] \quad (2)$$

where E_{FEA} is the Young's modulus of the shell material of microcapsules, F is the experimental compression force measured by micromanipulation, δ is the compressive axial displacement during compression, h/r is the ratio of shell thickness to the microcapsule initial radius, and f_1, f_2, and f_3 are polynomial functions of h/r detailed in Table 1 [5].

A typical example of FEA model fitting to the force versus fractional deformation (ε) data of an individual microcapsule is presented in Figure 3B (R^2 = 0.98). Figure 4A-ii displays the FEA-derived shell Young's modulus of microcapsules versus their diameter, which did not seem to change with diameter significantly. When analysing the data, several non-aligned points were obvious at a given diameter. Specifically, at a diameter of 18 μm and 24 μm, the data scattered vertically between 0.9–1.5 GPa and 0.4–1.6 GPa, respectively. However, the most significant variability in the Young's modulus values was found at a diameter similar to the mean size of microcapsules (~28 μm). Interestingly, the data ranged vertically from 0.15 GPa to 1.9 GPa, highlighting a variability of more than one order of magnitude. These findings likely suggest that the coacervate matter may form inhomogeneously around the oil droplets during microencapsulation, hence a direct effect on the intrinsic mechanical properties of the resulting shell material may be plausible (Figure 2A,B). When compared to the Young's modulus values determined by the Hertz model, there appears to be an overall upward shift of the data points obtained via FEA, including their vertical scattering at each given diameter. Clearly there is a significant difference in the mean value of the two Young's moduli determined by the two approaches. As mentioned, the Hertz model includes no h/r parameter, thus any physical difference between thin- and thick-shell microcapsule is difficult to predict. On average, the Young's modulus by FEA from thirty HS-microcapsules was determined to be E_{FEA} = 1.02 ± 0.13 GPa (with a corresponding unique $(h/r)_{FEA}$ = 0.132 ± 0.009), which is in line with that reported by Mercadé-Prieto et al. [5] for MF-based microcapsules via FEA (1.6 ± 0.3 GPa) [5]. Although these values appear to be statistically similar, the slight discrepancy is likely ascribable to the nature of the different shell materials. Specifically, the coacervate shell made of natural biopolymers (fChGA) may possibly form non-uniform and microporous structures, as also described by Espinosa-Andrews et al. [58]. In contrast, MF can form thin and highly smooth shells ($(h/r)_{MF}$~0.02 ± 0.01 [5]), as with many other plastic materials [46]. Furthermore, this value (E_{FEA} = 1.02 ± 0.13 GPa) is also in line with that of other polymeric microspheres for pharmaceutical applications (1.6 ± 0.2 GPa) investigated by Yap et al. [49]. In addition, similar values of the Young's modulus were also obtained from other types of microcapsule shells utilising AFM, which is a different technique from micromanipulation. Specifically, core–shell microcapsules formulated with aminoplast [59], poly(styrene sulfonate)/poly(allylamine) [60], and poly(D,L-lactide-co-glycolide) [36] exhibited a Young's modulus of to 1.7 GPa, 1.3 ± 0.15–1.9 ± 0.2 GPa, and 0.1–3.0 GPa, respectively. As discussed above, the Young's modulus value from AFM, although it is also determined by the Hertz model, tends to represent the local stiffness of the test material near the surface (due to a typical indentation depth ~20–200 nm), the resulting Young's modulus value obtained should be close to the intrinsic Young's modulus of the shell. Notwithstanding, it may be difficult to predict any effect of the liquid reservoir (beneath the shell) on the AFM measurements when the shell is particularly thin, as with synthetic shells whose thickness can be as low as 20–70 nm [5]. Tan et al. [61] conducted AFM measurements on oil-laden microcapsules made of thiolated chitosan tentatively resulting in an apparent Young's modulus of 1.44 MPa, which is surprisingly around three orders of magnitude lower than ours (1.02 ± 0.13 GPa). This major discrepancy was probably due to the combination of several effects, including the processing conditions (i.e., ultrasonic synthesis), and the chemistry of chitosan employed which had been grafted with thiol groups using DL-N-acetylhomocysteine thiolactone [61]. Moreover, a crucial role may also be played by the size of the spheres (<10 μm) and their extremely thin shells (<180 nm) leading to a h/r~0.018 which is one order of magnitude lower than our $(h/r)_{FEA}$ = 0.132 ± 0.009. Figure 4B displays the FEA-predicted h/r of microcapsules versus their diameter from compression experiments. It is found that h/r did not vary significantly with the diameter with 95% confidence.

3.3.3. Interrelationship between the Apparent Young's Modulus (EA) of Whole Microcapsules Determined Using the Hertz Model and That of the Shell Material Using FEA

As shown in Table 2, E_{FEA} is clearly higher (approximately by one order of magnitude) than the corresponding E_A. Such discrepancy is not surprising as the latter is based on treating a microcapsule as a homogenous microsphere and the Young's modulus of a liquid core (or a fluid) is nil. By considering the two Young's moduli can be determined using the same set of experimental data, it may be possible to establish their relationship mathematically. Given that the fractional deformation is $\varepsilon = \delta/(2r)$, the model proposed by Mercadé-Prieto et al. [5] (Equation (2)) can also be expressed as a dimensionless force $\widetilde{F} = F/E_{FEA}rh$:

$$\frac{F}{E_{FEA}rh} = f_1\varepsilon^2 + f_2\varepsilon + f_3, \qquad \{0.03 < \varepsilon < 0.1\} \tag{3}$$

where f_1, f_2 and f_3 are the polynomial functions of (h/r), see Table 1 [5]. For a given h/r (e.g., h/r =0.05; 0 < h/r $\leq$ 0.14), Equation (3) can be used to generate the $\widetilde{F}$ data corresponding to fractional deformations ε from 0.03 to 0.1 at progressively high steps (e.g., step-by-step threshold of 0.001). Five examples of the generated dimensionless force versus fractional deformation data (up to 10% deformation) for a given h/r are shown in Figure 5A. If the data is plotted in terms of the dimensional force $\widetilde{F}$ versus $\varepsilon^{3/2}$, see Figure 5B, their relationship looks approximately linear, and the slope increases with h/r, which is similar to Equation (1). Therefore, Equation (3) can also be expressed as follows:

$$\frac{F}{E_{FEA}rh} = f_4\varepsilon^{3/2} \tag{4}$$

where f_4 is a function of h/r only. For a given h/r of 0.05, the value of f_4 can be determined by linearly fitting the corresponding curve using Equation (4), which gives 0.549 with a coefficient of determination R^2 = 0.997 (See the supplementary material Figure S2). By comparing Equation (1) with Equation (4), the ratio of E_A/E_{FEA} defined by φ is given by:

$$\varphi = f_4 \frac{1-\nu^2}{\psi} \frac{h}{r} \tag{5}$$

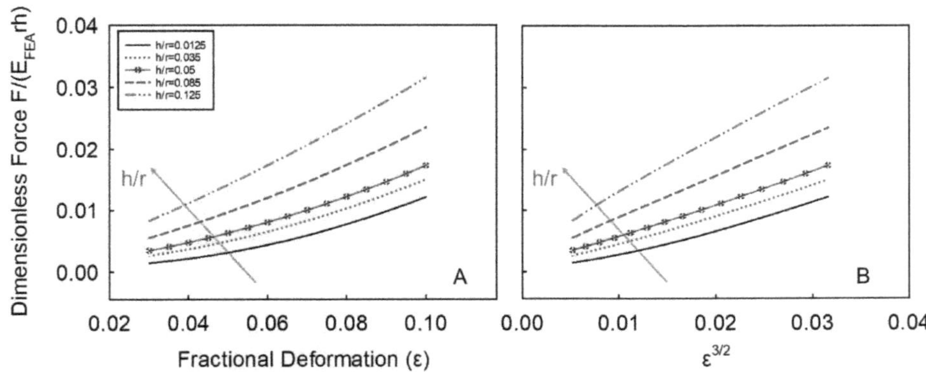

Figure 5. (**A**) Dimensionless force versus fractional deformation data (up to 10% deformation) for a given h/r; (**B**) dimensional force versus fractional deformation to a power of 3/2. The simulated data are based on 5 example values of h/r, namely 0.0125, 0.035, 0.05, 0.085, 0.125.

The corresponding φ value can therefore be determined to be 0.0154 for this particular value of h/r, which includes the Poisson ratio (ν) and the spherical shape factor (ψ).

Accordingly, increasingly high h/r values ($(h/r)_k$ = 0.01, 0.02, ... , 0.14) were chosen to generate the series of the corresponding $f_{4,k}$ values and therefore φ_k (Figure 6).

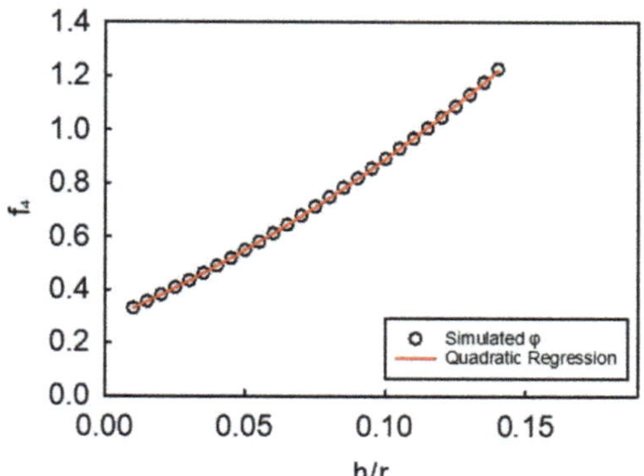

Figure 6. Simulated f_4 function (directly proportional to the scope function φ) generated using a series h/r values ($(h/r)_k$ = 0.01, 0.02, ... , 0.14) within the applicable domain of the model ($h/r \leq 0.14$). The circles (o) represent the simulated points as a function of h/r, whereas the red line (–) is the corresponding second-degree polynomial regression.

Interestingly, f_4 increases with h/r monotonically, and the resulting relationship is therefore fitted by a second-degree polynomial:

$$f_4 = k_1\left(\frac{h}{r}\right)^2 + k_2\left(\frac{h}{r}\right) + k_3 \tag{6}$$

where k_1 = 8.4673, k_2 = 2.5728, and k_3 = 0.1597 are the dimensionless constants of the polynomial, with their mean coefficient of determination of 0.999. Thus, the dimensionless force can be expressed as:

$$\widetilde{F} = \left[k_1\left(\frac{h}{r}\right)^2 + k_2\left(\frac{h}{r}\right) + k_3\right]\varepsilon^{3/2} \tag{7}$$

which leads to the following explicit equation between force (F) and displacement (δ) by substituting $\varepsilon^{3/2} = (\delta/2r)^{3/2}$ and $\widetilde{F} = F/(E_{FEA}rh)$:

$$F = E_{FEA}\left[k_1\left(\frac{h}{r}\right)^3 + k_2\left(\frac{h}{r}\right)^2 + k_3\left(\frac{h}{r}\right)\right]\left(\frac{\delta}{2}\right)^{3/2} r^{1/2} \tag{8}$$

Since the compression force F is equal in both models (i.e., the Hertz and FEA) for a given fractional deformation, the combination of Equations (1), (5) and (8) thus leads to:

$$\frac{E_A}{E_{FEA}} = k_1\left(\frac{h}{r}\right)^3 + k_2\left(\frac{h}{r}\right)^2 + k_3\left(\frac{h}{r}\right). \tag{9}$$

which is an explicitly third-degree polynomial function of h/r with $0 < h/r \leq 0.14$. Applying Equation (9), E_A/E_{FEA} was determined to be $0.085 \pm 2 \times 10^{-3}$ for HS laden microcapsules with a fCh-GA shell. For given values of E_A and h/r, the predicted intrinsic E_{FEA^*} was 1.09 ± 0.14 GPa, which have no significant difference from those calculated directly via FEA ($E_{FEA} = 1.02 \pm 0.13$ GPa by Equation (1)) with 95% confidence. These results confirm the effectiveness of the model herein developed. Nonetheless, a general validation of the model is required in order to verify its prediction capability independently of the chemistry of the microcapsules. Having said that, the newly established Equation (9) was also applied to the micromanipulation data obtained from core–shell melamine-gluteraldehyde-formaldehyde (MGF) microcapsules with a core of perfume oil (i.e., lily oil) fabricated via in situ polymerisation by Luo et al. [62]. Interestingly, it was found that the intrinsic Young's modulus of the shell material from FEA simulation (Equation (2)) and the predicted E_{FEA^*} (Equation (9)) were 2.92 ± 0.29 GPa and 2.91 ± 0.29 GPa, respectively (validation data shown in Supplementary Figure S3). The excellent agreement of the two values demonstrates the large applicability of the model to synthetic microparticles with a spherical morphology and a core–shell configuration. Moreover, the average number based diameter and h/r of thirty MGF microcapsules were 10.0 ± 0.8 μm and 0.067 ± 0.007, respectively, suggesting a good narrow dispersity and homogeneity of the sample. Similar results were also documented by Mercadé-Prieto et al. [5] for simple MF microcapsules.

In addition, Equation (6) was further validated against composite double coated microcapsules, reported in our previous studies [55]. For fCh-GA microcapsules with an additional maltodextrin based coating, it was determined that the intrinsic Young's modulus of the overall shell material obtained from FEA simulation was 2.59 ± 0.83 GPa, whereas the E_{FEA^*} predicted from the apparent Young's modulus (Hertz) was 2.61 ± 0.84 GPa (validation data shown in Supplementary Figure S4). As anticipated, the two values independently generated by FEA and φ were in total agreement, demonstrating the effectiveness of the model on double coated microcapsules, which had been fabricated using a two-stage chemical-physical approach (complex coacervation followed by spray drying) [55]. Based on the above, these findings confirm the applicability of the novel model to different types of core–shell microcapsules, independently of the microencapsulation process and shell chemistry. Notwithstanding, when compared to HS laden microcapsules made with a single fCh-GA shell, the mean Young's modulus value from maltodextrin coated microcapsules is significantly greater. This result indicates the extra maltodextrin coating provided additional stiffness to the shell effectively, thereby enhancing the overall mechanical properties of coacervate shell microcapsules [55]. Overall, these findings not only have demonstrated the superb prediction capability of the newly developed model but have also elucidated its broad spectrum applicability to synthetic, non-synthetic, and composite microcapsules.

4. Conclusions

Microcapsules with a liquid perfume core and a fungal chitosan–gum Arabic shell produced via CC were assessed for their intrinsic mechanical properties insightfully. The compression force dataset obtained from micromanipulation measurements based on compression of single microparticles to different deformations was utilised to determine the apparent Young's modulus of the whole microcapsule (including the liquid core) and the intrinsic Young's modulus of shell material by Hertz and FEA models, respectively. It was found that the apparent Young's modulus of whole microcapsules was 0.095 ± 0.014 GPa, whilst the intrinsic E_{FEA} of the shell of microcapsules was 1.02 ± 0.13 GPa (yielding $(h/r)_{FEA}$ = 0.132 ± 0.009) on average, which differ by about one order of magnitude. The mathematical interrelationship between E_A and E_{FEA} was determined by numerical simulations resulting in E_{FEA}/E_A as a third degree of polynomial function of h/r. The prediction capability of the newly developed model was validated against a broad spectrum of microcapsules (i.e., synthetic, non-synthetic, double coated) thereby elucidating its general validity regardless of the nature (chemistry) of microcapsules. The new model therefore represents a powerful and rapid tool to determine the intrinsic Young's modulus of the microcapsule shell material when only the apparent Young's modulus of whole microcapsules is known, or no FEA simulating tool is available. Moreover, micromanipulation provides a rapid pathway to investigate the mechanical rupture of microcapsules with accuracy and dexterity, which is crucial to ensuring their functionalities at end-use applications. Future work can be directed at developing rigorous software to help predict both structural and mechanical property parameters of microparticles, which may pave an avenue to facilitate the development, engineering, and functionalisation of microcapsules for a wide range of applications.

Supplementary Materials: The following supporting information can be downloaded at: https://www.mdpi.com/article/10.3390/mi14010123/s1, Figure S1: Particle Size Distribution of HS laden microcapsules (Master Sizer 2000, Malvern Instruments Ltd., Malvern, UK); Figure S2: The Hertz model fitting results of dimensionless force versus fractional deformation up to 10% deformation (h/r = 0.05, from Figure 5B) where the slope is f_4 = 0.549; Figure S3: Model application example: synthetic microcapsules by Luo et al. [62]; linear relationship between the FEA derived and model φ-predicted Young's moduli, with their average values of the shell material being 2.92 ± 0.29 GPa and 2.91 ± 0.29 GPa, respectively; Figure S4: Model application example: double coated microcapsules by Baiocco et al. [55]; linear relationship between the FEA derived and model φ-predicted Young's moduli, with their average values of the shell material being 2.59 ± 0.83 GPa and 2.61 ± 0.84 GPa, respectively.

Author Contributions: Author Contributions: Conceptualisation, D.B. and Z.Z. (Zhibing Zhang); methodology, D.B.; software, Z.Z. (Zhihua Zhang) and D.B.; validation, D.B. and Z.Z. (Zhihua Zhang); formal analysis, D.B.; investigation, D.B.; resources, Z.Z. (Zhibing Zhang), Y.H.; data curation, D.B.; writing—original draft preparation, D.B.; writing—review and editing, Z.Z. (Zhibing Zhang); visualization, D.B.; supervision, Z.Z. (Zhibing Zhang); project administration, Z.Z. (Zhibing Zhang); funding acquisition, Z.Z. (Zhibing Zhang) All authors have read and agreed to the published version of the manuscript.

Funding: This research was funded by EPSRC, UK, grant number EP/V027654/1-EP/R512436/1.

Data Availability Statement: Not applicable.

Acknowledgments: The authors wish to express their appreciation to the Engineering and Physical Sciences Research Council (EPSRC) and the National Productivity Investment Fund (NPIF) in the UK for the financial support.

Conflicts of Interest: The authors declare no conflict of interest.

References

1. Teixeira, M.A.; Rodríguez, O.; Rodrigues, S.; Martins, I.; Rodrigues, A.E. A case study of product engineering: Performance of microencapsulated perfumes on textile applications. *AIChE J.* **2012**, *58*, 1939–1950. [CrossRef]
2. Baiocco, D.; Preece, J.A.; Zhang, Z. Encapsulation of Hexylsalicylate in an Animal-Free Chitosan-Gum Arabic Shell by Complex Coacervation. *Colloids Surf. A Physicochem. Eng. Asp.* **2021**, *625*, 126861. [CrossRef]
3. Mercade-Prieto, R.; Allen, R.; York, D.; Preece, J.A.; Goodwin, T.E.; Zhang, Z. Determination of the shell permeability of microcapsules with a core of oil-based active ingredient. *J. Microencapsul.* **2012**, *29*, 463–474. [CrossRef] [PubMed]
4. Peña, B.; Panisello, C.; Aresté, G.; Garcia-Valls, R.; Gumí, T. Preparation and characterization of polysulfone microcapsules for perfume release. *Chem. Eng. J.* **2012**, *179*, 394–403. [CrossRef]
5. Mercadé-Prieto, R.; Nguyen, B.; Allen, R.; York, D.; Preece, J.A.; Goodwin, T.E.; Zhang, Z. Determination of the elastic properties of single microcapsules using micromanipulation and finite element modeling. *Chem. Eng. Sci.* **2011**, *66*, 2042–2049. [CrossRef]
6. Pan, X.; Mercadé-Prieto, R.; York, D.; Preece, J.A.; Zhang, Z. Structure and Mechanical Properties of Consumer-Friendly PMMA Microcapsules. *Ind. Eng. Chem. Res.* **2013**, *52*, 11253–11265. [CrossRef]
7. Guinebretiere, S.J.S.; Smets, J.; Sands, P.D.; Pintens, A.; Dihora, J.O. Benefit Agent-Containing Delivery Particle. European Patent 2,087,089,B1 (07862210.7), 20 November 2007.
8. Miró Specos, M.M.; Escobar, G.; Marino, P.; Puggia, C.; Defain Tesoriero, M.V.; Hermida, L. Aroma Finishing of Cotton Fabrics by Means of Microencapsulation Techniques. *J. Ind. Text.* **2010**, *40*, 13–32. [CrossRef]
9. Cole, P.; Adami, H.-O.; Trichopoulos, D.; Mandel, J. Formaldehyde and lymphohematopoietic cancers: A review of two recent studies. *Regul. Toxicol. Pharmacol.* **2010**, *58*, 161–166. [CrossRef]
10. Lazenby, V.; Hinwood, A.; Callan, A.; Franklin, P. Formaldehyde personal exposure measurements and time weighted exposure estimates in children. *Chemosphere* **2012**, *88*, 966–973. [CrossRef]
11. Tang, X.; Bai, Y.; Duong, A.; Smith, M.T.; Li, L.; Zhang, L. Formaldehyde in China: Production, consumption, exposure levels, and health effects. *Environ. Int.* **2009**, *35*, 1210–1224. [CrossRef]
12. Rodrigues, S.N.; Martins, I.M.; Fernandes, I.P.; Gomes, P.B.; Mata, V.G.; Barreiro, M.F.; Rodrigues, A.E. Scentfashion®: Microencapsulated perfumes for textile application. *Chem. Eng. J.* **2009**, *149*, 463–472. [CrossRef]

13. Alongi, J.; Ciobanu, M.; Tata, J.; Carosio, F.; Malucelli, G. Thermal stability and flame retardancy of polyester, cotton, and relative blend textile fabrics subjected to sol–gel treatments. *J. Appl. Polym. Sci.* **2011**, *119*, 1961–1969. [CrossRef]
14. Karlsson, L.E.; Jannasch, P. Polysulfone ionomers for proton-conducting fuel cell membranes: Sulfoalkylated polysulfones. *J. Membr. Sci.* **2004**, *230*, 61–70. [CrossRef]
15. Piirilä, P.L.; Meuronen, A.; Majuri, M.L.; Luukkonen, R.; Mäntylä, T.; Wolff, H.J.; Nordman, H.; Alenius, H.; Laitinen, A. Inflammation and functional outcome in diisocyanate-induced asthma after cessation of exposure. *Allergy* **2008**, *63*, 583–591. [CrossRef]
16. Bruyninckx, K.; Dusselier, M. Sustainable Chemistry Considerations for the Encapsulation of Volatile Compounds in Laundry-Type Applications. *ACS Sustain. Chem. Eng.* **2019**, *7*, 8041–8054. [CrossRef]
17. Leclercq, S.; Harlander, K.R.; Reineccius, G.A. Formation and characterization of microcapsules by complex coacervation with liquid or solid aroma cores. *Flavour Fragr. J.* **2009**, *24*, 17–24. [CrossRef]
18. Butstraen, C.; Salaün, F. Preparation of microcapsules by complex coacervation of gum Arabic and chitosan. *Carbohydr. Polym.* **2014**, *99*, 608–616. [CrossRef]
19. Lin, L.; Regenstein, J.M.; Lv, S.; Lu, J.; Jiang, S. An overview of gelatin derived from aquatic animals: Properties and modification. *Trends Food Sci. Technol.* **2017**, *68*, 102–112. [CrossRef]
20. Mercadé-Prieto, R.; Zhang, Z. Mechanical characterization of microspheres—Capsules, cells and beads: A review. *J. Microencapsul.* **2012**, *29*, 277–285. [CrossRef]
21. Gray, A.; Egan, S.; Bakalis, S.; Zhang, Z. Determination of microcapsule physicochemical, structural, and mechanical properties. *Particuology* **2016**, *24*, 32–43. [CrossRef]
22. Ahearne, M.; Yang, Y.; Liu, K.-K. Mechanical Characterisation of Hydrogels for Tissue Engineering Applications. *Top. Tissue Eng.* **2008**, *4*, 1–16.
23. Hochmuth, R.M. Micropipette aspiration of living cells. *J. Biomech.* **2000**, *33*, 15–22. [CrossRef] [PubMed]
24. Tan, Y.; Sun, D.; Huang, W.; Cheng, S.H. Mechanical Modeling of Biological Cells in Microinjection. *IEEE Trans. NanoBiosci.* **2008**, *7*, 257–266. [CrossRef] [PubMed]
25. Tan, Y.; Sun, D.; Wang, J.; Huang, W. Mechanical Characterization of Human Red Blood Cells Under Different Osmotic Conditions by Robotic Manipulation With Optical Tweezers. *IEEE Trans. Biomed. Eng.* **2010**, *57*, 1816–1825. [CrossRef] [PubMed]
26. Dols-Perez, A.; Marin, V.; Amador, G.J.; Kieffer, R.; Tam, D.; Aubin-Tam, M.-E. Artificial Cell Membranes Interfaced with Optical Tweezers: A Versatile Microfluidics Platform for Nanomanipulation and Mechanical Characterization. *ACS Appl. Mater. Interfaces* **2019**, *11*, 33620–33627. [CrossRef] [PubMed]
27. Corbin, E.A.; Kong, F.; Teck Lim, C.; King, W.P.; Bashir, R. Biophysical properties of human breast cancercells measured using silicon MEMS resonatorsand atomic force microscopy. *Lab Chip* **2015**, *15*, 839–847. [CrossRef]
28. Lefebvre, Y.; Leclerc, E.; Barthès-Biesel, D.; Walter, J.; Edwards-Lévy, F. Flow of artificial microcapsules in microfluidic channels: A method for determining the elastic properties of the membrane. *Phys. Fluids* **2008**, *20*, 123102. [CrossRef]
29. Zhang, L.; D'Acunzi, M.; Kappl, M.; Imhof, A.; van Blaaderen, A.; Butt, H.-J.; Graf, R.; Vollmer, D. Tuning the mechanical properties of silica microcapsules. *Phys. Chem. Chem. Phys.* **2010**, *12*, 15392–15398. [CrossRef]
30. Liu, S.; Wang, Y. A Review of the Application of Atomic Force Microscopy (AFM) in Food Science and Technology. *Adv. Food Nutr. Res.* **2011**, *62*, 201–240.
31. Dufrêne, Y.F. Sticky microbes: Forces in microbial cell adhesion. *Trends Microbiol.* **2015**, *23*, 376–382. [CrossRef]
32. Wright, C.J.; Shah, M.K.; Powell, L.C.; Armstrong, I. Application of AFM from microbial cell to biofilm. *Scanning* **2010**, *32*, 134–149. [CrossRef] [PubMed]
33. Whited, A.M.; Park, P.S.H. Atomic force microscopy: A multifaceted tool to study membrane proteins and their interactions with ligands. *Biochim. Biophys. Acta (BBA)-Biomembr.* **2014**, *1838*, 56–68. [CrossRef] [PubMed]
34. Chen, S.-w.W.; Odorico, M.; Meillan, M.; Vellutini, L.; Teulon, J.-M.; Parot, P.; Bennetau, B.; Pellequer, J.-L. Nanoscale structural features determined by AFM for single virus particles. *Nanoscale* **2013**, *5*, 10877–10886. [CrossRef] [PubMed]
35. Giro-Paloma, J.; Konuklu, Y.; Fernández, A.I. Preparation and exhaustive characterization of paraffin or palmitic acid microcapsules as novel phase change material. *Sol. Energy* **2015**, *112*, 300–309. [CrossRef]
36. Sarrazin, B.; Tsapis, N.; Mousnier, L.; Taulier, N.; Urbach, W.; Guenoun, P. AFM Investigation of Liquid-Filled Polymer Microcapsules Elasticity. *Langmuir* **2016**, *32*, 4610–4618. [CrossRef]
37. Ghorbanzadeh Ahangari, M.; Fereidoon, A.; Jahanshahi, M.; Sharifi, N. Effect of nanoparticles on the micromechanical and surface properties of poly(urea–formaldehyde) composite microcapsules. *Compos. Part B Eng.* **2014**, *56*, 450–455. [CrossRef]
38. Zhang, Y.; Baiocco, D.; Mustapha, A.N.; Zhang, X.; Yu, Q.; Wellio, G.; Zhang, Z.; Li, Y. Hydrocolloids: Nova materials assisting encapsulation of volatile phase change materials for cryogenic energy transport and storage. *Chem. Eng. J.* **2020**, *382*, 123028. [CrossRef]
39. Zhang, Z.; Stenson, J.D.; Thomas, C.R. Chapter 2 Micromanipulation in Mechanical Characterisation of Single Particles. In *Advances in Chemical Engineering*; Li, J., Ed.; Academic Press: Cambridge, MA, USA, 2009; Volume 37, pp. 29–85.
40. Zhang, Z.; He, Y.; Zhang, Z. Micromanipulation and Automatic Data Analysis to Determine the Mechanical Strength of Microparticles. *Micromachines* **2022**, *13*, 751. [CrossRef]

41. Mercadé-Prieto, R.; Allen, R.; York, D.; Preece, J.A.; Goodwin, T.E.; Zhang, Z. Compression of elastic–perfectly plastic microcapsules using micromanipulation and finite element modelling: Determination of the yield stress. *Chem. Eng. Sci.* **2011**, *66*, 1835–1843. [CrossRef]
42. Xue, J.; Zhang, Z. Physical, structural, and mechanical characterization of calcium–shellac microspheres as a carrier of carbamide peroxide. *J. Appl. Polym. Sci.* **2009**, *113*, 1619–1625. [CrossRef]
43. Zhang, Y.; Mustapha, A.N.; Zhang, X.; Baiocco, D.; Wellio, G.; Davies, T.; Zhang, Z.; Li, Y. Improved volatile cargo retention and mechanical properties of capsules via sediment-free in situ polymerization with cross-linked poly(vinyl alcohol) as an emulsifier. *J. Colloid Interface Sci.* **2020**, *568*, 155–164. [CrossRef] [PubMed]
44. Smith, A.E.; Moxham, K.E.; Middelberg, A.P.J. On uniquely determining cell–wall material properties with the compression experiment. *Chem. Eng. Sci.* **1998**, *53*, 3913–3922. [CrossRef]
45. Mercadé-Prieto, R.; Allen, R.; Zhang, Z.; York, D.; Preece, J.A.; Goodwin, T.E. Failure of elastic-plastic core–shell microcapsules under compression. *AIChE J.* **2012**, *58*, 2674–2681. [CrossRef]
46. Sun, G.; Zhang, Z. Mechanical properties of melamine-formaldehyde microcapsules. *J. Microencapsul.* **2001**, *18*, 593–602.
47. Baiocco, D. *Fabrication and Characterisation of Vegetable Chitosan Derived Microcapsules*; University of Birmingham: Birmingham, UK, 2021.
48. Baiocco, D.; Zhang, Z. Microplastic-Free Microcapsules to Encapsulate Health-Promoting Limonene Oil. *Molecules* **2022**, *27*, 7215. [CrossRef]
49. Yap, S.F.; Adams, M.J.; Seville, J.P.K.; Zhang, Z. Single and bulk compression of pharmaceutical excipients: Evaluation of mechanical properties. *Powder Technol.* **2008**, *185*, 1–10. [CrossRef]
50. Hu, J.; Chen, H.-Q.; Zhang, Z. Mechanical properties of melamine formaldehyde microcapsules for self-healing materials. *Mater. Chem. Phys.* **2009**, *118*, 63–70. [CrossRef]
51. Singh, A.; Geveke, D.J.; Yadav, M.P. Improvement of rheological, thermal and functional properties of tapioca starch by using gum arabic. *LWT* **2017**, *80*, 155–162. [CrossRef]
52. Cho, J.; Heuzey, M.-C.; Bégin, A.; Carreau, P.J. Viscoelastic properties of chitosan solutions: Effect of concentration and ionic strength. *J. Food Eng.* **2006**, *74*, 500–515. [CrossRef]
53. Martínez-Ruvalcaba, A.; Chornet, E.; Rodrigue, D. Viscoelastic properties of dispersed chitosan/xanthan hydrogels. *Carbohydr. Polym.* **2007**, *67*, 586–595. [CrossRef]
54. Farshchi, A.; Hassanpour, A.; Bayly, A.E. The structure of spray-dried detergent powders. *Powder Technol.* **2019**, *355*, 738–754. [CrossRef]
55. Baiocco, D.; Preece, J.A.; Zhang, Z. Microcapsules with a Fungal Chitosan-gum Arabic-maltodextrin Shell to Encapsulate Health-Beneficial Peppermint Oil. *Food Hydrocoll. Health* **2021**, *1*, 100016. [CrossRef]
56. Espinosa-Andrews, H.; Baez-Gonzalez, J.G.; Cruz-Sosa, F.; Vernon-Carter, E.J. Gum Arabic-Chitosan Complex Coacervation. *Biomacromolecules* **2007**, *8*, 1313–1318. [CrossRef] [PubMed]
57. Yu, F.; Xue, C.; Zhang, Z. Mechanical characterization of fish oil microcapsules by a micromanipulation technique. *LWT* **2021**, *144*, 111194. [CrossRef]
58. Espinosa-Andrews, H.; Sandoval-Castilla, O.; Vázquez-Torres, H.; Vernon-Carter, E.J.; Lobato-Calleros, C. Determination of the gum Arabic–chitosan interactions by Fourier Transform Infrared Spectroscopy and characterization of the microstructure and rheological features of their coacervates. *Carbohydr. Polym.* **2010**, *79*, 541–546. [CrossRef]
59. Pretzl, M.; Neubauer, M.; Tekaat, M.; Kunert, C.; Kuttner, C.; Leon, G.; Berthier, D.; Erni, P.; Ouali, L.; Fery, A. Formation and Mechanical Characterization of Aminoplast Core/Shell Microcapsules. *ACS Appl. Mater. Interfaces* **2012**, *4*, 2940–2948. [CrossRef]
60. Dubreuil, F.; Elsner, N.; Fery, A. Elastic properties of polyelectrolyte capsules studied by atomic-force microscopy and RICM. *Eur. Phys. J. E* **2003**, *12*, 215–221. [CrossRef]
61. Tan, S.; Mettu, S.; Biviano, M.D.; Zhou, M.; Babgi, B.; White, J.; Dagastine, R.R.; Ashokkumar, M. Ultrasonic synthesis of stable oil filled microcapsules using thiolated chitosan and their characterization by AFM and numerical simulations. *Soft Matter* **2016**, *12*, 7212–7222. [CrossRef]
62. Luo, S.; Gao, M.; Pan, X.; Wang, Y.; He, Y.; Zhu, L.; Si, T.; Sun, Y. Fragrance oil microcapsules with low content of formaldehyde: Preparation and characterization. *Colloids Surf. A Physicochem. Eng. Asp.* **2022**, *648*, 129019. [CrossRef]

 micromachines

Article

Direct Kinetostatic Analysis of a Gripper with Curved Flexures

Alessandro Cammarata [1,*], Pietro Davide Maddio [1], Rosario Sinatra [1] and Nicola Pio Belfiore [2]

1 Department of Civil Engineering and Architecture, University of Catania, Viale Andrea Doria, 6, 95123 Catania, Italy
2 Department of Industrial, Electronic and Mechanical Engineering, University of Roma Tre, Via Vito Volterra, 62, 00146 Rome, Italy
* Correspondence: alessandro.cammarata@unict.it; Tel.: +39-095-738-2403

Abstract: Micro-electro-mechanical-systems (MEMS) extensively employed planar mechanisms with elastic curved beams. However, using a curved circular beam as a flexure hinge, in most cases, needs a more sophisticated kinetostatic model than the conventional planar flexures. An elastic curved beam generally allows its outer sections to experience full plane mobility with three degrees of freedom, making complex non-linear models necessary to predict their behavior. This paper describes the direct kinetostatic analysis of a planar gripper with an elastic curved beam is described and then solved by calculating the tangent stiffness matrix in closed form. Two simplified models and different contributions to derive their tangent stiffness matrices are considered. Then, the Newton–Raphson iterative method solves the non-linear direct kinetostatic problem. The technique, which appears particularly useful for real-time applications, is finally applied to a case study consisting of a four-bar linkage gripper with elastic curved beam joints that can be used in real-time grasping operations at the microscale.

Keywords: micro-grippers; micro-manipulators; flexure; CFSH; compliant mechanism; the tangent stiffness matrix

Citation: Cammarata, A.; Maddio, P.D., Sinatra, R.; Belfiore, N.P. Direct Kinetostatic Analysis of a Gripper with Curved Flexures. *Micromachines* **2022**, *13*, 2172. https://doi.org/10.3390/mi13122172

Academic Editors: Michael Gauthier and Eui-Hyeok Yang

Received: 3 November 2022
Accepted: 3 December 2022
Published: 8 December 2022

Publisher's Note: MDPI stays neutral with regard to jurisdictional claims in published maps and institutional affiliations.

1. Introduction

Compliant mechanisms have been adopted in several applications for centuries, because of several well-known advantages, such as the absence of sliding friction, backlash, and significant wear, with the advantage of requiring minimum effort assembly. The first complete and systematic description of their characteristics appeared in the literature about thirty years ago, around 1994, when Midha and Howell introduced a classification [1] that has been used until today. Another significant step forward in developing the compliant mechanisms was taken a couple of years later, when the Pseudo-Rigid-Body equivalent Model (PRBM) was introduced to evaluate the elasticity of compliant mechanisms with significant deflection capabilities [2]. More materials concerning the compliant mechanisms were presented in 2001 [3].

Compliant mechanisms are nowadays used in many fields [4], as, for instance, for the development of aerospace [5] and biomedical [6,7] devices, compliant bistable mechanisms [8–11], grasping and releasing micro-objects devices [12], precision engineering [13], MEMS [14–19], polishing and deburring [20–22], lab-on-chip micro and nanosystems [23], and automotive devices [24,25].

The actual configuration of a compliant mechanism depends not only on the applied forces and torques but also on its geometric characteristics, with kinematic and mechanical coupling and non-linearity problems that especially arise in case of large deformations.

A comprehensive survey on many recent techniques for modeling the kinetostatic and dynamic behavior of flexure-based compliant mechanisms has been recently presented [26].

Kinetostatic models of complex plane compliant mechanisms have been developed in both micro and macro scale devices by using a wide variety of different linear and non-linear methods, such as, for only representative example, those based on: Loop-closure equations and the static equilibrium conditions for multi-loops compliant mechanisms [27]; chained beam constraint model and geometric parameter optimization, specially conceived for translational motion [28]; a combination of a beam constraint model, load equilibrium conditions, and geometric compatibility equations, specially conceived for 3-PPR compliant parallel mechanisms [29]; a kinetostatic modeling approach that integrates the screw theory with the energy method, with consequent avoidance of the problem of finding a solution to the equilibrium equations of nodal forces and the possibility of taking into account the parasitic deformations in space [30]; an extension of the chained beam constraint model specially revisited to analyze flexible beams of effective variable length [31]; a mathematical formulation of the compliance matrix method, combined with the inverse kinematic, specially introduced for modeling the flexure-based parallel compliant mechanisms with multiple actuation forces [32]; the adoption of three representations of multiple segments 2D beam models, namely, beam constraint model, linear Euler–Bernoulli beam, and PRBM [33]; the adoption of a new two-colored digraph representation of planar flexure-based compliant mechanisms for the automatic generation of the kinetostatic equations [34]; the introduction of virtual flexure hinges, link-flexure incidence matrices, and path matrices to generate automatically the formulation of the kinetostatic equations [35].

A more specific contribution has been dedicated in 2013 [36] to the solution of the problem of inverse kinetostatic analysis of a compliant four-bar linkage with flexible circular joints and pseudo-rigid bodies. This problem was attacked by extensively applying the theory of curved beams to the flexible parts, which gave rise to the closed-form symbolic expression of the compliance matrix, and by applying the static balance equations to both the elastic and pseudo-rigid parts.

The present investigation explores the opposite problem of direct kinetostatic analysis of a planar gripper with circular flexures. Two possible models based on the static balance of flexures are provided. The first linear model considers the static equilibrium in the undeformed configuration, while the second considers the balance in the deformed configuration. Both models simplify the fully non-linear model by exploiting a constant stiffness matrix of the undeformed curved beam element. These models allow us to find the tangent stiffness matrix in closed form as the sum of different contributions. Furthermore, dividing the tangent stiffness matrix into its contributions allows for evaluating each term's importance and setting strategies to speed up convergence. Furthermore, through a validation process of the fully non-linear model results performed on a case study, it will be possible to ascertain how the simplifications still provide accurate values in almost the entire mechanism's range of motion. As known, the tangent stiffness matrix is the heart of an iterative solving method. It is the basis of many implicit integrators widely used for the study of flexible mechanisms, such as the generalized α-method [37] or the HHT-method [38].

The main target of this article is to create two simplified models:

- Solving the problem of the direct kinetostatic analysis of planar grippers with curved beams;
- Being reliable in terms of motion accuracy and actuation forces;
- Being computationally efficient to extend the formulation for real-time applications.

Any Finite Element Analysis (FEA) or Multibody Dynamics Simulation (MBDS) package is very reliable for solving any general problem in kinetostatic analysis numerically. Despite this, the availability of a ready-to-use independent algorithm to solve the direct kinetostatic problem gives rise to the possibility of implementing it in any real-time applications. Nevertheless, the MBDS Adams software has been used herein for validation purposes.

The paper is divided into the following sections. Section 2 gives the fundamentals of the curved beam model. Section 3 outlines the kinetostatic analysis. Two simplified linear and partial non-linear models are developed, and their tangent stiffness matrices are obtained in closed form. Section 4 includes a detailed case study description. Section 5 compares and validates the two models and gives essential insights into convergence and computational burden. Finally, Section 6 gives the concluding remarks.

2. The Adopted Curved Beam Model

Flexures employed in this context are curved beams. It has been demonstrated that curved beams can provide large rotations while maintaining small errors in terms of displacements of its center [39], as it is typical for classic revolute pairs. This feature is important to guarantee finite rotations of the end-effector in monolithic structures such as MEMS-based grippers. Furthermore, a linear model is capable of faithfully reproducing the displacements and in-plane rotation of the curved beam tip up to rotations of approximately $\pm 20°$. This feature has the considerable advantage of using a constant stiffness matrix, as will be recalled below.

In the following, the curved beam compliance matrix, and its inverse stiffness matrix, will be recalled from [36]. Let us consider a curved beam with a circular profile of radius r_f and beam characteristic angle θ_f, as displayed in Figure 1. First, let us consider the generalized displacement array $\boldsymbol{\psi}_f = [\hat{\boldsymbol{\xi}}_f^T, \phi_f]^T$ containing the displacement $\hat{\boldsymbol{\xi}}_f$ and the rotation angle ϕ_f of the end section due to the deformation. Then, introducing the generalized wrench array $\mathbf{w}_f = [\mathbf{F}_f^T, M_f]^T$ containing the force vector and the torque applied to the end section, the compliance matrix $\mathbf{C}_f$, derived in [36], follows from

$$\boldsymbol{\psi}_f = \mathbf{C}_f \mathbf{w}_f \tag{1}$$

and depends only on the geometric and structural parameters of the curved beam, i.e.,

$$\mathbf{C}_f = \frac{1}{EI} \begin{bmatrix} \frac{r_f^3}{4}(6\theta_f + s(2\theta_f) - 8s(\theta_f)) & \frac{r_f^3}{2}(c^2(\theta_f) - 2c(\theta_f) + 1) & r_f^2(\theta_f - s(\theta_f)) \\ \cdots & \frac{r_f^3}{4}(2\theta_f - s(2\theta_f)) & r_f^2(1 - c(\theta_f)) \\ \text{(sym)} & \cdots & r_f\theta_f \end{bmatrix} \tag{2}$$

where E is Young's modulus and I is the area moment of inertia, assumed both constant for the circular profile. In the following sections, the inverse of the compliance matrix $\mathbf{C}_f$, i.e., the stiffness matrix $\mathbf{K}_f$ will be employed to write the kinetostatic equations of planar mechanisms with curved beams. Furthermore, the stiffness matrix will be expressed in its locale frame $\hat{\mathcal{S}}_f$ attached to the end-section of the curved beam in the undeformed configuration, as shown in Figure 1.

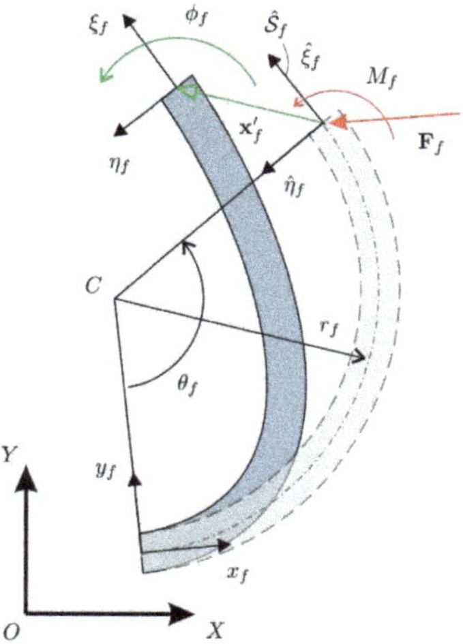

Figure 1. Curved beam in its initial (dashed line) and deformed configuration (solid line).

3. Kinetostatic Analysis

Hereafter, all vectors denoted with the *hat* will refer to the undeformed configuration, while the same vectors will indicate the deformed configuration without the *hat*. Position vectors of local frames as well as rotation matrices of frames describing the orientation of bodies in the undeformed configuration are constant.

A curved beam links two components, as it happens for the two bodies displayed in Figure 2. First, consider the undeformed system composed of two rigid bodies, identified by the reference frames $\hat{\mathcal{S}}_i$ and $\hat{\mathcal{S}}_j$, and by the flexure $\hat{f}$. The vectors $\hat{\mathbf{r}}_i$ and $\hat{\mathbf{r}}_j$ denote the positions of the body reference frame origins with respect to the fixed reference frame $\boldsymbol{\Sigma}$. In contrast, the vectors $\hat{\mathbf{s}}_{if}$ and $\hat{\mathbf{s}}_{jf}$, respectively, indicate the distance vectors going from the body-reference frame origins to the attachment points of the curved beam to the bodies.

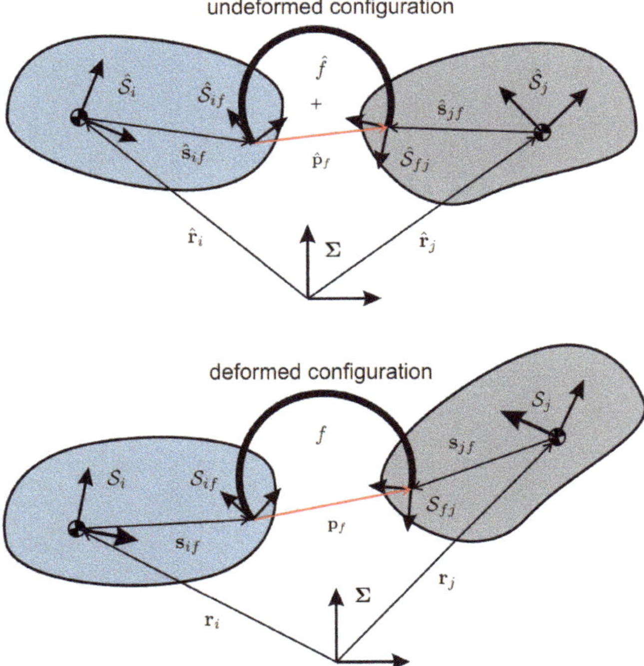

Figure 2. Deformation of a curved beam due to the relative motions of the bodies connected to its extremities.

In the undeformed configuration, the position vector $\hat{\mathbf{p}}_f$ going from the attachment point on body i to that on body j is obtained through the following expression

$$\hat{\mathbf{p}}_f = \hat{\mathbf{r}}_j + \hat{\mathbf{s}}_{jf} - \hat{\mathbf{r}}_i - \hat{\mathbf{s}}_{if} \tag{3}$$

Then, consider a generic configuration in which the bodies undergo finite displacements and rotations, and the flexure is deformed. For the assumption of rigid bodies, it follows that $\hat{\mathbf{s}}_{if}^{(\hat{\mathcal{S}}_i)} \equiv \mathbf{s}_{if}^{(\mathcal{S}_i)} \equiv \bar{\mathbf{s}}_{if}$ and $\hat{\mathbf{s}}_{jf}^{(\hat{\mathcal{S}}_j)} \equiv \mathbf{s}_{jf}^{(\mathcal{S}_j)} \equiv \bar{\mathbf{s}}_{jf}$ where the superscript denotes the reference frame in which the vector is expressed. In the previous expressions, $\bar{\mathbf{s}}_{if}$ and $\bar{\mathbf{s}}_{jf}$ have been introduced to simplify the notation. Then, the following closure equation stands,

$$\mathbf{r}_i + \mathbf{A}_i\bar{\mathbf{s}}_{if} + \mathbf{p}_f - \mathbf{r}_j - \mathbf{A}_j\bar{\mathbf{s}}_{jf} = \mathbf{0} \tag{4}$$

where $\mathbf{A}_i$ and $\mathbf{A}_j$ are the rotation matrices mapping $\mathcal{S}_i$ and $\mathcal{S}_j$ into $\boldsymbol{\Sigma}$ and $\mathbf{p}_f$ is the distance vector between the two flexure extremities in the deformed configuration.

Let us introduce the deformation vector $\mathbf{x}'_f$ containing the deformations of the flexure due to the displaced configuration described in Section 2. As known from the continuum mechanics, this vector can be represented using either the material or the spatial description of motion. In the following, only the material description is implemented. Therefore, expressing $\mathbf{x}'_f$ in the frame $\hat{\mathcal{S}}_{if}$ of Figure 2, it follows

$$\mathbf{x}'_f = \hat{\mathbf{A}}_i \mathbf{A}_i^T \mathbf{p}_f - \hat{\mathbf{p}}_f \tag{5}$$

where $\mathbf{p}_f$ has been pulled back to the undeformed configuration as required in the material description of motion.

Then, as recalled in Section 2, the circular flexure model requires $\mathbf{x}'_f$ to be expressed in the frame $\hat{\mathcal{S}}_{fj}$ instead of $\hat{\mathcal{S}}_{if}$, therefore

$$\mathbf{x}_f^{\prime(\hat{\mathcal{S}}_{fj})} = \hat{\mathbf{A}}_{fj}^T \hat{\mathbf{A}}_j^T \mathbf{x}'_f \tag{6}$$

where $\hat{\mathbf{A}}_{fj}$ is the constant rotation matrix mapping $\hat{\mathcal{S}}_{fj}$ into $\hat{\mathcal{S}}_j$. The rotation angle ϕ_f due to the flexure deformation reads

$$\phi_f = \theta_j - \theta_i - \hat{\theta}_j + \hat{\theta}_i \equiv \Delta\theta_{ij} - \Delta\hat{\theta}_{ij} \tag{7}$$

where θ and $\hat{\theta}$, respectively, are the rotation angles of the bodies in the spatial and material configurations and $\Delta\theta$, $\Delta\hat{\theta}$ denote the corresponding relative rotation angles. The generalized displacement array of the curved beam f, already introduced in Section 2, becomes

$$\boldsymbol{\psi}_f^{(\hat{\mathcal{S}}_{fj})} = \begin{bmatrix} \mathbf{x}_f^{\prime(\hat{\mathcal{S}}_{fj})} \\ \phi_f \end{bmatrix} \tag{8}$$

3.1. *Jacobian of the Deformation Vector*

Since the direct kinetostatic analysis will be solved using an iterative procedure, the variation of the generalized displacement array must be calculated. Using Equation (5), the variation $\delta\mathbf{x}'_f$ is

$$\delta\mathbf{x}'_f = \hat{\mathbf{A}}_i \bar{\mathbf{A}}_i^T \mathbf{p}_f \delta\theta_i + \hat{\mathbf{A}}_i \mathbf{A}_i^T (\delta\mathbf{r}_j - \delta\mathbf{r}_i + \bar{\mathbf{A}}_j \bar{\mathbf{s}}_{jf} \delta\theta_j - \bar{\mathbf{A}}_i \bar{\mathbf{s}}_{if} \delta\theta_i) \tag{9}$$

where $\bar{\mathbf{A}} = \partial\mathbf{A}/\partial\theta$ while $\delta\mathbf{r}$, $\delta\theta$ are the variations of the body coordinates in the deformed configuration. The variation $\delta\mathbf{x}'_f$ is evaluated in the reference frame $\boldsymbol{\Sigma}$ but can be easily expressed in $\hat{\mathcal{S}}_{fj}$ remembering that $\hat{\mathbf{A}}_{fj}$ and $\hat{\mathbf{A}}_j$ in Equation (6) are constant, i.e.,

$$\delta\mathbf{x}_f^{\prime(\hat{\mathcal{S}}_{fj})} = \hat{\mathbf{A}}_{fj}^T \hat{\mathbf{A}}_j^T \delta\mathbf{x}'_f \tag{10}$$

Considering the angles, the variation of Equation (7), leads to

$$\delta\phi_f = \delta\theta_j - \delta\theta_i \equiv \delta\Delta\theta_{ij} \tag{11}$$

Finally, the variations can be combined to form the variation of the generalized deformation vector $\delta\boldsymbol{\psi}_f$. The latter satisfies the following expression

$$\delta\boldsymbol{\psi}_f = \check{\mathbf{J}}_f \delta\mathbf{q}_{ij} \tag{12}$$

where $\delta\mathbf{q}_{ij} = [\delta\mathbf{q}_i^T \delta\mathbf{q}_j^T]^T$ is a 6-dimensional vector containing the variations of the body coordinates of bodies i and j and $\check{\mathbf{J}}_f$ is the (3×6) Jacobian matrix, defined as

$$\breve{\mathbf{J}}_f = \left[\ \breve{\mathbf{J}}_f^i \mid \breve{\mathbf{J}}_f^j\ \right] = \left[\begin{array}{cc|cc} -\hat{\mathbf{A}}_i\mathbf{A}_i^T & \hat{\mathbf{A}}_i(\bar{\mathbf{A}}_i^T\mathbf{p}_f - \tilde{\mathbf{1}}\bar{\mathbf{s}}_{if}) & \hat{\mathbf{A}}_i\mathbf{A}_i^T & \hat{\mathbf{A}}_i\mathbf{A}_i^T\bar{\mathbf{A}}_j\bar{\mathbf{s}}_{jf} \\ \mathbf{0}^T & -1 & \mathbf{0}^T & 1 \end{array} \right] \tag{13}$$

In deriving $\breve{\mathbf{J}}_f$, the property $\mathbf{A}_i^T\bar{\mathbf{A}}_i = \tilde{\mathbf{1}}$ has been employed, being

$$\tilde{\mathbf{1}} = \begin{bmatrix} 0 & -1 \\ 1 & 0 \end{bmatrix} \tag{14}$$

a particular skew-symmetric matrix used to define the cross-product in the planar case.

3.2. Kinetostatic Equations

The kinetostatic equations of the system require the static equilibrium of a curved beam. Consider the layout of Figure 3 showing the static balance of a curved beam connecting the bodies i and j. The deformation of the beam yields force and moment applied on the section $\hat{\mathcal{S}}_{fj}$ that must be equilibrated at section $\hat{\mathcal{S}}_{if}$. A first simplified model, hereafter referred to as the *linear model*, performs the balance in the undeformed configuration and leads to the following expressions

$$\begin{bmatrix} \mathbf{F}_{if}^{(\hat{\mathcal{S}}_{if})} \\ M_{if} \end{bmatrix} = -\underbrace{\begin{bmatrix} \mathbf{A}_f & \mathbf{0} \\ -(\mathbf{A}_f^T\mathbf{d}_f^{(\hat{\mathcal{S}}_{if})})^T\tilde{\mathbf{1}} & 1 \end{bmatrix}}_{\mathbf{T}_f} \begin{bmatrix} \mathbf{F}_{fj}^{(\hat{\mathcal{S}}_{fj})} \\ M_{fj} \end{bmatrix} \tag{15}$$

where $\mathbf{A}_f$ is the rotation matrix mapping $\hat{\mathcal{S}}_{fj}$ to $\hat{\mathcal{S}}_{if}$ and $\mathbf{d}_f$ is the distance vector between the two sections, respectively, defined as

$$\mathbf{A}_f = \begin{bmatrix} \cos(\theta_f) & -\sin(\theta_f) \\ \sin(\theta_f) & \cos(\theta_f) \end{bmatrix}, \quad \mathbf{d}_f^{(\hat{\mathcal{S}}_{if})} = \begin{bmatrix} r_f\sin(\theta_f) \\ r_f(1-\cos(\theta_f)) \end{bmatrix} \tag{16}$$

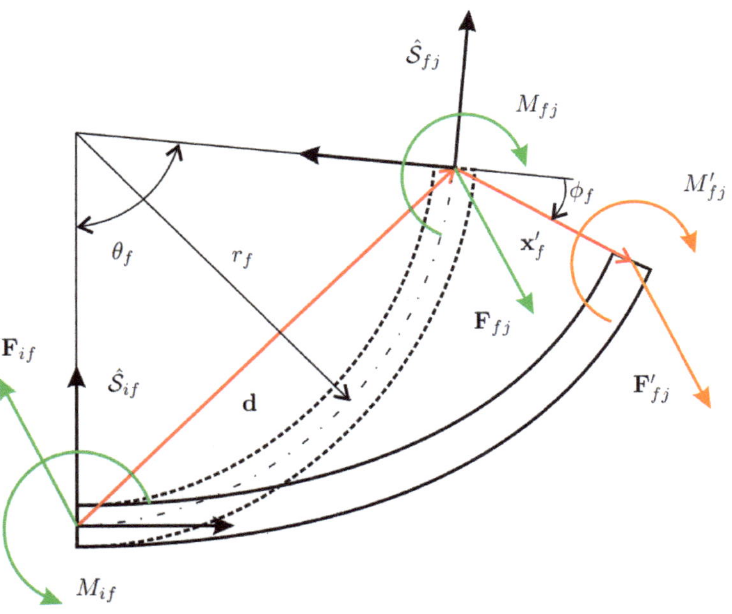

Figure 3. Static balance of a curved beam. Dashed line for the undeformed beam and a solid line for the deformed beam.

In Section 2, the stiffness matrix of the curved beam has been derived considering the undeformed configuration, meaning that the stiffness model is linear and cannot capture the geometrical non-linearity coming from the change of configuration during the beam deformation. Despite this, the balance in the deformed configuration can be modified including the tip displacement due to deformation. Referring to Figure 3, the tip displacement can be included in deriving the moment M_{if} at the first section $\hat{\mathcal{S}}_{if}$, therefore modifying the previous Equation (15) into

$$\begin{bmatrix} \mathbf{F}_{if}^{(\hat{\mathcal{S}}_{if})} \\ M_{if} \end{bmatrix} = -\underbrace{\begin{bmatrix} \mathbf{A}_f & \mathbf{0} \\ -(\mathbf{A}_f^T \mathbf{d}_f^{(\hat{\mathcal{S}}_{if})} + \mathbf{x}_f^{\prime(\hat{\mathcal{S}}_{fj})})^T \tilde{\mathbf{1}} & 1 \end{bmatrix}}_{\mathbf{T}'_f} \begin{bmatrix} \mathbf{F}_{fj}^{\prime(\hat{\mathcal{S}}_{fj})} \\ M'_{fj} \end{bmatrix} \tag{17}$$

where $\mathbf{F}'_{fj}$ and M'_{fj} are referred to the deformed configuration. It is noteworthy that this *partial non-linear model* is not the geometrically exact fully non-linear model of the curved beam since the forces and moment at section $\hat{\mathcal{S}}_{fj}$ are still obtained using a linear stiffness model for the curved beam.

In the following, either the linear or the partial non-linear model will be included to derive the kinetostatic equations of a planar mechanism. Let us consider the body i in its deformed configuration, as displayed in Figure 4. The flexures have been removed and replaced with their reaction forces and torques where the minus signs come from Newton's third law. The static balance of body i requires that the following system be satisfied

$$\mathbf{F}_i - \mathbf{F}_{1i} - \mathbf{F}_{i2} = \mathbf{0} \tag{18a}$$

$$M_i - M_{1i} - M_{i2} + \mathbf{s}_{i1}^T \tilde{\mathbf{1}} \mathbf{F}_{1i} + \mathbf{s}_{i2}^T \tilde{\mathbf{1}} \mathbf{F}_{i2} = 0 \tag{18b}$$

where $\mathbf{F}_i$ and M_i are the external force and torque applied to body i, respectively. From the balance Equations (15) and (17), the forces and moments coming from the flexures have been expressed in the undeformed configuration and are now turned into the deformed one. From Figure 4, it can be found that $\mathbf{A}_{1i}\mathbf{F}_{1i}^{(\mathcal{S}_{1i})} \equiv \hat{\mathbf{A}}_{1i}\hat{\mathbf{F}}_{1i}^{(\hat{\mathcal{S}}_{1i})}$ and $\mathbf{A}_{i2}\mathbf{F}_{i2}^{(\mathcal{S}_{i2})} \equiv \hat{\mathbf{A}}_{i2}\hat{\mathbf{F}}_{i2}^{(\hat{\mathcal{S}}_{i2})}$, hence it follows that

$$\mathbf{F}_i - \mathbf{A}_i\hat{\mathbf{A}}_{1i}\hat{\mathbf{F}}_{1i}^{(\hat{\mathcal{S}}_{1i})} - \mathbf{A}_i\hat{\mathbf{A}}_{i2}\hat{\mathbf{F}}_{i2}^{(\hat{\mathcal{S}}_{i2})} = \mathbf{0} \tag{19a}$$

$$M_i - M_{1i} - M_{i2} + \mathbf{s}_{i1}^T \tilde{\mathbf{1}} \mathbf{F}_{1i} + \mathbf{s}_{i2}^T \tilde{\mathbf{1}} \mathbf{F}_{i2} = 0 \tag{19b}$$

Considering the frame invariance of the scalar equation of moments, the final system reads

$$\mathbf{F}_i - \mathbf{A}_i\hat{\mathbf{A}}_{1i}\hat{\mathbf{F}}_{1i}^{(\hat{\mathcal{S}}_{1i})} - \mathbf{A}_i\hat{\mathbf{A}}_{i2}\hat{\mathbf{F}}_{i2}^{(\hat{\mathcal{S}}_{i2})} = \mathbf{0} \tag{20a}$$

$$M_i - \hat{M}_{1i} - \hat{M}_{i2} + \hat{\mathbf{s}}_{i1}^T \tilde{\mathbf{1}} \hat{\mathbf{A}}_{1i}\hat{\mathbf{F}}_{1i}^{(\hat{\mathcal{S}}_{1i})} + \hat{\mathbf{s}}_{i2}^T \tilde{\mathbf{1}} \hat{\mathbf{A}}_{i2}\hat{\mathbf{F}}_{i2}^{(\hat{\mathcal{S}}_{i2})} = 0 \tag{20b}$$

or in matrix form

$$\begin{bmatrix} \mathbf{F}_i \\ M_i \end{bmatrix} - \begin{bmatrix} \mathbf{A}_i\hat{\mathbf{A}}_{1i} & \mathbf{0} \\ -\hat{\mathbf{s}}_{i1}^T \tilde{\mathbf{1}} \hat{\mathbf{A}}_{1i} & 1 \end{bmatrix} \begin{bmatrix} \hat{\mathbf{F}}_{1i}^{(\hat{\mathcal{S}}_{1i})} \\ \hat{M}_{1i} \end{bmatrix} - \begin{bmatrix} \mathbf{A}_i\hat{\mathbf{A}}_{i2} & \mathbf{0} \\ -\hat{\mathbf{s}}_{i2}^T \tilde{\mathbf{1}} \hat{\mathbf{A}}_{i2} & 1 \end{bmatrix} \begin{bmatrix} \hat{\mathbf{F}}_{i2}^{(\hat{\mathcal{S}}_{i2})} \\ \hat{M}_{i2} \end{bmatrix} = \mathbf{0} \tag{21}$$

Then, denoting with

$$\mathbf{N}_{1i} = \begin{bmatrix} \mathbf{A}_i\hat{\mathbf{A}}_{1i} & \mathbf{0} \\ -\hat{\mathbf{s}}_{i1}^T \tilde{\mathbf{1}} \hat{\mathbf{A}}_{1i} & 1 \end{bmatrix}, \quad \mathbf{N}_{i2} = \begin{bmatrix} \mathbf{A}_i\hat{\mathbf{A}}_{i2} & \mathbf{0} \\ -\hat{\mathbf{s}}_{i2}^T \tilde{\mathbf{1}} \hat{\mathbf{A}}_{i2} & 1 \end{bmatrix} \tag{22}$$

and with $\mathbf{R}_i$ the residual vector of body i, the final kinetostatic model is

$$\mathbf{R}_i \equiv -\mathbf{w}_i + \mathbf{N}_{1i}\mathbf{K}_1^{(\hat{\mathcal{S}}_{1i})}\boldsymbol{\psi}_1^{(\hat{\mathcal{S}}_{1i})} + \mathbf{N}_{i2}\mathbf{T}'_2\mathbf{K}_2^{(\hat{\mathcal{S}}_{2j})}\boldsymbol{\psi}_2^{(\hat{\mathcal{S}}_{2j})} = 0 \tag{23}$$

where $\mathbf{w}_i = [\mathbf{F}_i^T, M_i]^T$ is the generalized force vector or wrench acting on body i, $\mathbf{K}_1$ and $\mathbf{K}_2$ are the stiffness matrices of the curved beams connected to the body, and $\mathbf{T}_2'$ is the matrix defined in Equation (17). If the latter is replaced by the matrix $\mathbf{T}_2$ of the linear model in Equation (15), the kinetostatic model turns into

$$\mathbf{R}_i \equiv -\mathbf{w}_i + \mathbf{N}_{1i}\mathbf{K}_1^{(\hat{\mathcal{S}}_{1i})}\boldsymbol{\psi}_1^{(\hat{\mathcal{S}}_{1i})} + \mathbf{N}_{i2}\mathbf{T}_2\mathbf{K}_2^{(\hat{\mathcal{S}}_{2j})}\boldsymbol{\psi}_2^{(\hat{\mathcal{S}}_{2j})} = \mathbf{0} \tag{24}$$

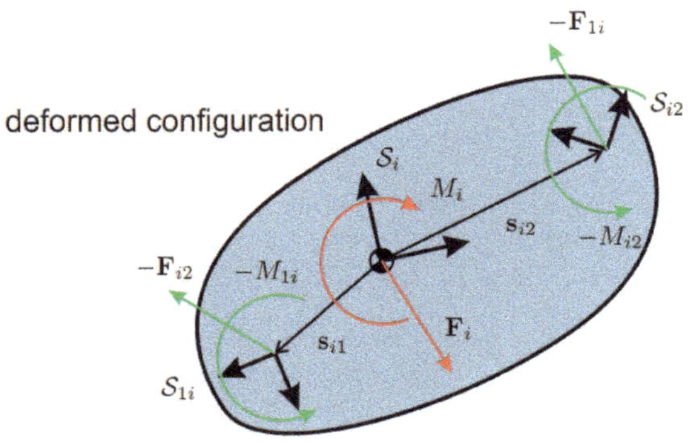

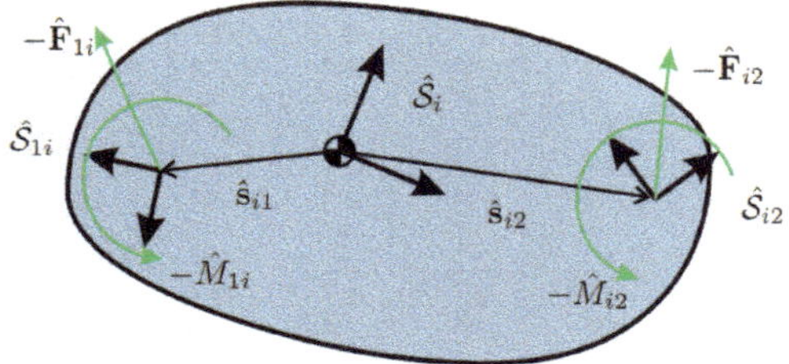

Figure 4. Kinetostatic balance of a rigid-body.

The kinetostatic equations of a complex multibody system are derived by assembling the residual vectors of all rigid bodies. The final system can be cast in the form

$$\mathbf{R}(\mathbf{w}, \mathbf{q}) = \mathbf{0}, \tag{25}$$

where $\mathbf{R}$ indicates the global residual vector, $\mathbf{w}$ is the global wrench of external forces and torques, and $\mathbf{q}$ is the global vector of generalized coordinates. The elastostatics model offers two types of analyses: the inverse and the direct kinetostatic analysis. The deformed configuration is the input, and the global wrench is the output in the inverse analysis, as has been already described in [36]. The solution of the inverse kinetostatic analysis is straightforward and does not require an iterative procedure. In the direct kinetostatic analysis, the forces and moments applied to the system are known, while the final configuration of the deformed mechanism is sought. This highly non-linear problem can be solved using an iterative procedure such as the Newton–Raphson method described in Algorithm 1.

Algorithm 1 : Newton–Raphson iterative method

1: $\epsilon \leftarrow$ given threshold to terminate iterations
2: $k = 1 \leftarrow$ iteration number
3: $\mathbf{q}^{(k)} = \hat{\mathbf{q}} \leftarrow$ set the solution guess value
4: **procedure** NEWTON–RAPHSON ITERATIVE METHOD($\mathbf{q}^{(k)}$,$\hat{\mathbf{q}}$,$\mathbf{w}$)
5:     $\mathbf{R}(\mathbf{w}, \mathbf{q}^{(k)}) \leftarrow$ calculate the residuals as in Equation (24)
6:     $\mathbf{K}_T(\mathbf{q}^{(k)}) \leftarrow$ calculate the tangent stiffness matrix as in Equation (26)
7:     $\mathbf{q}^{(k+1)} = \mathbf{q}^{(k)} - \mathbf{K}_T^{(k)^{-1}} \mathbf{R}(\mathbf{w}, \mathbf{q}^{(k)}) \leftarrow$ update the solution
8:     **if** $\|\mathbf{q}^{(k+1)} - \mathbf{q}^{(k)}\| < \epsilon$ **then**
9:         $\mathbf{q} = \mathbf{q}^{(k+1)} \leftarrow$ DKP solution
10:         exit **procedure**
11:     **else**
12:         $k \leftarrow k + 1$
13:         **goto** step 4
14:     **end if**
15: **end procedure**

3.3. *Tangent Stiffness Matrix Determination*

Suppose that the external forces and moments are fixed in space. Then, considering the residual of the partial non-linear model of Equation (23), the tangent stiffness matrix is

$$\mathbf{K}_{Ti} = \frac{\partial \mathbf{R}_i}{\partial \mathbf{q}} \equiv \mathbf{K}_{Ti}^{I} + \mathbf{K}_{Ti}^{II} + \mathbf{K}_{Ti}^{III} \tag{26}$$

where

$$\mathbf{K}_{Ti}^{I} = \mathbf{N}_{1i}\mathbf{K}_1^{(\hat{\mathcal{S}}_{1i})} \frac{\partial \boldsymbol{\psi}_1^{(\hat{\mathcal{S}}_{1i})}}{\partial \mathbf{q}} + \mathbf{N}_{i2}\mathbf{T}_2'\mathbf{K}_2^{(\hat{\mathcal{S}}_{2j})} \frac{\partial \boldsymbol{\psi}_2^{(\hat{\mathcal{S}}_{2j})}}{\partial \mathbf{q}} \tag{27a}$$

$$\mathbf{K}_{Ti}^{II} = \frac{\partial \mathbf{N}_{1i}}{\partial \mathbf{q}}\mathbf{K}_1^{(\hat{\mathcal{S}}_{1i})} \boldsymbol{\psi}_1^{(\hat{\mathcal{S}}_{1i})} + \frac{\partial \mathbf{N}_{i2}}{\partial \mathbf{q}}\mathbf{T}_2'\mathbf{K}_2^{(\hat{\mathcal{S}}_{2j})} \boldsymbol{\psi}_2^{(\hat{\mathcal{S}}_{2j})} \tag{27b}$$

$$\mathbf{K}_{Ti}^{III} = \mathbf{N}_{i2}\frac{\partial \mathbf{T}_2'}{\partial \mathbf{q}}\mathbf{K}_2^{(\hat{\mathcal{S}}_{2j})} \boldsymbol{\psi}_2^{(\hat{\mathcal{S}}_{2j})} \tag{27c}$$

If the residual of the linear model of Equation (24) is used instead, the tangent stiffness matrix turns into

$$\mathbf{K}_{Ti} = \frac{\partial \mathbf{R}_i}{\partial \mathbf{q}} \equiv \mathbf{K}_{Ti}^{I} + \mathbf{K}_{Ti}^{II} \tag{28}$$

The Appendix A reports the expressions for the terms of $\mathbf{K}_{Ti}$.

4. Case Study: The Four-Bar Linkage

In this section, a compliant gripper with curved beams, shown in Figure 5a, is studied. The monolithic structure of the mechanism can be reduced to two in-parallel four-bar linkages, as revealed in Figure 5b. For symmetry along the vertical axis, only half a mechanism will be analyzed, i.e., the right side of Figure 5b. The layout of Figure 6 has been plotted following the notation presented in Section 3. Body 1 is the frame, here considered fixed, while bodies 2, 3, and 4 are moving rigid bodies. Considering grippers with CSFH joints, one link is a part of the monolithic structure enclosed between two circular beams. In an MEMS device, the entire monolithic structure can deform. However, the assumption of rigid links coupled to flexible circular beams has been verified using FE models. It is fully justified since the links are at least one order of magnitude stiffer than the circular beam flexures.

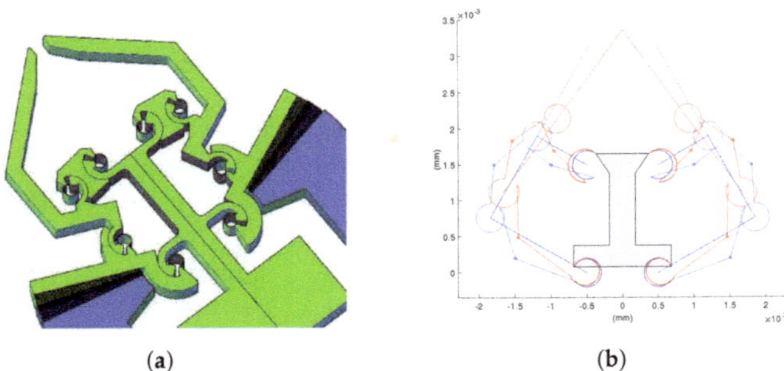

(**a**) (**b**)

Figure 5. MEMS-based gripper with curved beams. (**a**) CAD layout. (**b**) Matlab model: (red) deformed configuration, (blue) undeformed configuration.

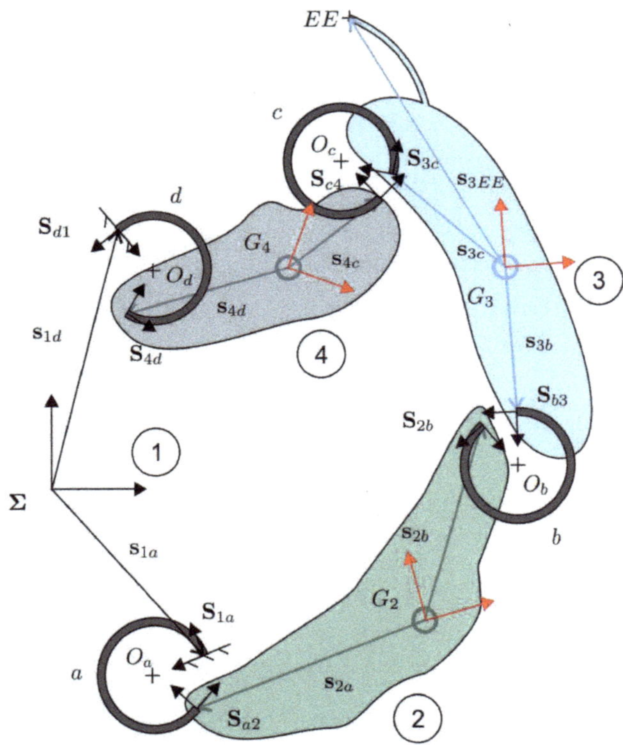

Figure 6. Layout of a four-bar linkage with curved beams.

First, vectors $\hat{\mathbf{p}}_f$ can be calculated knowing the undeformed configuration, i.e., $\hat{\mathbf{q}}$, therefore

$$\hat{\mathbf{p}}_a = \hat{\mathbf{r}}_2 - \hat{\mathbf{r}}_1 + \hat{\mathbf{A}}_2\hat{\mathbf{s}}_{2a} - \hat{\mathbf{A}}_1\hat{\mathbf{s}}_{1a} \tag{29a}$$

$$\hat{\mathbf{p}}_b = \hat{\mathbf{r}}_3 - \hat{\mathbf{r}}_2 + \hat{\mathbf{A}}_3\hat{\mathbf{s}}_{3b} - \hat{\mathbf{A}}_2\hat{\mathbf{s}}_{2b} \tag{29b}$$

$$\hat{\mathbf{p}}_c = \hat{\mathbf{r}}_4 - \hat{\mathbf{r}}_3 + \hat{\mathbf{A}}_4\hat{\mathbf{s}}_{4c} - \hat{\mathbf{A}}_3\hat{\mathbf{s}}_{3c} \tag{29c}$$

$$\hat{\mathbf{p}}_d = \hat{\mathbf{r}}_1 - \hat{\mathbf{r}}_4 + \hat{\mathbf{A}}_1\hat{\mathbf{s}}_{1d} - \hat{\mathbf{A}}_4\hat{\mathbf{s}}_{4d} \tag{29d}$$

Similarly, the vectors $\mathbf{p}_f$ of the flexures in the deformed configuration, i.e., $\mathbf{q}$, are

$$\mathbf{p}_a = \mathbf{r}_2 - \mathbf{r}_1 + \mathbf{A}_2\mathbf{s}_{2a} - \mathbf{A}_1\mathbf{s}_{1a} \tag{30a}$$
$$\mathbf{p}_b = \mathbf{r}_3 - \mathbf{r}_2 + \mathbf{A}_3\mathbf{s}_{3b} - \mathbf{A}_2\mathbf{s}_{2b} \tag{30b}$$
$$\mathbf{p}_c = \mathbf{r}_4 - \mathbf{r}_3 + \mathbf{A}_4\mathbf{s}_{4c} - \mathbf{A}_3\mathbf{s}_{3c} \tag{30c}$$
$$\mathbf{p}_d = \mathbf{r}_1 - \mathbf{r}_4 + \mathbf{A}_1\mathbf{s}_{1d} - \mathbf{A}_4\mathbf{s}_{4d} \tag{30d}$$

Since body 1 is the frame, $\mathbf{r}_1 = \hat{\mathbf{r}}_1$ and $\mathbf{A}_1 = \hat{\mathbf{A}}_1$. Furthermore, if the frame $\mathbf{S}_1$ of body 1 is coincident with $\boldsymbol{\Sigma}$, it follows that $\mathbf{r}_1 = \mathbf{0}$ and $\mathbf{A}_1 = \mathbf{1}$. Here, these matrices are written for the sake of completeness.

Notice that $\mathbf{q}$ could be one of the iterative solutions $\mathbf{q}^{(k)}$ employed in the Newton–Raphson algorithm. The deformation vectors $\mathbf{x}'_f$ in the material description and expressed in the local frames of the undeformed flexures are

$$\mathbf{x}_a'^{(\hat{\mathcal{S}}_{a2})} = \hat{\mathbf{A}}_{a2}^T\hat{\mathbf{A}}_2^T(\hat{\mathbf{A}}_1\mathbf{A}_1^T\mathbf{p}_a - \hat{\mathbf{p}}_a) \tag{31a}$$
$$\mathbf{x}_b'^{(\hat{\mathcal{S}}_{b3})} = \hat{\mathbf{A}}_{b3}^T\hat{\mathbf{A}}_3^T(\hat{\mathbf{A}}_2\mathbf{A}_2^T\mathbf{p}_b - \hat{\mathbf{p}}_b) \tag{31b}$$
$$\mathbf{x}_c'^{(\hat{\mathcal{S}}_{c4})} = \hat{\mathbf{A}}_{c4}^T\hat{\mathbf{A}}_4^T(\hat{\mathbf{A}}_3\mathbf{A}_3^T\mathbf{p}_c - \hat{\mathbf{p}}_c) \tag{31c}$$
$$\mathbf{x}_d'^{(\hat{\mathcal{S}}_{d1})} = \hat{\mathbf{A}}_{d1}^T\hat{\mathbf{A}}_1^T(\hat{\mathbf{A}}_4\mathbf{A}_4^T\mathbf{p}_d - \hat{\mathbf{p}}_d) \tag{31d}$$

The angular deformation ϕ_f is obtained as

$$\phi_a = \theta_2 - \theta_1 - \hat{\theta}_2 + \hat{\theta}_1 \tag{32a}$$
$$\phi_b = \theta_3 - \theta_2 - \hat{\theta}_3 + \hat{\theta}_2 \tag{32b}$$
$$\phi_c = \theta_4 - \theta_3 - \hat{\theta}_4 + \hat{\theta}_3 \tag{32c}$$
$$\phi_d = \theta_1 - \theta_4 - \hat{\theta}_1 + \hat{\theta}_4 \tag{32d}$$

The expressions (31) and (32) allows for determining the flexure generalized deformations $\boldsymbol{\psi}_a^{(\hat{\mathcal{S}}_{a2})}$, $\boldsymbol{\psi}_b^{(\hat{\mathcal{S}}_{b3})}$, $\boldsymbol{\psi}_c^{(\hat{\mathcal{S}}_{c4})}$, and $\boldsymbol{\psi}_d^{(\hat{\mathcal{S}}_{d1})}$.

The transformation matrices $\mathbf{T}'_f$ of Equation (17) are defined as

$$\mathbf{T}'_a = -\begin{bmatrix} \mathbf{A}_a & \mathbf{0} \\ -(\mathbf{A}_a^T\mathbf{d}_a^{(\hat{\mathcal{S}}_{1a})} + \mathbf{x}_a'^{(\hat{\mathcal{S}}_{a2})})^T\tilde{\mathbf{1}} & 0 \end{bmatrix} \tag{33a}$$
$$\mathbf{T}'_b = -\begin{bmatrix} \mathbf{A}_b & \mathbf{0} \\ -(\mathbf{A}_b^T\mathbf{d}_b^{(\hat{\mathcal{S}}_{2b})} + \mathbf{x}_b'^{(\hat{\mathcal{S}}_{b3})})^T\tilde{\mathbf{1}} & 0 \end{bmatrix} \tag{33b}$$
$$\mathbf{T}'_c = -\begin{bmatrix} \mathbf{A}_c & \mathbf{0} \\ -(\mathbf{A}_c^T\mathbf{d}_c^{(\hat{\mathcal{S}}_{3c})} + \mathbf{x}_c'^{(\hat{\mathcal{S}}_{c4})})^T\tilde{\mathbf{1}} & 0 \end{bmatrix} \tag{33c}$$
$$\mathbf{T}'_d = -\begin{bmatrix} \mathbf{A}_d & \mathbf{0} \\ -(\mathbf{A}_d^T\mathbf{d}_d^{(\hat{\mathcal{S}}_{4d})} + \mathbf{x}_d'^{(\hat{\mathcal{S}}_{d1})})^T\tilde{\mathbf{1}} & 0 \end{bmatrix} \tag{33d}$$

where $\mathbf{A}_f$ and $\mathbf{d}_f$ can be found for each curved beam using Equation (16). Expressions similar to $\mathbf{T}'_f$, not reported for brevity, can be written to determine $\mathbf{T}_f$ of Equation (15).

Then, the matrices $\mathbf{N}$ of Equation (22) are

$$\mathbf{N}_{d1} = \begin{bmatrix} \mathbf{A}_1\hat{\mathbf{A}}_{d1} & \mathbf{0} \\ -\hat{\mathbf{s}}_{1d}^T\tilde{\mathbf{1}}\hat{\mathbf{A}}_{d1} & 1 \end{bmatrix}, \quad \mathbf{N}_{1a} = \begin{bmatrix} \mathbf{A}_1\hat{\mathbf{A}}_{1a} & \mathbf{0} \\ -\hat{\mathbf{s}}_{1a}^T\tilde{\mathbf{1}}\hat{\mathbf{A}}_{1a} & 1 \end{bmatrix} \tag{34a}$$
$$\mathbf{N}_{a2} = \begin{bmatrix} \mathbf{A}_2\hat{\mathbf{A}}_{a2} & \mathbf{0} \\ -\hat{\mathbf{s}}_{2a}^T\tilde{\mathbf{1}}\hat{\mathbf{A}}_{a2} & 1 \end{bmatrix}, \quad \mathbf{N}_{2b} = \begin{bmatrix} \mathbf{A}_2\hat{\mathbf{A}}_{2b} & \mathbf{0} \\ -\hat{\mathbf{s}}_{2b}^T\tilde{\mathbf{1}}\hat{\mathbf{A}}_{2b} & 1 \end{bmatrix} \tag{34b}$$
$$\mathbf{N}_{b3} = \begin{bmatrix} \mathbf{A}_3\hat{\mathbf{A}}_{b3} & \mathbf{0} \\ -\hat{\mathbf{s}}_{3b}^T\tilde{\mathbf{1}}\hat{\mathbf{A}}_{b3} & 1 \end{bmatrix}, \quad \mathbf{N}_{3c} = \begin{bmatrix} \mathbf{A}_3\hat{\mathbf{A}}_{3c} & \mathbf{0} \\ -\hat{\mathbf{s}}_{3c}^T\tilde{\mathbf{1}}\hat{\mathbf{A}}_{3c} & 1 \end{bmatrix} \tag{34c}$$
$$\mathbf{N}_{c4} = \begin{bmatrix} \mathbf{A}_4\hat{\mathbf{A}}_{c4} & \mathbf{0} \\ -\hat{\mathbf{s}}_{4c}^T\tilde{\mathbf{1}}\hat{\mathbf{A}}_{c4} & 1 \end{bmatrix}, \quad \mathbf{N}_{4d} = \begin{bmatrix} \mathbf{A}_4\hat{\mathbf{A}}_{4d} & \mathbf{0} \\ -\hat{\mathbf{s}}_{4d}^T\tilde{\mathbf{1}}\hat{\mathbf{A}}_{4d} & 1 \end{bmatrix} \tag{34d}$$

Setting the external wrenches $\mathbf{w}_i$ applied at the mass centers G_i of the rigid bodies, the four residual vectors $\mathbf{R}_i$, $i = 1, \ldots, 4$, are

$$\mathbf{R}_1 \equiv -\mathbf{w}_1 + \mathbf{N}_{d1}\mathbf{K}_d^{(\hat{\mathcal{S}}_{d1})}\boldsymbol{\psi}_d^{(\hat{\mathcal{S}}_{d1})} + \mathbf{N}_{1a}\mathbf{T}'_a\mathbf{K}_a^{(\hat{\mathcal{S}}_{a2})}\boldsymbol{\psi}_a^{(\hat{\mathcal{S}}_{a2})} \tag{35a}$$

$$\mathbf{R}_2 \equiv -\mathbf{w}_2 + \mathbf{N}_{a2}\mathbf{K}_a^{(\hat{\mathcal{S}}_{a2})}\boldsymbol{\psi}_a^{(\hat{\mathcal{S}}_{a2})} + \mathbf{N}_{2b}\mathbf{T}'_b\mathbf{K}_b^{(\hat{\mathcal{S}}_{b3})}\boldsymbol{\psi}_b^{(\hat{\mathcal{S}}_{b3})} \tag{35b}$$

$$\mathbf{R}_3 \equiv -\mathbf{w}_3 + \mathbf{N}_{b3}\mathbf{K}_b^{(\hat{\mathcal{S}}_{b3})}\boldsymbol{\psi}_b^{(\hat{\mathcal{S}}_{b3})} + \mathbf{N}_{3c}\mathbf{T}'_c\mathbf{K}_c^{(\hat{\mathcal{S}}_{c4})}\boldsymbol{\psi}_c^{(\hat{\mathcal{S}}_{c4})} \tag{35c}$$

$$\mathbf{R}_4 \equiv -\mathbf{w}_4 + \mathbf{N}_{c4}\mathbf{K}_c^{(\hat{\mathcal{S}}_{c4})}\boldsymbol{\psi}_c^{(\hat{\mathcal{S}}_{c4})} + \mathbf{N}_{4d}\mathbf{T}'_d\mathbf{K}_d^{(\hat{\mathcal{S}}_{d1})}\boldsymbol{\psi}_d^{(\hat{\mathcal{S}}_{d1})} \tag{35d}$$

The residuals are employed to form a system of 12 non-linear kinetostatic equations, i.e.,

$$\mathbf{R} \equiv [\mathbf{R}_1^T, \mathbf{R}_2^T, \mathbf{R}_3^T, \mathbf{R}_4^T]^T = \mathbf{0} \tag{36}$$

that must be solved using an iterative procedure. To calculate the tangent stiffness matrix necessary to apply the Newton–Raphson algorithm, let us define the Jacobians $\check{\mathbf{J}}_f^{(\hat{\mathcal{S}}_{fj})}$, i.e.,

$$\check{\mathbf{J}}_a^{(\hat{\mathcal{S}}_{a2})} \equiv \left[\ \check{\mathbf{J}}_a^{1(\hat{\mathcal{S}}_{a2})} \mid \check{\mathbf{J}}_a^{2(\hat{\mathcal{S}}_{a2})}\ \right] = \begin{bmatrix} \hat{\mathbf{A}}_{a2}^T\hat{\mathbf{A}}_2^T & \mathbf{0} \\ \mathbf{0}^T & 1 \end{bmatrix}\check{\mathbf{J}}_a \tag{37a}$$

$$\check{\mathbf{J}}_b^{(\hat{\mathcal{S}}_{b3})} \equiv \left[\ \check{\mathbf{J}}_b^{2(\hat{\mathcal{S}}_{b3})} \mid \check{\mathbf{J}}_b^{3(\hat{\mathcal{S}}_{b3})}\ \right] = \begin{bmatrix} \hat{\mathbf{A}}_{b3}^T\hat{\mathbf{A}}_3^T & \mathbf{0} \\ \mathbf{0}^T & 1 \end{bmatrix}\check{\mathbf{J}}_b \tag{37b}$$

$$\check{\mathbf{J}}_c^{(\hat{\mathcal{S}}_{c4})} \equiv \left[\ \check{\mathbf{J}}_c^{3(\hat{\mathcal{S}}_{c4})} \mid \check{\mathbf{J}}_c^{4(\hat{\mathcal{S}}_{c4})}\ \right] = \begin{bmatrix} \hat{\mathbf{A}}_{c4}^T\hat{\mathbf{A}}_4^T & \mathbf{0} \\ \mathbf{0}^T & 1 \end{bmatrix}\check{\mathbf{J}}_c \tag{37c}$$

$$\check{\mathbf{J}}_d^{(\hat{\mathcal{S}}_{d1})} \equiv \left[\ \check{\mathbf{J}}_d^{4(\hat{\mathcal{S}}_{d1})} \mid \check{\mathbf{J}}_d^{1(\hat{\mathcal{S}}_{d1})}\ \right] = \begin{bmatrix} \hat{\mathbf{A}}_{d1}^T\hat{\mathbf{A}}_1^T & \mathbf{0} \\ \mathbf{0}^T & 1 \end{bmatrix}\check{\mathbf{J}}_d \tag{37d}$$

where

$$\check{\mathbf{J}}_a = \left[\begin{array}{cc|cc} -\hat{\mathbf{A}}_1\mathbf{A}_1^T & \hat{\mathbf{A}}_1(\bar{\mathbf{A}}_1^T\mathbf{p}_a - \tilde{\mathbf{1}}\bar{\mathbf{s}}_{1a}) & \hat{\mathbf{A}}_1\mathbf{A}_1^T & \hat{\mathbf{A}}_1\mathbf{A}_1^T\bar{\mathbf{A}}_2\bar{\mathbf{s}}_{2a} \\ \mathbf{0}^T & -1 & \mathbf{0}^T & 1 \end{array}\right] \tag{38a}$$

$$\check{\mathbf{J}}_b = \left[\begin{array}{cc|cc} -\hat{\mathbf{A}}_2\mathbf{A}_2^T & \hat{\mathbf{A}}_2(\bar{\mathbf{A}}_2^T\mathbf{p}_b - \tilde{\mathbf{1}}\bar{\mathbf{s}}_{2b}) & \hat{\mathbf{A}}_2\mathbf{A}_2^T & \hat{\mathbf{A}}_2\mathbf{A}_2^T\bar{\mathbf{A}}_3\bar{\mathbf{s}}_{3b} \\ \mathbf{0}^T & -1 & \mathbf{0}^T & 1 \end{array}\right] \tag{38b}$$

$$\check{\mathbf{J}}_c = \left[\begin{array}{cc|cc} -\hat{\mathbf{A}}_3\mathbf{A}_3^T & \hat{\mathbf{A}}_3(\bar{\mathbf{A}}_3^T\mathbf{p}_c - \tilde{\mathbf{1}}\bar{\mathbf{s}}_{3c}) & \hat{\mathbf{A}}_3\mathbf{A}_3^T & \hat{\mathbf{A}}_3\mathbf{A}_3^T\bar{\mathbf{A}}_4\bar{\mathbf{s}}_{4c} \\ \mathbf{0}^T & -1 & \mathbf{0}^T & 1 \end{array}\right] \tag{38c}$$

$$\check{\mathbf{J}}_d = \left[\begin{array}{cc|cc} -\hat{\mathbf{A}}_4\mathbf{A}_4^T & \hat{\mathbf{A}}_4(\bar{\mathbf{A}}_4^T\mathbf{p}_d - \tilde{\mathbf{1}}\bar{\mathbf{s}}_{4d}) & \hat{\mathbf{A}}_4\mathbf{A}_4^T & \hat{\mathbf{A}}_4\mathbf{A}_4^T\bar{\mathbf{A}}_1\bar{\mathbf{s}}_{1d} \\ \mathbf{0}^T & -1 & \mathbf{0}^T & 1 \end{array}\right] \tag{38d}$$

The Jacobians $\check{\mathbf{J}}_f^{(\hat{\mathcal{S}}_{fj})}$ can be mapped using Boolean matrices to the final dimension of the system, i.e.,

$$\mathbf{J}_a^{(\hat{\mathcal{S}}_{a2})} = \left[\ \check{\mathbf{J}}_a^{1(\hat{\mathcal{S}}_{a2})} \mid \check{\mathbf{J}}_a^{2(\hat{\mathcal{S}}_{a2})} \mid \mathbf{O} \mid \mathbf{O}\ \right] \tag{39a}$$

$$\mathbf{J}_b^{(\hat{\mathcal{S}}_{b3})} = \left[\ \mathbf{O} \mid \check{\mathbf{J}}_b^{2(\hat{\mathcal{S}}_{b3})} \mid \check{\mathbf{J}}_b^{3(\hat{\mathcal{S}}_{b3})} \mid \mathbf{O}\ \right] \tag{39b}$$

$$\mathbf{J}_c^{(\hat{\mathcal{S}}_{c4})} = \left[\ \mathbf{O} \mid \mathbf{O} \mid \check{\mathbf{J}}_c^{3(\hat{\mathcal{S}}_{c4})} \mid \check{\mathbf{J}}_c^{4(\hat{\mathcal{S}}_{c4})}\ \right] \tag{39c}$$

$$\mathbf{J}_d^{(\hat{\mathcal{S}}_{d1})} = \left[\ \check{\mathbf{J}}_d^{1(\hat{\mathcal{S}}_{d1})} \mid \mathbf{O} \mid \mathbf{O} \mid \check{\mathbf{J}}_d^{4(\hat{\mathcal{S}}_{d1})}\ \right] \tag{39d}$$

Then, following the expression (A2) of $\mathbf{K}_{Ti}^{I}$ reported in the Appendix, it yields

$$\mathbf{K}^{I}_{T1} = \mathbf{N}_{d1}\mathbf{K}_{d}^{(\hat{\mathcal{S}}_{d1})}\mathbf{J}_{d}^{(\hat{\mathcal{S}}_{d1})} + \mathbf{N}_{1a}\mathbf{T}'_{a}\mathbf{K}_{a}^{(\hat{\mathcal{S}}_{a2})}\mathbf{J}_{a}^{(\hat{\mathcal{S}}_{a2})} \tag{40a}$$

$$\mathbf{K}^{I}_{T2} = \mathbf{N}_{a2}\mathbf{K}_{a}^{(\hat{\mathcal{S}}_{a2})}\mathbf{J}_{a}^{(\hat{\mathcal{S}}_{a2})} + \mathbf{N}_{2b}\mathbf{T}'_{b}\mathbf{K}_{b}^{(\hat{\mathcal{S}}_{b3})}\mathbf{J}_{b}^{(\hat{\mathcal{S}}_{b3})} \tag{40b}$$

$$\mathbf{K}^{I}_{T3} = \mathbf{N}_{b3}\mathbf{K}_{b}^{(\hat{\mathcal{S}}_{b3})}\mathbf{J}_{b}^{(\hat{\mathcal{S}}_{b3})} + \mathbf{N}_{3c}\mathbf{T}'_{c}\mathbf{K}_{c}^{(\hat{\mathcal{S}}_{c4})}\mathbf{J}_{c}^{(\hat{\mathcal{S}}_{c4})} \tag{40c}$$

$$\mathbf{K}^{I}_{T4} = \mathbf{N}_{c4}\mathbf{K}_{c}^{(\hat{\mathcal{S}}_{c4})}\mathbf{J}_{c}^{(\hat{\mathcal{S}}_{c4})} + \mathbf{N}_{4d}\mathbf{T}'_{d}\mathbf{K}_{d}^{(\hat{\mathcal{S}}_{d1})}\mathbf{J}_{d}^{(\hat{\mathcal{S}}_{d1})} \tag{40d}$$

The final expression for the first part of the tangent stiffness matrix is

$$\mathbf{K}^{I}_{T} = \begin{bmatrix} \mathbf{K}_{T1} \\ \mathbf{K}^{I}_{T2} \\ \mathbf{K}^{I}_{T3} \\ \mathbf{K}^{I}_{T4} \end{bmatrix} \tag{41}$$

The expressions for $\mathbf{K}^{II}_{Ti}$ can be obtained starting from $\mathbf{z}_i$ in Equation (A7), i.e.,

$$\mathbf{z}_1 = \mathbf{G}_{d1}\mathbf{K}_{d}^{(\hat{\mathcal{S}}_{d1})}\boldsymbol{\psi}_{d}^{(\hat{\mathcal{S}}_{d1})} + \mathbf{G}_{1a}\mathbf{T}'_{a}\mathbf{K}_{a}^{(\hat{\mathcal{S}}_{a2})}\boldsymbol{\psi}_{a}^{(\hat{\mathcal{S}}_{a2})} \tag{42a}$$

$$\mathbf{z}_2 = \mathbf{G}_{a2}\mathbf{K}_{a}^{(\hat{\mathcal{S}}_{a2})}\boldsymbol{\psi}_{a}^{(\hat{\mathcal{S}}_{a2})} + \mathbf{G}_{2b}\mathbf{T}'_{b}\mathbf{K}_{b}^{(\hat{\mathcal{S}}_{b3})}\boldsymbol{\psi}_{b}^{(\hat{\mathcal{S}}_{b3})} \tag{42b}$$

$$\mathbf{z}_3 = \mathbf{G}_{b3}\mathbf{K}_{b}^{(\hat{\mathcal{S}}_{b3})}\boldsymbol{\psi}_{b}^{(\hat{\mathcal{S}}_{b3})} + \mathbf{G}_{3c}\mathbf{T}'_{c}\mathbf{K}_{c}^{(\hat{\mathcal{S}}_{c3})}\boldsymbol{\psi}_{c}^{(\hat{\mathcal{S}}_{c4})} \tag{42c}$$

$$\mathbf{z}_4 = \mathbf{G}_{c4}\mathbf{K}_{c}^{(\hat{\mathcal{S}}_{c4})}\boldsymbol{\psi}_{c}^{(\hat{\mathcal{S}}_{c4})} + \mathbf{G}_{2b}\mathbf{T}'_{d}\mathbf{K}_{d}^{(\hat{\mathcal{S}}_{d4})}\boldsymbol{\psi}_{d}^{(\hat{\mathcal{S}}_{d4})} \tag{42d}$$

where the matrices $\mathbf{G}$ of Equation (A4) are defined as

$$\mathbf{G}_{d1} = \begin{bmatrix} \bar{\mathbf{A}}_1\hat{\mathbf{A}}_{d1} & \mathbf{0} \\ \mathbf{0}^T & 1 \end{bmatrix}, \quad \mathbf{G}_{1a} = \begin{bmatrix} \bar{\mathbf{A}}_1\hat{\mathbf{A}}_{1a} & \mathbf{0} \\ \mathbf{0}^T & 1 \end{bmatrix} \tag{43a}$$

$$\mathbf{G}_{a2} = \begin{bmatrix} \bar{\mathbf{A}}_2\hat{\mathbf{A}}_{a2} & \mathbf{0} \\ \mathbf{0}^T & 1 \end{bmatrix}, \quad \mathbf{G}_{2b} = \begin{bmatrix} \bar{\mathbf{A}}_2\hat{\mathbf{A}}_{2b} & \mathbf{0} \\ \mathbf{0}^T & 1 \end{bmatrix} \tag{43b}$$

$$\mathbf{G}_{b3} = \begin{bmatrix} \bar{\mathbf{A}}_3\hat{\mathbf{A}}_{b3} & \mathbf{0} \\ \mathbf{0}^T & 1 \end{bmatrix}, \quad \mathbf{G}_{3c} = \begin{bmatrix} \bar{\mathbf{A}}_3\hat{\mathbf{A}}_{3c} & \mathbf{0} \\ \mathbf{0}^T & 1 \end{bmatrix} \tag{43c}$$

$$\mathbf{G}_{c4} = \begin{bmatrix} \bar{\mathbf{A}}_4\hat{\mathbf{A}}_{c4} & \mathbf{0} \\ \mathbf{0}^T & 1 \end{bmatrix}, \quad \mathbf{G}_{4d} = \begin{bmatrix} \bar{\mathbf{A}}_4\hat{\mathbf{A}}_{4d} & \mathbf{0} \\ \mathbf{0}^T & 1 \end{bmatrix} \tag{43d}$$

Therefore, $\mathbf{K}^{II}_{T}$ becomes

$$\mathbf{K}^{II}_{T} = \left[\begin{array}{ccc|ccc|ccc|ccc} \mathbf{0} & \mathbf{0} & \mathbf{z}_1 & \mathbf{0} & \mathbf{0} & \mathbf{0} & \mathbf{0} & \mathbf{0} & \mathbf{0} & \mathbf{0} & \mathbf{0} & \mathbf{0} \\ \mathbf{0} & \mathbf{0} & \mathbf{0} & \mathbf{0} & \mathbf{0} & \mathbf{z}_2 & \mathbf{0} & \mathbf{0} & \mathbf{0} & \mathbf{0} & \mathbf{0} & \mathbf{0} \\ \mathbf{0} & \mathbf{0} & \mathbf{0} & \mathbf{0} & \mathbf{0} & \mathbf{0} & \mathbf{0} & \mathbf{0} & \mathbf{z}_3 & \mathbf{0} & \mathbf{0} & \mathbf{0} \\ \mathbf{0} & \mathbf{0} & \mathbf{0} & \mathbf{0} & \mathbf{0} & \mathbf{0} & \mathbf{0} & \mathbf{0} & \mathbf{0} & \mathbf{0} & \mathbf{0} & \mathbf{z}_4 \end{array}\right] \tag{44}$$

Finally, the third part of the tangent stiffness matrix $\mathbf{K}^{III}_{T}$ is derived through the 6-dimensional vectors $\mathbf{v}_f$ of Equation (A11), i.e.,

$$\mathbf{v}_a = \check{\mathbf{J}}^{T}_{ax}\hat{\mathbf{A}}_2\hat{\mathbf{A}}_{a2}\tilde{\mathbf{1}}\mathbf{F}_{a}^{(\hat{\mathcal{S}}_{a2})} \equiv \mathbf{J}_{ax}^{(\hat{\mathcal{S}}_{a2})T}\tilde{\mathbf{1}}\mathbf{F}_{a}^{(\hat{\mathcal{S}}_{a2})} \tag{45a}$$

$$\mathbf{v}_b = \check{\mathbf{J}}^{T}_{bx}\hat{\mathbf{A}}_3\hat{\mathbf{A}}_{b3}\tilde{\mathbf{1}}\mathbf{F}_{b}^{(\hat{\mathcal{S}}_{b3})} \equiv \mathbf{J}_{bx}^{(\hat{\mathcal{S}}_{b3})T}\tilde{\mathbf{1}}\mathbf{F}_{b}^{(\hat{\mathcal{S}}_{b3})} \tag{45b}$$

$$\mathbf{v}_c = \check{\mathbf{J}}^{T}_{cx}\hat{\mathbf{A}}_4\hat{\mathbf{A}}_{c4}\tilde{\mathbf{1}}\mathbf{F}_{c}^{(\hat{\mathcal{S}}_{c4})} \equiv \mathbf{J}_{cx}^{(\hat{\mathcal{S}}_{c4})T}\tilde{\mathbf{1}}\mathbf{F}_{c}^{(\hat{\mathcal{S}}_{c4})} \tag{45c}$$

$$\mathbf{v}_d = \check{\mathbf{J}}^{T}_{dx}\hat{\mathbf{A}}_1\hat{\mathbf{A}}_{d1}\tilde{\mathbf{1}}\mathbf{F}_{d}^{(\hat{\mathcal{S}}_{d1})} \equiv \mathbf{J}_{dx}^{(\hat{\mathcal{S}}_{d1})T}\tilde{\mathbf{1}}\mathbf{F}_{d}^{(\hat{\mathcal{S}}_{d1})} \tag{45d}$$

where $\mathbf{F}_{f}^{(\hat{\mathcal{S}}_{fj})}$ is obtained taking the force vector from the flexure wrench $\mathbf{w}_{f}^{(\hat{\mathcal{S}}_{fj})} = \mathbf{K}_{f}^{(\hat{\mathcal{S}}_{fj})}\boldsymbol{\psi}_{f}^{(\hat{\mathcal{S}}_{fj})}$. The block-matrices $\check{\mathbf{J}}_{fx}$, or $\mathbf{J}_{fx}^{(\hat{\mathcal{S}}_{fj})}$, are derived taking only the first two rows, i.e., the block of

$\mathbf{x}'_f$, from the corresponding Jacobian matrices. Using the equations of (A12), the matrices $\mathbf{V}_f = \left[\begin{array}{c|c} \mathbf{V}_{fi} & \mathbf{V}_{fj} \end{array}\right]$ can be obtained and therefore, the matrix $\mathbf{K}_T^{III}$ becomes

$$\mathbf{K}_T^{III} = \left[\begin{array}{c|c|c|c} \mathbf{N}_{1a}\mathbf{V}_{a1} & \mathbf{N}_{1a}\mathbf{V}_{a2} & \mathbf{O} & \mathbf{O} \\ \mathbf{O} & \mathbf{N}_{2b}\mathbf{V}_{b2} & \mathbf{N}_{2b}\mathbf{V}_{b3} & \mathbf{O} \\ \mathbf{O} & \mathbf{O} & \mathbf{N}_{3c}\mathbf{V}_{c3} & \mathbf{N}_{3c}\mathbf{V}_{c4} \\ \mathbf{N}_{4d}\mathbf{V}_{d1} & \mathbf{O} & \mathbf{O} & \mathbf{N}_{4d}\mathbf{V}_{d4} \end{array}\right] \tag{46}$$

The final expression for the tangent stiffness matrix is $\mathbf{K}_T = \mathbf{K}_T^I + \mathbf{K}_T^{II} + \mathbf{K}_T^{III}$. If the DOFs of the first body, i.e., the fixed frame, are removed by imposing fixed boundary conditions, the final form of $\mathbf{K}_T$ will be a (9×9) matrix with the following pattern

$$\mathbf{K}_T = \left[\begin{array}{c|c|c} \bullet & \bullet & \\ \hline & \bullet & \bullet \\ \hline & & \bullet \end{array}\right] \tag{47}$$

5. Numerical Application

Referring to Figure 6, body 1 is fixed while body 2 is actuated through a vertical force applied at its center of mass. Finally, the end-effector is attached to body 3. Following the layout of Figure 6, all geometric and structural parameters necessary for direct kinetostatic analysis of the case study are reported in Table 1.

Table 1. Geometric and structural parameters of the case study.

Four-Bar Mechanism		
mass center G_2	[1.5000, 0.2000]	(mm)
mass center G_3	[1.8000, 1.5000]	(mm)
mass center G_4	[1.0000, 1.5000]	(mm)
hinge center O_a	[0.5000, 0.0000]	(mm)
hinge center O_b	[1.8478, 0.7654]	(mm)
hinge center O_c	[1.1928, 1.9000]	(mm)
hinge center O_d	[0.5000, 1.5000]	(mm)
local base vector $\bar{\mathbf{s}}_{1a}$	[0.6848, 0.0765]	(mm)
local base vector $\bar{\mathbf{s}}_{2a}$	[−0.8413, −0.3218]	(mm)
local base vector $\bar{\mathbf{s}}_{2b}$	[0.2139, 0.7140]	(mm)
local base vector $\bar{\mathbf{s}}_{3b}$	[0.0408, −0.5348]	(mm)
local base vector $\bar{\mathbf{s}}_{3c}$	[−0.4140, 0.3482]	(mm)
local base vector $\bar{\mathbf{s}}_{3EE}$	[−1.3571, 1.6991]	(mm)
local base vector $\bar{\mathbf{s}}_{4c}$	[0.3342, 0.2586]	(mm)
local base vector $\bar{\mathbf{s}}_{4d}$	[−0.6000, −0.1732]	(mm)
local base vector $\bar{\mathbf{s}}_{1d}$	[0.3714, 1.6532]	(mm)
curved beam		
beam radius r_f, ($f \in [a, b, c, d]$)	0.2	(mm)
beam characteristic angle $[\theta_a, \theta_b, \theta_c, \theta_d]$	[300, 320, 330, 250]	(°)
Young modulus E_y	100	(GPa)
cross-section base b	25	(μm)
cross-section height h	5	(μm)

5.1. Comparison and Validation

First, let us consider the linear flexure model expressed through Equation (24) of Section 3. Considering an initial actuation force $F_{2y} = -60$ (μN), the latter is increased until the x-coordinate of the end-effector becomes zero, i.e., the gripper clamp is completely closed. The gripper deforms as displayed on the left side of Figure 7, wherein the limit cases and the undeformed configuration are reported. Then, let us consider the partial non-linear flexure model expressed through the Equation (23) of Section 3. Performing the same simulations, the workspace becomes that of the right side of Figure 7. The values of F_{2y} for which the clamp is closed, respectively, are $F_{2y} = 63$ (μN) for the linear model and $F_{2y} = 77$ (μN) for the partial non-linear model.

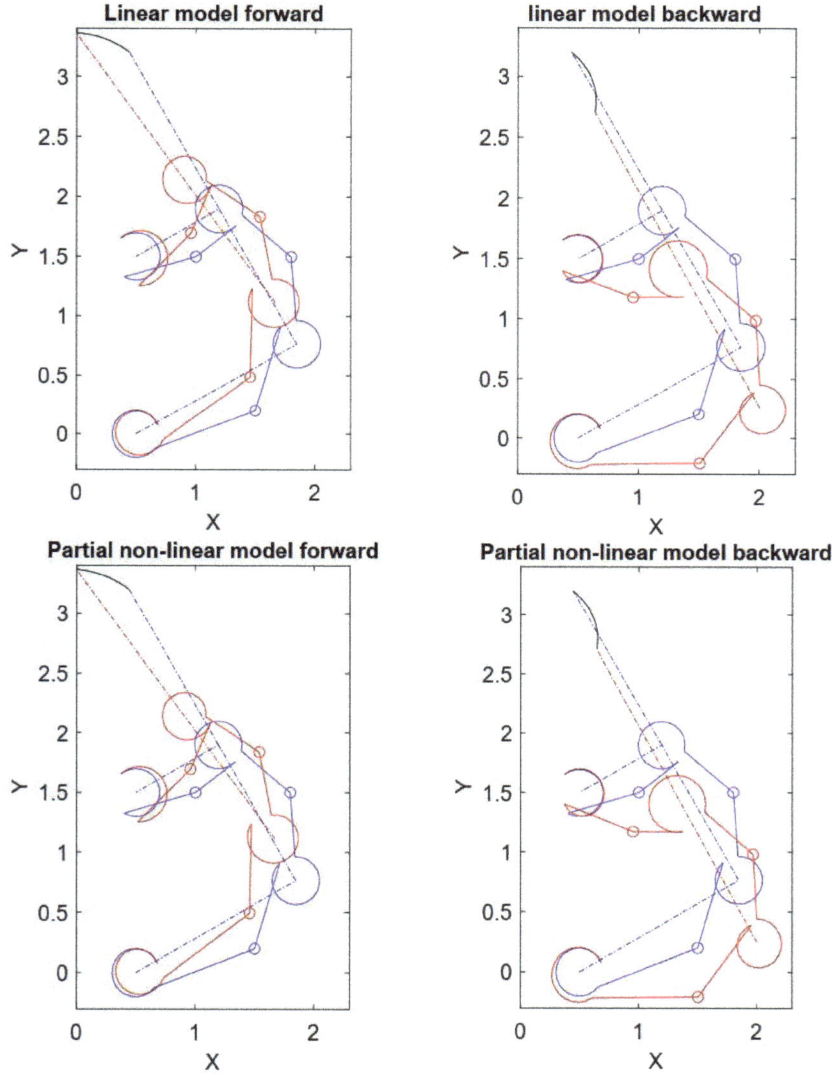

Figure 7. Workspace of the gripper and end-effector trajectory for the linear and partial non-linear models during the opening/closing maneuver: (red) deformed configuration, (blue) undeformed configuration, (black) end-effector trajectory. Units in millimeters.

The two models have been compared to a model implemented using the commercial multibody software MSC Adams©. The model includes rigid links and flexible circular beams. The latter are combined to the links' endpoints through fixed connections. The model has been developed to be comparable with the Matlab model. The only difference pertains to the flexures since the curved beams have been modeled using a two-dimensional geometrical non-linear representation for beam-like structures. Compared to the previous Matlab models, this representation is fully non-linear, as the stiffness matrix of each flexure is updated during deformation. The actuation force has been applied through step functions, and two simulations have been carried out to achieve convergency; one for the forward movement of closing and the other for the backward movement of opening. Figure 8 shows the two simulations and the trajectories accomplished by the end-effector point. As for Figure 7, the deformed configuration has also been plotted to understand the deformation process better.

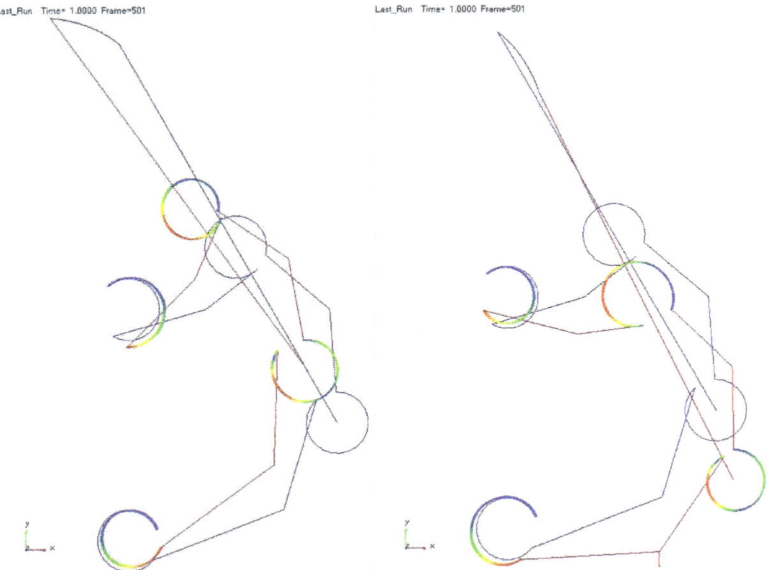

Figure 8. Adams results: (**left**) forward movement; (**right**) backward movement. The deformed gripper is in red, while the undeformed configuration is represented in blue color. The trajectory of the end-effector for the two movements is displayed in black color.

The three models have been compared in Figure 9 in terms of the end-effector trajectory, also referred to as the mechanism's workspace (left subplot) and actuation forces (right subplot).

Two materials have been modeled for the curved beams: silicon with Young modulus $E_y = 100$ (GPa) and nylon with Young modulus $E_y = 3.84$ (GPa). Considering the same rectangular cross-section, whose dimensions are reported in Table 1, the silicon bending stiffness is $EI = 26.0417$ (μNμm^2) while the nylon bending stiffness is $EI = 1$ (μNμm^2). The two materials have different properties in terms of elasticity. Silicon has a brittle behavior, while nylon has an anisotropic hyperelastic or visco-hyperelastic behavior. The two bending stiffness values should be seen as extreme cases to test the proposed method. With this premise in mind, the two materials will be assumed to have both isotropic linear behavior, while the different degrees of non-linearity of the models will only concern the geometric stiffness.

The first row of plots in Figure 9 pertains to the silicon while the second one is the nylon. First, let us consider silicon. Observing the top-left subplot, the three arc-shaped trajectories of the end-effector reveal relevant differences only in the final part of the path. The influence of the fully non-linear flexures becomes more evident in the opening movement, where the trajectories become more distant. Compared to the linear model, the top-right subplot reveals that the flexure non-linearities introduce a stiffening effect, and the actuation force required to produce the same displacement grows. It can be observed that the partial non-linear model is stiffer than the fully non-linear model in the forward closing movement and softer in the opening movement. The three models have equal stiffness only at the undeformed configuration where the actuation force is zero.

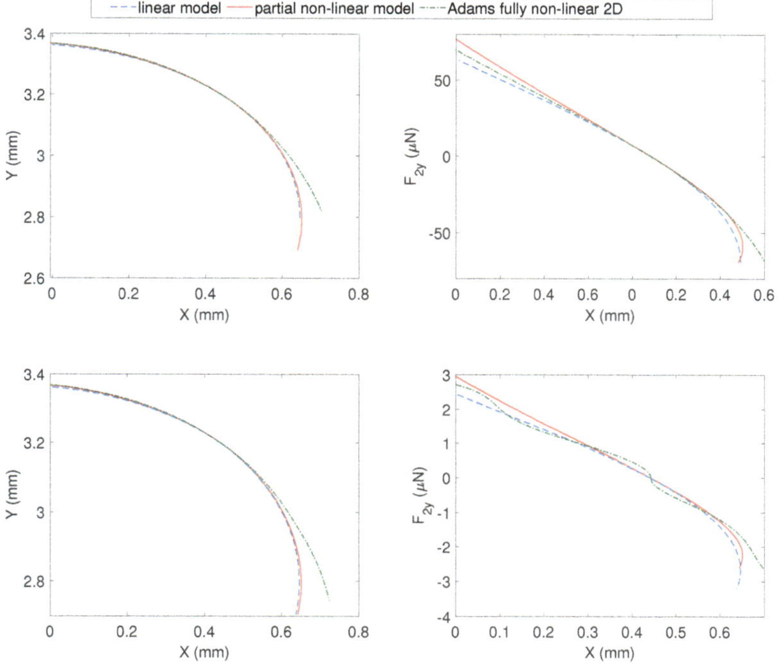

Figure 9. Model comparison in terms of workspace and actuation forces. Left subplot: end-effector trajectory in an opening–closing movement; right subplot: actuation force vs. x-coordinate of the end-effector during the opening–closing movement.

Then, let us consider nylon. Observing the bottom-left subplot, the end-effector trajectory follows a trend similar to the previous case. The bottom-right plot reveals differences in force range, as could be expected considering the lower bending stiffness of the nylon. Now, the non-linearities are more pronounced, and three inflection points appear for the fully non-linear flexures, which are totally absent in the remaining cases. Despite this, the plot is similar to the simplified cases.

Excluding the limit points of the workspace during the opening movement, the simplified models provide excellent results. Added to this is that the proper workspace is usually limited by other constraints such as the electrical interfaces or the maximum stress in the material, thus making the three models closer than they might appear in Figure 9.

For example, the Conjugate Surfaces Flexure Hinges (CSFH) employed in MEMS micro-grippers have a rotation range limited to $\pm 20^\circ$ to prevent the silicon from breaking [39,40]. This range is displayed in the opening–closing movement of Figure 9. However, this range is further limited to about $\pm 2^\circ$ by other phenomena coming from the electrostatic actuation such as sticking-friction anomalies, the *pull-in* or the impossibility to generating high actuation forces [41].

5.2. Shape Optimization

The simplified models have been employed to perform the cross-section optimization of the curved beams, as shown in Figure 10. These plots can be used in various ways; for example, knowing the maximum actuation force the electrical interface can produce, the section parameters can be chosen to cope with this value. Another example could be related to the choice of the section parameters based on the maximum allowable stress of the material or its fatigue limit. Likewise, the optimization could affect other structural parameters or mechanism lengths.

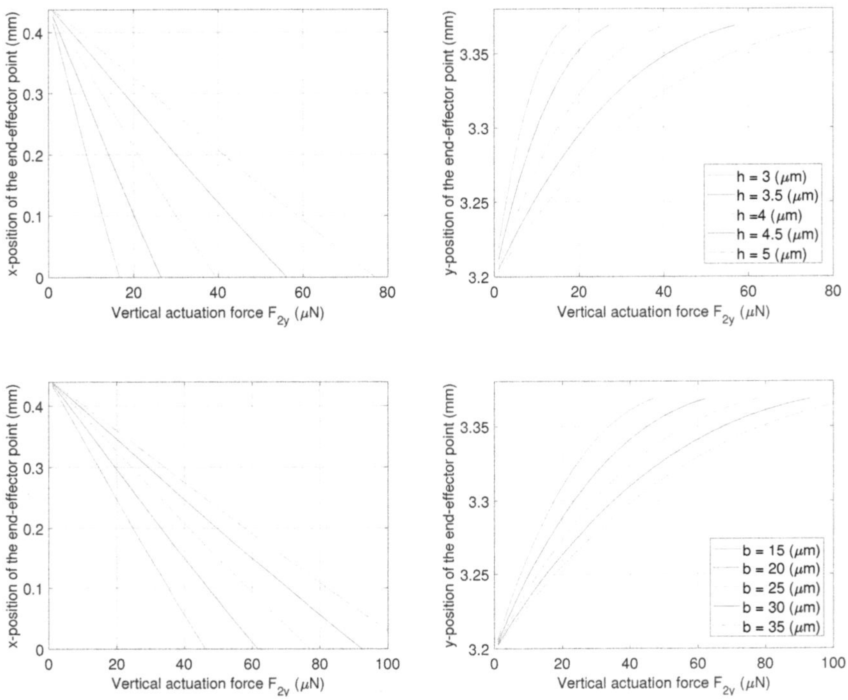

Figure 10. Cross-section optimization of the flexures.

5.3. *Tangent Stiffness Matrix*

Since the tangent stiffness matrix is the key element of the direct kinetostatic analysis, in the following, a detailed analysis of the tangent stiffness matrix and its role in the Newton–Raphson algorithm convergence is detailed. As already described in Section 4, after imposing fixed boundary conditions on body 1, the tangent stiffness matrix turns into a 9 × 9 symmetric matrix. Referring to the partial non-linear model, $\mathbf{K}_T$ has the expression reported in Table 2 in the undeformed configuration.

Table 2. Tangent stiffness matrix of the partial non-linear model in the undeformed configuration.

2.40×10^3	1.84×10^2	-5.42×10^2	-1.32×10^3	-1.11×10^2	-1.01×10^3			
1.84×10^2	2.71×10^3	-1.34×10^3	-1.11×10^2	-1.09×10^3	-1.46×10^2			
-5.42×10^2	-1.34×10^3	2.29×10^3	6.76×10^2	-3.29×10^2	4.70×10^2			
-1.32×10^3	-1.11×10^2	6.76×10^2	2.41×10^3	2.06×10^2	4.98×10^2	-1.10×10^3	-9.53×10^1	4.31×10^2
-1.11×10^2	-1.09×10^3	-3.29×10^2	2.06×10^2	2.30×10^3	-6.44×10^2	-9.53×10^1	-1.21×10^3	-1.75×10^2
-1.01×10^3	-1.46×10^2	4.70×10^2	4.98×10^2	-6.44×10^2	1.51×10^3	5.07×10^2	7.89×10^2	-9.01×10^1
			-1.10×10^3	-9.53×10^1	5.07×10^2	2.34×10^3	-5.58×10^1	-3.75×10^2
			-9.53×10^1	-1.21×10^3	7.89×10^2	-5.58×10^1	4.16×10^3	-1.08×10^3
			4.31×10^2	-1.75×10^2	-9.01×10^1	-3.75×10^2	-1.08×10^3	7.93×10^2

Let us consider the mechanism in the final deformed configuration obtained by applying $F_{2y} = 50$ (μN). In this case, the expression of $\mathbf{K}_T$ is reported in Table 3, while the percentage difference between the deformed and undeformed case is provided in Table 4. It can be observed that relevant differences appear during the deformation process, especially in the off-diagonal components. Furthermore, the matrix $\mathbf{K}_T$ is no longer symmetrical in the deformed configuration.

Table 3. Tangent stiffness matrix of the partial non-linear model in the final deformed configuration obtained applying $F_{2y} = 50$ (μN).

2.31×10^3	-9.15×10^1	-1.66×10^2	-1.27×10^3	-1.45×10^2	-9.67×10^2			
4.22×10^2	2.75×10^3	-1.20×10^3	-1.45×10^2	-1.14×10^3	-2.65×10^2			
-5.90×10^2	-1.45×10^3	2.20×10^3	7.24×10^2	-2.25×10^2	4.81×10^2			
-1.28×10^3	-2.79×10^2	6.74×10^2	2.36×10^3	3.65×10^2	5.15×10^2	-1.08×10^3	-8.63×10^1	5.32×10^2
6.01×10^0	-1.11×10^3	-3.28×10^2	8.03×10^1	2.33×10^3	-6.07×10^2	-8.63×10^1	-1.22×10^3	-2.72×10^1
-9.59×10^2	-3.34×10^2	4.46×10^2	4.64×10^2	-4.42×10^2	1.46×10^3	4.95×10^2	7.76×10^2	-2.18×10^2
			-1.02×10^3	2.52×10^2	2.50×10^2	2.23×10^3	-1.41×10^3	-1.90×10^2
			-3.80×10^2	-1.20×10^3	8.58×10^2	6.65×10^2	3.92×10^3	-1.12×10^3
			4.43×10^2	-1.43×10^2	-1.12×10^2	-4.97×10^2	-9.28×10^2	6.76×10^2

Table 4. Percentage difference of the tangent stiffness matrices of the partial non-linear model between the final deformed configuration of Table 3 and the undeformed configuration of Table 2.

−3.5	−149.8	−69.4	−3.8	31.4	−3.8			
129.6	1.2	−10.4	31.4	4.5	82.0			
8.8	7.8	−4.1	7.1	−31.6	2.3			
−3.1	152.5	−0.4	−2.3	77.6	3.5	−1.2	−9.4	23.4
−105.4	2.2	−0.2	−61.0	1.6	−5.6	−9.4	1.1	−84.4
−4.6	129.3	−5.1	−6.7	−31.4	−3.9	−2.5	−1.7	141.6
			−7.2	−364.2	−50.7	−4.4	2424.5	−49.4
			299.1	−0.8	8.7	−1290.9	−5.8	3.7
			2.7	−18.2	23.9	32.7	−14.1	−14.7

Since $\mathbf{K}_T$ is composed of three terms, it is legitimate to ask what the contribution of each term is. As reported in Table 5, the first term $\mathbf{K}_T^I$ is the closest to the final expression of $\mathbf{K}_T$.

Table 5. The first term $\mathbf{K}_T^I$ in the deformed configuration obtained applying $F_{2y} = 50$ (μN).

2.31×10^3	-9.15×10^1	-1.16×10^2	-1.27×10^3	-1.45×10^2	-9.67×10^2			
4.22×10^2	2.75×10^3	-1.20×10^3	-1.45×10^2	-1.14×10^3	-2.65×10^2			
-5.74×10^2	-1.42×10^3	2.19×10^3	7.08×10^2	-2.46×10^2	4.71×10^2			
-1.28×10^3	-2.79×10^2	6.74×10^2	2.36×10^3	3.65×10^2	5.15×10^2	-1.08×10^3	-8.63×10^1	5.32×10^2
6.01×10^0	-1.11×10^3	-3.28×10^2	8.03×10^1	2.33×10^3	-6.07×10^2	-8.63×10^1	-1.22×10^3	-2.72×10^1
-9.59×10^2	-3.34×10^2	4.46×10^2	4.83×10^2	-4.22×10^2	1.44×10^3	4.76×10^2	7.56×10^2	-2.15×10^2
			-1.02×10^3	2.52×10^2	2.50×10^2	2.23×10^3	-1.41×10^3	-1.90×10^2
			-3.80×10^2	-1.20×10^3	8.58×10^2	6.65×10^2	3.92×10^3	-1.12×10^3
			4.43×10^2	-1.43×10^2	-1.12×10^2	-4.77×10^2	-9.10×10^2	6.62×10^2

The second term $\mathbf{K}_T^{II}$ is reported in Table 6. Remembering the expression for $\mathbf{z}_i$ in Equation (A7) and the form of $\mathbf{K}_T^{II}$ in Equation (A8), observing Table 6, it can be found that at the equilibrium, i.e., if and only if the residuals are zero, the following expression stands

$$\mathbf{z}_i = \begin{bmatrix} \tilde{\mathbf{1}} & \mathbf{0} \\ \mathbf{0}^T & 0 \end{bmatrix}, \; \mathbf{w}_i \equiv \begin{bmatrix} -F_{iy} \\ +F_{ix} \\ 0 \end{bmatrix} \tag{48}$$

For the case study, only body 2, and therefore $\mathbf{z}_2$, has components different from zero.

Finally, the third term $\mathbf{K}_T^{III}$ is reported in Table 7. It can be noticed that only the components of the inner moments, i.e., due to the flexures, are different from zero. This feature comes from the particular form of $\mathbf{V}_{fi}$, or $\mathbf{V}_{fj}$, in Equation (A12).

Table 6. Second term $\mathbf{K}_T^{II}$ in the deformed configuration obtained applying $F_{2y} = 50$ (μN).

0	0	-5.00×10^{1}	0	0	0			
0	0	-6.25×10^{-12}	0	0	0			
0	0	0	0	0	0			
0	0	0	0	0	-6.75×10^{-13}	0	0	0
0	0	0	0	0	4.73×10^{-13}	0	0	0
0	0	0	0	0	0	0	0	0
			0	0	0	0	0	-1.95×10^{-11}
			0	0	0	0	0	7.21×10^{-11}
			0	0	0	0	0	0

Table 7. Third term $\mathbf{K}_T^{III}$ in the deformed configuration obtained applying $F_{2y} = 50$ (μN).

0	0	0	0	0	0			
0	0	0	0	0	0			
-1.62×10^{1}	-2.14×10^{1}	8.03×10^{0}	1.62×10^{1}	-2.14×10^{1}	-1.03×10^{1}			
0	0	0	0	0	0	0	0	0
0	0	0	0	0	0	0	0	0
0	0	0	-1.86×10^{1}	-1.94×10^{1}	1.30×10^{1}	1.86×10^{1}	1.94×10^{1}	-2.29×10^{0}
			0	0	0	0	0	0
			0	0	0	0	0	0
			0	0	0	-1.99×10^{1}	-1.80×10^{1}	1.45×10^{1}

To better focus on the importance of the tangent stiffness matrix in achieving solution convergence, let us keep the tangent stiffness matrix constant and equal to that obtained in the undeformed configuration for the entire simulation, i.e., $\mathbf{K}_T = \mathbf{K}_T(\hat{\mathbf{q}})$. The results of Figure 11 reveal that the number of iterations necessary to achieve convergence grows exponentially as the input load increases and no longer converges beyond $F_{2y} = 76$ (μN).

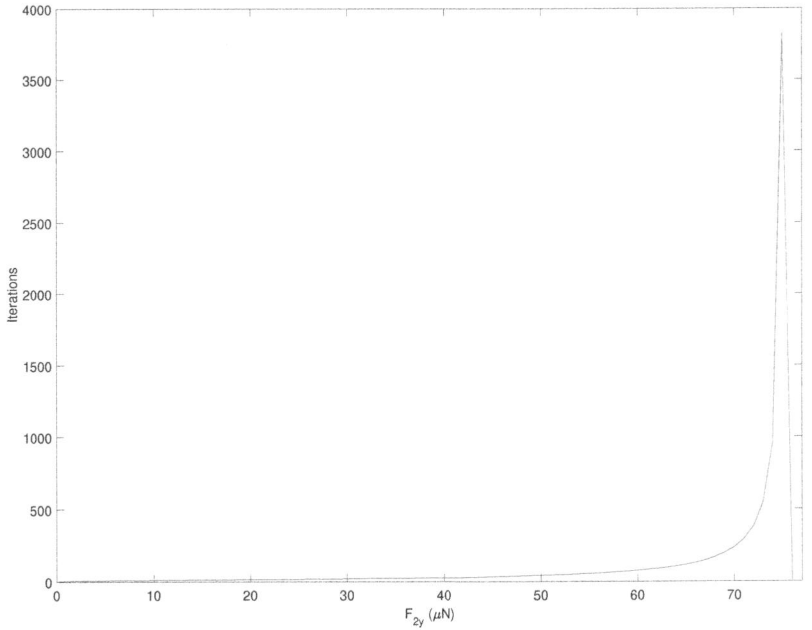

Figure 11. Number of iterations to achieve convergence of the direct kinetostatic analysis varying the input force. The simulation employs the constant tangent stiffness matrix obtained in the undeformed configuration.

The same simulation has been repeated by updating the tangent stiffness matrix following four strategies based on:

- The complete matrix $\mathbf{K}_T$ starting each simulation from the undeformed solution;
- The first term $\mathbf{K}_T^I$ starting each simulation from the undeformed solution;
- The complete matrix $\mathbf{K}_T$ starting each simulation from the previous converged solution;
- The first term $\mathbf{K}_T^I$ starting each simulation from the previous converged solution.

It can be seen that the approaches using the previous converged solution reduce the number of iterations. Similarly, using the complete matrix instead of the first-term approximated matrix results in fewer iterations. The results are displayed in Figure 12.

It is interesting to observe how these trends translate into a computational burden. The reduced number of iterations provided by the strategies based on the previous converged solution translates directly into savings in computation time. The two strategies based on the first term of the tangent stiffness matrix lead to a higher number of iterations but, simultaneously, require a smaller number of variables to be determined and save on calculation times, as shown in Figure 13. The peaks observed in Figure 13 come from memory allocation and other inner processes of Matlab during the first computation. On the other hand, it can be observed from Figure 12 that a higher number of iterations do not correspond to these CPU times. The results have been obtained using an HP workstation equipped with an Intel Xeon CPU @3.20GHz with 32 GB of RAM.

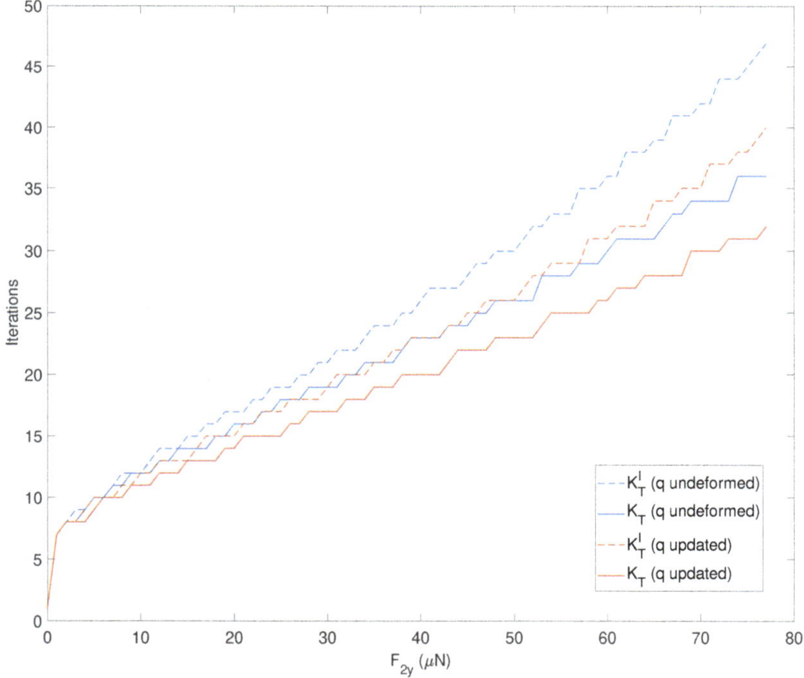

Figure 12. Number of iterations to achieve convergency of the direct kinetostatic analysis varying the input force. The simulation compares four strategies to upload the tangent stiffness matrix: considering the complete matrix $\mathbf{K}_T$ or only the first term $\mathbf{K}_T^I$ starting each simulation from the undeformed solution, considering the complete matrix $\mathbf{K}_T$ or only the first term $\mathbf{K}_T^I$ starting each simulation from the previous converged solution.

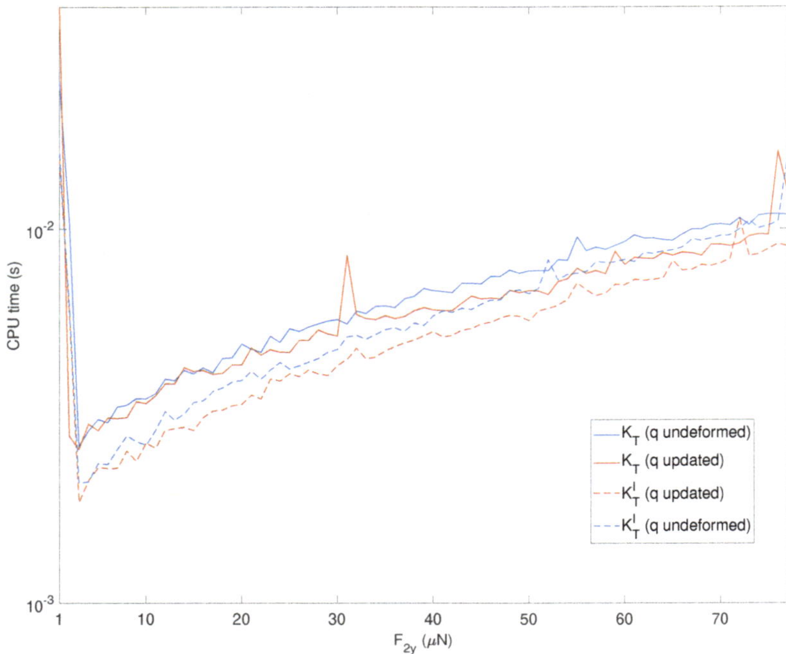

Figure 13. Computation time of the four strategies of Figure 12.

This section is concluded by giving some insights into the computational time obtained using Adams. In Table 8, the CPU times obtained in the opening–closing movement for the Adams fully non-linear model, the Matlab linear model, and the Matlab partial non-linear model are compared. Adams simulations have been performed by disabling the graphic display. The comparison has been carried out for both silicon and nylon. As can be observed, the CPU time of the simplified methods is from 30 to 500 times faster than Adams. It is noteworthy that the simulation time for nylon is ten times faster than silicon for the Adams fully non-linear model.

Table 8. CPU-time comparison in seconds for the opening–closing movement.

	Silicon	
	opening	closing
Adams fully non-linear model	41.00	50.00
Matlab linear model	0.150	0.100
Matlab partial non-linear model	0.150	0.150
	Nylon	
	opening	closing
Adams Fully non-linear model	4.500	4.500
Matlab linear model	0.135	0.100
Matlab partial non-linear model	0.145	0.135

6. Conclusions and Discussion

The tangent stiffness matrix has been used as a conceptual base to solve the direct kinetostatic problem of planar grippers with curved beams. Two models have been presented to cope with flexure deformations. The first linear model considers the flexure equilibrium in the initial undeformed configuration, while the second partial non-linear model considers the equilibrium in the deformed configuration. Both methods do not include a fully non-linear geometric description of the curved beam flexure whose stiffness matrix is kept constant and equal to that obtained in the undeformed state. The tangent stiffness matrix has been divided into sub-parts to facilitate both the theoretical treatment and the numerical implementation. The linear model led to two sub-parts, while the partial non-linear model introduced a further third sub-part. Both models were tested and compared with a fully non-linear model obtained using the commercial software MSC Adams. The results proved to be in good agreement on most of the mechanism's workspace, except for the extreme areas wherein the geometric non-linear effects become relevant. The same case study was used to show the method's potential; for example, in conducting a shape optimization of the flexure cross-section. Finally, the importance of each term of the tangent stiffness matrix in the convergence process was detailed in terms of the number of iterations required to achieve convergence and computational load.

From what has been outlined, the proposed method offers various advantages:

1. The results of Figure 13 suggest a possible extension to real-time applications of micro and nano-grippers. It is known that the control often requires simplified models to be executed quickly by the control unit. Often these models are obtained by linearizing the equilibrium equations around one or more operating points. Using models with reduced complexity would allow more efficient control strategies such as control in the operating space, inverse dynamics control, pre-calculated torque control. Furthermore, the closed form helps creating more efficient reduced order models [42–44].
2. The tangent stiffness matrix is obtained in closed form. This feature prevents the use of numeric differentiation, making the convergence process of the direct kinetostatic solution more robust. Furthermore, splitting the expression of $\mathbf{K}_T$ allowed for identifying its most basic terms and calibrating the compromise between the number of iterations and calculation time. The calculation times are considerably reduced by using only the first term of the tangent stiffness matrix and recalculating it at each iteration of the Newton–Raphson algorithm described in Algorithm 1.
3. The tangent stiffness matrix can be employed to develop a dynamics model to study vibrations. The tangent stiffness matrix is the core of implicit time integration methods primarily employed in flexible multibody dynamics [45]. Shape optimization takes further advantage of the closed form of $\mathbf{K}_T$ opening scenarios to gradient-based constrained optimization problems based on the kinetostatic analysis.
4. Both the two simplified models employ curved beams modeled by a constant stiffness matrix. Despite this, the curved beams guarantee finite displacements/rotations in the mechanism, allowing for the expansion of the reachable workspace. The model remains reliable for most of the mechanism's workspace. The results are accurate in the functional area except for the limit zones of the workspace in which physical constraints usually prevent motion. When the maximum rotations of the curved beams exceed $\pm 20^\circ$ the constant stiffness hypothesis can no longer capture the geometric nonlinearities, and the results deviate from the actual case.
5. Although the proposed method is valid only for planar cases, it can be extended to other compliant mechanisms with constant stiffness flexures without changing the mathematical background.

Author Contributions: Conceptualization, A.C. and N.P.B.; methodology, A.C. and N.P.B.; software, A.C.; validation, A.C.; formal analysis, A.C. and N.P.B.; investigation, P.D.M.; resources, R.S.; data curation, P.D.M.; writing—original draft preparation, A.C.; writing—review and editing, R.S.; visualization, A.C.; supervision, R.S. All authors have read and agreed to the published version of the manuscript.

Funding: This research received no external funding.

Data Availability Statement: Not applicable.

Conflicts of Interest: The authors declare no conflict of interest.

Appendix A

The first term $\mathbf{K}_{Ti}^{I}$ of the tangent stiffness matrix in Equation (26) can be derived by mapping the Jacobian of Equation (13) to the whole n-dimensional vector of generalized coordinates $\mathbf{q}$, therefore obtaining the new $(3 \times n)$ extended Jacobian $\mathbf{J}_f$

$$\mathbf{J}_f = \begin{bmatrix} \mathbf{O} & \cdots & \check{\mathbf{J}}_f^{i} & \cdots & \check{\mathbf{J}}_f^{j} & \cdots \mathbf{O} \end{bmatrix} \tag{A1}$$

where $\mathbf{O}$ is the 3×3 zero matrix. By substituting Equation (A1) into Equation (27a), the $(3 \times n)$ matrix $\mathbf{K}_{Ti}^{I}$ becomes

$$\mathbf{K}_{Ti}^{I} = \mathbf{N}_{1i}\mathbf{K}_1^{(\hat{\mathcal{S}}_{1i})}\mathbf{J}_1^{(\hat{\mathcal{S}}_{1i})} + \mathbf{N}_{i2}\mathbf{T}_2'\mathbf{K}_2^{(\hat{\mathcal{S}}_{2j})}\mathbf{J}_2^{(\hat{\mathcal{S}}_{2j})} \tag{A2}$$

with $\mathbf{J}_1$ and $\mathbf{J}_2$ referred to the curved beams connected to body i.

To find the second term $\mathbf{K}_{Ti}^{II}$, let us consider the variation of matrices $\mathbf{N}_{1i}$ and $\mathbf{N}_{i2}$ of Equation (22), i.e.,

$$\delta\mathbf{N}_{1i} = \delta\theta_i\mathbf{G}_{1i}, \quad \delta\mathbf{N}_{i2} = \delta\theta_i\mathbf{G}_{i2} \tag{A3}$$

where

$$\mathbf{G}_{1i} = \begin{bmatrix} \bar{\mathbf{A}}_i\hat{\mathbf{A}}_{1i} & \mathbf{0} \\ \mathbf{0}^T & 0 \end{bmatrix}, \quad \mathbf{G}_{i2} = \begin{bmatrix} \bar{\mathbf{A}}_i\hat{\mathbf{A}}_{i2} & \mathbf{0} \\ \mathbf{0}^T & 0 \end{bmatrix} \tag{A4}$$

From this expression, it follows that

$$\delta\mathbf{N}_{1i}\mathbf{K}_1^{(\hat{\mathcal{S}}_{1i})}\boldsymbol{\psi}_1^{(\hat{\mathcal{S}}_{1i})} + \delta\mathbf{N}_{i2}\mathbf{T}_2'\mathbf{K}_2^{(\hat{\mathcal{S}}_{2j})}\boldsymbol{\psi}_2^{(\hat{\mathcal{S}}_{2j})} \equiv \delta\theta_i\mathbf{z}_i \tag{A5}$$

where $\mathbf{z}_i$ is a 3-dimensional vector defined as

$$\mathbf{z}_i = \frac{\partial\mathbf{N}_{1i}}{\partial\theta_i}\mathbf{K}_1^{(\hat{\mathcal{S}}_{1i})}\boldsymbol{\psi}_1^{(\hat{\mathcal{S}}_{1i})} + \frac{\partial\mathbf{N}_{i2}}{\partial\theta_i}\mathbf{T}_2'\mathbf{K}_2^{(\hat{\mathcal{S}}_{2j})}\boldsymbol{\psi}_2^{(\hat{\mathcal{S}}_{2j})} \tag{A6}$$

Using Equation (A3), $\mathbf{z}_i$ becomes

$$\mathbf{z}_i = \mathbf{G}_{1i}\mathbf{K}_1^{(\hat{\mathcal{S}}_{1i})}\boldsymbol{\psi}_1^{(\hat{\mathcal{S}}_{1i})} + \mathbf{G}_{i2}\mathbf{T}_2'\mathbf{K}_2^{(\hat{\mathcal{S}}_{2j})}\boldsymbol{\psi}_2^{(\hat{\mathcal{S}}_{2j})} \tag{A7}$$

Comparing Equation (A5) with (27b), the $(3 \times n)$ matrix $\mathbf{K}_{Ti}^{II}$ reads

$$\mathbf{K}_{Ti}^{II} = [\ \mathbf{O} \mid \cdots \mid \mathbf{0} \ \ \mathbf{0} \ \ \mathbf{z}_i \mid \cdots \mid \mathbf{O}\ ] \tag{A8}$$

where $\mathbf{z}_i$ is mapped in the column corresponding to the angle θ_i while all the other entries are zero.

Finally, starting from the expression of $\mathbf{T}_f'$ in Equation (17) and calculating its variation provides

$$\delta\mathbf{T}_f' = \begin{bmatrix} \mathbf{O} & \mathbf{0} \\ (\delta\mathbf{x}_f^{\prime(\hat{\mathcal{S}}_{f2})})^T\tilde{\mathbf{1}} & 0 \end{bmatrix} \equiv \begin{bmatrix} \mathbf{O} & \mathbf{0} \\ -(\hat{\mathbf{A}}_{fj}^T\hat{\mathbf{A}}_j^T\check{\mathbf{J}}_{fx}\delta\mathbf{q}_{ij})^T\tilde{\mathbf{1}} & 0 \end{bmatrix} \tag{A9}$$

where Equation (10) has been employed to pass from the global to the local system, and $\check{\mathbf{J}}_{fx}$ is the block matrix obtained considering the first two rows of $\check{\mathbf{J}}_f$ pertaining $\mathbf{x}_f'$ only. Multiplying $\delta\mathbf{T}_f'$ by the wrench $\mathbf{w}_{fj}^{\hat{\mathcal{S}}_{fj}} = \mathbf{K}_f^{\hat{\mathcal{S}}_{fj}}\boldsymbol{\psi}_f^{\hat{\mathcal{S}}_{fj}}$ at section $\hat{\mathcal{S}}_{fj}$, the following expression is obtained

$$\delta\mathbf{T}_f'\mathbf{w}_{fj}^{(\hat{\mathcal{S}}_{fj})} = \begin{bmatrix} \mathbf{0} \\ \mathbf{v}_f^T\delta\mathbf{q}_{ij} \end{bmatrix} = \begin{bmatrix} \mathbf{O}_{2,6} \\ \mathbf{v}_f^T \end{bmatrix}\delta\mathbf{q}_{ij} = \mathbf{V}_f\delta\mathbf{q}_{ij} \tag{A10}$$

where $\mathbf{V}_f$ is a (3×6) matrix while the 6-dimensional vector $\mathbf{v}_f = [\mathbf{v}_{fi}^T, \mathbf{v}_{fj}^T]^T$ is defined as

$$\mathbf{v}_f = \check{\mathbf{J}}_{fx}^T \hat{\mathbf{A}}_j \hat{\mathbf{A}}_{fj} \tilde{\mathbf{I}} \mathbf{F}_f^{\hat{\mathcal{S}}_{fj}} \tag{A11}$$

being $\mathbf{F}_f^{(\hat{\mathcal{S}}_{fj})}$ the force composing the wrench $\mathbf{w}_f^{(\hat{\mathcal{S}}_{fj})} = [\mathbf{F}_f^{(\hat{\mathcal{S}}_{fj})^T}, M_f]^T$. The matrix $\mathbf{V}_f$ can also be written as

$$\mathbf{V}_f = \left[\begin{array}{c|c} \mathbf{V}_{fi} & \mathbf{V}_{fj} \end{array} \right], \quad \mathbf{V}_{fi} = \begin{bmatrix} \mathbf{0}^T \\ \mathbf{0}^T \\ \mathbf{v}_{fi}^T \end{bmatrix}, \quad \mathbf{V}_{fj} = \begin{bmatrix} \mathbf{0}^T \\ \mathbf{0}^T \\ \mathbf{v}_{fj}^T \end{bmatrix} \tag{A12}$$

Using these expressions, the $(3 \times n)$ matrix $\mathbf{K}_{Ti}^{III}$ becomes

$$\mathbf{K}_{Ti}^{III} = \left[\begin{array}{c|c|c|c|c|c|c} \mathbf{O} & \cdots & \mathbf{N}_{i2}\mathbf{V}_{2i} & \cdots & \mathbf{N}_{i2}\mathbf{V}_{2j} & \cdots & \mathbf{O} \end{array} \right] \tag{A13}$$

in which the matrices are mapped in correspondence with the dofs of the body i and j, respectively.

References

1. Midha, A.; Norton, T.W.; Howell, L.L. On the nomenclature, classification and abstractions of compliant mechanisms. *J. Mech. Des.* **1994**, *116*, 270–279. [CrossRef]
2. Howell, L.L.; Midha, A.; Norton, T.W. Evaluation of Equivalent Spring Stiffness for Use in a Pseudo-Rigid-Body Model of Large-Deflection Compliant Mechanisms. *J. Mech. Des. Trans. ASME* **1996**, *118*, 126–131. [CrossRef]
3. Howell, L.L. *Compliant Mechanisms*; John Wiley & Sons: Hoboken, NJ, USA, 2001; p. 459.
4. Shuib, S.; Ridzwan, M.I.Z.; Kadarman, A.H. Methodology of compliant mechanisms and its current developments in applications: A review. *Am. J. Appl. Sci.* **2007**, *4*, 160–167. [CrossRef]
5. Mathew, B.C.; Bharatpatil, V.; Anilchamoli.; Raikwar, M.; Negi, M.S.; Singh, H. Compliant Mechanism and Origami usage in Aerospace and Space Application. In *IOP Conference Series: Earth and Environmental Science*; IOP Publishing: Bristol, UK, 2021; Volume 775.
6. Kota, S.; Lu, K.J.; Kreiner, Z.; Trease, B.; Arenas, J.; Geiger, J. Design and application of compliant mechanisms for surgical tools. *J. Biomech. Eng.* **2005**, *127*, 981–989. [CrossRef] [PubMed]
7. Peterlík, I.; Nouicer, M.; Duriez, C.; Cotin, S.; Kheddar, A. Constraint-based haptic rendering of multirate compliant mechanisms. *IEEE Trans. Haptics* **2011**, *4*, 175–187. [CrossRef]
8. Ma, F.; Chen, G. Influence of non-ideal fixed-end constraints on kinetostatic behaviors of compliant bistable mechanisms. *Mech. Mach. Theory* **2019**, *133*, 267–277. [CrossRef]
9. Li, G.; Chen, G. A function for characterizing complete kinetostatic behaviors of compliant bistable mechanisms. *Mech. Sci.* **2014**, *5*, 67–78. [CrossRef]
10. Han, Q.; Huang, X.; Shao, X. Nonlinear kinetostatic modeling of double-tensural fully-compliant bistable mechanisms. *Int. J. Non-Linear Mech.* **2017**, *93*, 41–46. [CrossRef]
11. Chen, G.; Ma, F. Kinetostatic Modeling of Fully Compliant Bistable Mechanisms Using Timoshenko Beam Constraint Model. *J. Mech. Des. Trans. ASME* **2015**, *137*,022301. [CrossRef]
12. Vurchio, F.; Ursi, P.; Buzzin, A.; Veroli, A.; Scorza, A.; Verotti, M.; Sciuto, S.A.; Belfiore, N.P. Grasping and releasing agarose micro beads in water drops. *Micromachines* **2019**, *10*, 436. [CrossRef]
13. Hartmann, L.; Zentner, L. A compliant mechanism as rocker arm with spring capability for precision engineering applications. *Mech. Mach. Sci.* **2015**, *30*, 1–7. . [CrossRef]
14. Kemeny, D.C.; Howell, L.L.; Magleby, S.P. Using compliant mechanisms to improve manufacturability in MEMS. In Proceedings of the ASME Design Engineering Technical Conference, Montreal, ON, Canada, 29 September–2 October 2002; Volume 3, pp. 247–254.
15. Belfiore, N.P.; Bagolini, A.; Rossi, A.; Bocchetta, G.; Vurchio, F.; Crescenzi, R.; Scorza, A.; Bellutti, P.; Sciuto, S.A. Design, fabrication, testing and simulation of a rotary double comb drives actuated microgripper. *Micromachines* **2021**, *12*, 1263. [CrossRef]
16. Kota, S. Design of compliant mechanisms: Applications to MEMS. In Proceedings of the SPIE—The International Society for Optical Engineering, San Jose, CA, USA, 28–29 January 1999; Volume 3673, pp. 45–54.
17. Kota, S.; Joo, J.; Li, Z.; Rodgers, S.; Sniegowski, J. Design of compliant mechanisms: Applications to MEMS. *Analog. Integr. Circuits Signal Process.* **2001**, *29*, 7–15. [CrossRef]
18. Tolou, N.; Henneken, V.A.; Herder, J.L. Statically balanced compliant micro mechanisms (sb-mems): Concepts and simulation. In Proceedings of the ASME Design Engineering Technical Conference, Montreal, QC, Canada, 15–18 August 2010; Volume 2, pp. 447–454.

19. Xi, X.; Clancy, T.; Wu, X.; Sun, Y.; Liu, X. A MEMS XY-stage integrating compliant mechanism for nanopositioning at sub-nanometer resolution. *J. Micromech. Microeng.* **2016**, *26*, 025014. [CrossRef]
20. Guo, T.; Li, J.; Wang, H. Application of compliant mechanism on the polishing robot. *Appl. Mech. Mater.* **2011**, *63–64*, 593–596.
21. Zhan, J.; Zhang, M. Study on edge-forming mechanism for aspheric surface compliant polishing. *Zhongguo Jixie Gongcheng/China Mech. Eng.* **2013**, *24*, 1587–1591.
22. Zhu, W.; Liu, J.; Li, H.; Gu, K. Design and analysis of a compliant polishing manipulator with tensegrity-based parallel mechanism. *Aust. J. Mech. Eng.* **2021**, *19*, 414–422.
23. Verotti, M.; Bagolini, A.; Bellutti, P.; Belfiore, N.P. Design and validation of a single-SOI-wafer 4-DOF crawling microgripper. *Micromachines* **2019**, *10*, 376. [CrossRef] [PubMed]
24. Pantano, A.; Minore, G.; Carollo, G. Design of a wiper as compliant mechanisms with a monolithic layout. *SAE Int. J. Passeng. Cars—Mech. Syst.* **2020**, *13*, 173–188. [CrossRef]
25. Shinde, S.; Bandi, P.; Detwiler, D.; Tovar, A. Structural Optimization of Thin-Walled Tubular Structures for Progressive Buckling Using Compliant Mechanism Approach. *SAE Int. J. Passeng. Cars—Mech. Syst.* **2013**, *6*, 109–120. [CrossRef]
26. Ling, M.; Howell, L.L.; Cao, J.; Chen, G. Kinetostatic and dynamic modeling of flexure-based compliant mechanisms: A survey. *Appl. Mech. Rev.* **2020**, *72*, 030802. [CrossRef]
27. Chi, I.T.; Chang, P.L.; Tran, N.; Wang, D.A. Kinetostatic Modeling of Planar Compliant Mechanisms with Flexible Beams, Linear Sliders, Multinary Rigid Links, and Multiple Loops. *J. Mech. Robot.* **2021**, *13* , 051002. [CrossRef]
28. Kong, K.; Chen, G.; Hao, G. Kinetostatic modeling and optimization of a novel horizontal-displacement compliant mechanism. *J. Mech. Robot.* **2019**, *11*, 064502. [CrossRef]
29. Zhang, C.; Yu, H.; Yang, M.; Chen, S.; Yang, G. Nonlinear kinetostatic modeling and analysis of a large range 3-PPR planar compliant parallel mechanism. *Precis. Eng.* **2022**, *74*, 264–277. [CrossRef]
30. Wu, S.; Shao, Z.; Su, H.; Fu, H. An energy-based approach for kinetostatic modeling of general compliant mechanisms. *Mech. Mach. Theory* **2019**, *142*, 103588. [CrossRef]
31. Ma, F.; Chen, G. Kinetostatic modeling and characterization of compliant mechanisms containing flexible beams of variable effective length. *Mech. Mach. Theory* **2020**, *147*, 103770. [CrossRef]
32. Arredondo-Soto, M.; Cuan-Urquizo, E.; Gómez-Espinosa, A. The compliance matrix method for the kinetostatic analysis of flexure-based compliant parallel mechanisms: Conventions and general force–displacement cases. *Mech. Mach. Theory* **2022**, *168*, 104583. [CrossRef]
33. Turkkan, O.; Su, H.J. A general and efficient multiple segment method for kinetostatic analysis of planar compliant mechanisms. *Mech. Mach. Theory* **2017**, *112*, 205–217. [CrossRef]
34. Li, J.; Liu, X.; Chen, G. Automatically generating kinetostatic model for planar flexure-based compliant mechanisms. *J. Xidian Univ.* **2016**, *43*, 74–79.
35. Li, J.; Chen, G. A general approach for generating kinetostatic models for planar flexure-based compliant mechanisms using matrix representation. *Mech. Mach. Theory* **2018**, *129*, 131–147. [CrossRef]
36. Belfiore, N.P.; Simeone, P. Inverse kinetostatic analysis of compliant four-bar linkages. *Mech. Mach. Theory* **2013**, *69*, 350–372. [CrossRef]
37. Chung, J.; Hulbert, G. A time integration algorithm for structural dynamics with improved numerical dissipation: The generalized-α method. *J. Appl. Mech.* **1993**, *60*, 371–375. [CrossRef]
38. Hilber, H.M.; Hughes, T.J.; Taylor, R.L. Improved numerical dissipation for time integration algorithms in structural dynamics. *Earthq. Eng. Struct. Dyn.* **1977**, *5*, 283–292. [CrossRef]
39. Verotti, M.; Crescenzi, R.; Balucani, M.; Belfiore, N.P. MEMS-based conjugate surfaces flexure hinge. *J. Mech. Des.* **2015**, *137*, 012301. [CrossRef]
40. Bagolini, A.; Ronchin, S.; Bellutti, P.; Chistè, M.; Verotti, M.; Belfiore, N.P. Fabrication of novel MEMS microgrippers by deep reactive ion etching with metal hard mask. *J. Microelectromech. Syst.* **2017**, *26*, 926–934. [CrossRef]
41. Buzzin, A.; Cupo, S.; Giovine, E.; de Cesare, G.; Belfiore, N.P. Compliant nano-pliers as a biomedical tool at the nanoscale: Design, simulation and fabrication. *Micromachines* **2020**, *11*, 1087. [CrossRef]
42. Cammarata, A.; Sinatra, R.; Maddìo, P.D. Static condensation method for the reduced dynamic modeling of mechanisms and structures. *Arch. Appl. Mech.* **2019**, *89*, 2033–2051. [CrossRef]
43. Cammarata, A.; Sinatra, R.; Maddìo, P.D. A system-based reduction method for spatial deformable multibody systems using global flexible modes. *J. Sound Vib.* **2021**, *504*, 116118. [CrossRef]
44. Cammarata, A.; Sinatra, R.; Maddìo, P.D. Interface reduction in flexible multibody systems using the Floating Frame of Reference Formulation. *J. Sound Vib.* **2022**, *523*, 116720. [CrossRef]
45. Bauchau, O.A. *Flexible Multibody Dynamics*; Springer: Dordrecht, The Netherlands; Heidelberg, Germany; London, UK; New York, NY, USA, 2011; Volume 176.

 *micromachines*

Article

A Substructure Condensed Approach for Kinetostatic Modeling of Compliant Mechanisms with Complex Topology

Shilei Wu [1,2], Zhongxi Shao [2,*] and Hongya Fu [2,*]

[1] School of Mechanical Engineering, Suzhou University of Science and Technology, Suzhou 215009, China
[2] School of Mechatronics Engineering, Harbin Institute of Technology, 92 West Dazhi Street, Harbin 150001, China
* Correspondence: shaozhongxi78@hit.edu.cn (Z.S.); hongyafu@hit.edu.cn (H.F.)

Abstract: Compliant mechanisms with complex topology have previously been employed in various precision devices due to the superiorities of high precision and compact size. In this paper, a substructure condensed approach for kinetostatic analysis of complex compliant mechanisms is proposed to provide concise solutions. In detail, the explicit relationships between the theoretical stiffness matrix, element stiffness matrix, and element transfer matrix for the common flexible beam element are first derived based on the energy conservation law. The transfer matrices for three types of serial–parallel substructures are then developed by combining the equilibrium equations of nodal forces with the transfer matrix approach, so that each branch chain can be condensed into an equivalent beam element. Based on the derived three types of transfer matrices, a kinetostatic model describing only the force-displacement relationship of the input/output nodes is established. Finally, two typical precision positioning platforms with complex topology are employed to demonstrate the conciseness and efficiency of this modeling approach. The superiority of this modeling approach is that the input/output stiffness, coupling stiffness, and input/output displacement relations of compliant mechanisms with multiple actuation forces and complex substructures can be simultaneously obtained in concise and explicit matrix forms, which is distinct from the traditional compliance matrix approach.

Keywords: compliant mechanisms; flexible elements; element transfer matrix; transfer matrix approach

Citation: Wu, S.; Shao, Z.; Fu, H. A Substructure Condensed Approach for Kinetostatic Modeling of Compliant Mechanisms with Complex Topology. *Micromachines* **2022**, *13*, 1734. https://doi.org/10.3390/mi13101734

Academic Editor: Alessandro Cammarata

Received: 22 September 2022
Accepted: 11 October 2022
Published: 13 October 2022

1. Introduction

With the superiorities of high precision, no clearance, no friction, and no assembly [1–4], compliant mechanisms have been increasingly employed in the fields of precision positioning [5–8], precision machining [9–12], and micro-electro mechanical systems (MEMS) [13,14], and so forth. Compared with conventional rigid-body mechanisms, compliant mechanisms convert energy, forces, and motion using the elastic deformations of flexible elements. Both kinematic and elasto-mechanical behaviors need to be considered simultaneously. Therefore, the design and analysis of compliant mechanisms are much more complicated and labor intensive than those of conventional rigid-body mechanisms.

In order to accurately analyze compliant mechanisms, it is crucial to establish a concise and efficient kinetostatic model for performance analysis, optimization design, and motion control. Finite element analysis is a common and effective approach for the kinetostatic analysis of compliant mechanisms with arbitrarily complex topology. In this approach, compliant mechanisms are first discretized into a large number of flexible elements. The kinetostatic model can then be derived based on the topology relationships of the flexible elements [15,16]. However, it fails to provide a concise and explicit motion control model and is also inapplicable for fast performance prediction due to the considerable degrees of freedom of the model. Hence, a number of studies focusing on this area have been carried out over the past few decades.

To develop the kinetostatic models for compliant mechanisms with small deflections, numerous modeling approaches have been proposed, for instance, the elastic beam theory [17,18], Castigliano's second theorem [19,20], the compliance matrix approach [21–23], and the transfer matrix approach [24–26]. According to the literatures, the elastic beam theory and Castigliano's second theorem are the most common ways for kinematic and static analyses of various flexure hinges and compliant mechanisms with relatively simple topology. Ma [27] simplified the parallel-type compliant amplifier into a rhombic-type amplifier and developed the analytical kinetostatic model for this amplifier by employing the elastic beam theory. Subsequently, the elastic beam theory was increasingly applied in the design and analysis of amplifying mechanisms. Lobontiu et al. [28,29] derived the stiffness models of various flexure hinges based on Castigliano's second theorem. Moreover, the theoretical models of input stiffness, amplification ratio, and the output stiffness of bridge-type amplifying mechanisms were established by this approach in [19]. However, the modeling approaches based on the elastic beam theory or Castigliano's second theorem require tedious internal force analysis, which is inapplicable for compliant mechanisms with complex topology. Compared with those two approaches, the compliance matrix approach can avoid the analysis of tedious internal forces. Therefore, the compliance matrix approach is particularly suitable for kinetostatic analysis of compliant mechanisms with complex topology. It was used to develop the stiffness models for compliant mechanisms with series and parallel topology in [30]. Additionally, Li and Xu [31,32] applied the compliance matrix approach to derive the analytical kinetostatic models for various compliant precision positioning manipulators. It is worth mentioning that the forces and corresponding displacements are at the same point. Lobontiu [33,34] established a series of kinetostatic models of serial–parallel compliant mechanisms with complex conditions, such as multi-point force loads and complex branched substructures, by employing the compliance matrix approach. However, the output stiffness, input stiffness, and coupling stiffness need to be respectively modeled [16]. As a result, the kinetostatic modeling procedure will become complicated for complex serial–parallel topology. In order to resolve this problem, a matrix displacement approach based on the nodal forces equilibrium equations was proposed to establish the kinetostatic models of compliant mechanisms with serial–parallel topology [24]. To avoid the solution of labor intensive equilibrium equations of nodal forces, an energy-based approach was proposed in our previous research [35]. However, this modeling approach is still cumbersome and has a large number of elements. In view of this, a two-port dynamic stiffness model for complex serial–parallel topology was developed based on the transfer matrix approach [25]. It is worth mentioning that this modeling approach is only suitable for specific topology and the element dynamic stiffness matrix is very difficult to formulate.

As a result, it is still necessary to further improve the conciseness and accuracy of the kinetostatic models for complex serial–parallel compliant mechanisms. This paper aims to extend our previous research [35] to develop a concise and efficient kinetostatic model with a low number of elements for compliant mechanisms with complex topology. The novelty of this paper is to propose a substructure condensed approach for the kinetostatic modeling of compliant mechanisms with complex topology and realize the explicit expression of input/output displacement relations with concise and matrix forms. The major contributions of this paper are summarized as follows. The explicit relationships between the theoretical stiffness matrix, element stiffness matrix, and element transfer matrix of the common flexible beam element are established via the energy conservation law. Similarly, the transfer matrices for three types of serial–parallel substructures are developed by combining the equilibrium equations of nodal forces with the transfer matrix approach. Additionally, kinematic and static performance analyses such as input/output stiffness, coupling stiffness, and the input/output displacement relation of complex compliant mechanisms can be simultaneously obtained.

This paper is organized as follows. Section 2 presents the general expression of the element transfer matrix for the common flexible beam element derived by employing the energy conservation law. The general kinetostatic model describing the force-displacement relationships of the input/output nodes for complex serial–parallel topology is proposed by combining the equilibrium equations of nodal forces with the transfer matrix approach in Section 3. Two typical precision positioning platforms are employed to validate the conciseness and accuracy of the proposed modeling approach in Section 4. Finally, conclusions are drawn in Section 5.

2. General Expression of Element Transfer Matrix for the Common Flexible Beam Element

As the basic element of compliant mechanisms, the compliance matrices of flexible beams with equal-section or variable-section have been extensively studied. For example, the compliance matrices of sheet, elliptic, and hyperbolic flexure hinges have been established based on the elastic beam theory and Castigliano's second theorem. As for flexible beams with equal-section, the element stiffness matrices can be calculated by employing the finite element approach. However, further investigation focusing on the general expression of the element transfer matrix for the common flexible element still needs to be carried out. As shown in Figure 1, the flexible beam element (i) has node j and node k, with six degrees of freedom per node. The force-displacement relationship of node j and node k in the local coordinate system is then obtained based on the elastic beam theory:

$$\begin{Bmatrix} \boldsymbol{F}_{i,j}^{e} \\ \boldsymbol{F}_{i,k}^{e} \end{Bmatrix} = \boldsymbol{K}_i^e \begin{Bmatrix} \boldsymbol{x}_{i,j}^{e} \\ \boldsymbol{x}_{i,k}^{e} \end{Bmatrix} \tag{1}$$

where $\boldsymbol{F}_{i,j}^{e} = [F_{uj}, F_{vj}, F_{wj}, M_{xj}, M_{yj}, M_{zj}]^{\mathrm{T}}$ and $\boldsymbol{F}_{i,k}^{e} = [F_{uk}, F_{vk}, F_{wk}, M_{xk}, M_{yk}, M_{zk}]^{\mathrm{T}}$ are, respectively, the nodal forces at the free end of the flexible beam element (i). $\boldsymbol{x}_{i,j}^{e} = [u_j, v_j, w_j, \theta_{xj}, \theta_{yj}, \theta_{zj}]^{\mathrm{T}}$ and $\boldsymbol{x}_{i,k}^{e} = [u_k, v_k, w_k, \theta_{xk}, \theta_{yk}, \theta_{zk}]^{\mathrm{T}}$ are the nodal displacements at the free the end of the flexible beam element (i). $\boldsymbol{K}_i^e$ is the 12×12 element stiffness matrix in the local coordinate system.

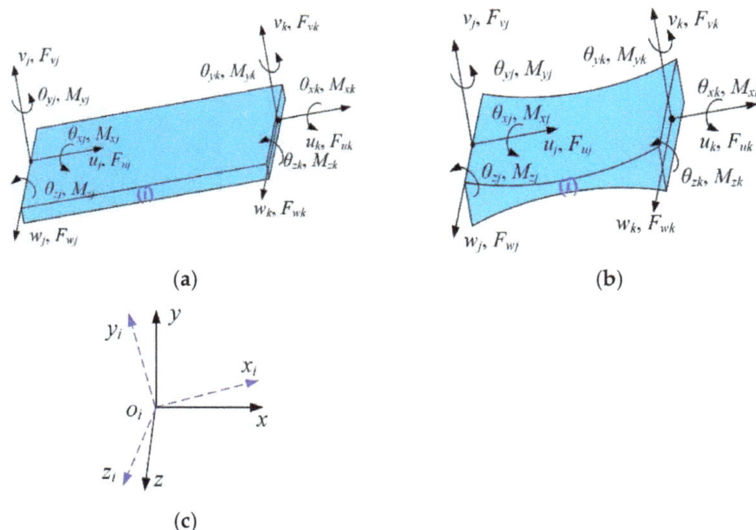

Figure 1. Nodal displacements and nodal forces of flexible elements: (**a**) the flexible beam with equal-section; (**b**) the flexible beam with variable-section; (**c**) the local and reference coordinate systems.

Rewriting the above equations into a block submatrices form, Equation (1) can be further expressed as:

$$\begin{Bmatrix} F^e_{i,j} \\ F^e_{i,k} \end{Bmatrix} = \begin{bmatrix} K^e_{i,1} & K^e_{i,2} \\ K^e_{i,3} & K^e_{i,4} \end{bmatrix} \begin{Bmatrix} x^e_{i,j} \\ x^e_{i,k} \end{Bmatrix} \tag{2}$$

where $K^e_{i,1}$, $K^e_{i,2}$, $K^e_{i,3}$, and $K^e_{i,4}$ are block submatrices of the element stiffness matrix in the local coordinate system.

This element can be seen as a flexible beam when the left end of the flexible beam element is clamped. The force-displacement relationship of node k can be obtained from Equation (2).

$$F^e_{i,k} = K^e_{i,4} x^e_{i,k} = K_i x^e_{i,k} \tag{3}$$

where $K_i = K^e_{i,4}$ is the theoretical stiffness matrix of this flexible beam.

Equation (3) can be further rewritten as below:

$$x^e_{i,k} = C_i F^e_{i,k} \tag{4}$$

where $C_i = K_i^{-1}$ is the theoretical compliance matrix of the flexible beam. The detailed expression of the compliance matrix can be derived from previous investigations [36,37] and expressed as below:

$$C_i = \begin{bmatrix} C_{u-F_u} & 0 & 0 & 0 & 0 & 0 \\ 0 & C_{v-F_v} & 0 & 0 & 0 & C_{v-M_z} \\ 0 & 0 & C_{w-F_w} & 0 & C_{w-M_y} & 0 \\ 0 & 0 & 0 & C_{\theta_x-M_x} & 0 & 0 \\ 0 & 0 & C_{\theta_y-F_w} & 0 & C_{\theta_y-M_y} & 0 \\ 0 & C_{\theta_z-F_v} & 0 & 0 & 0 & C_{\theta_z-M_z} \end{bmatrix} \tag{5}$$

For the flexible beam element (i), the displacement at node k equals the linear superposition of the elastic deformation under the external force and the displacement at node j. Hence, the nodal displacement of node k can be expressed by Equation (6).

$$x^e_{i,k} = \Delta x^e_i + [Ad] x^e_{i,j} \tag{6}$$

where Δx^e_i is the elastic deformation of the flexible beam element (i). $[Ad]$ is a 6×6 coordinate transformation matrix, which can transfer the nodal displacement at node j from the coordinate system $ox_jy_jz_j$ to the coordinate system $ox_ky_kz_k$, as shown in Figure 1, and can be expressed as:

$$[Ad] = \begin{bmatrix} R & D(d)R \\ 0 & R \end{bmatrix} \tag{7}$$

where R is a 3×3 rotation transformation matrix of the coordinate system $ox_jy_jz_j$ to the coordinate system $ox_ky_kz_k$. $d = [d_x, d_y, d_z]^T$ is the position vector of node j expressed in the coordinate system $ox_ky_kz_k$. $D(d)$ is a skew-symmetric matrix of the position vector d and can be obtained as:

$$D(d) = \begin{bmatrix} 0 & -d_z & d_y \\ d_z & 0 & -d_x \\ -d_y & d_x & 0 \end{bmatrix} \tag{8}$$

The flexible element can be considered as a cantilever beam when the left end is clamped while the right end is free, and the elastic deformation at the free end is Δx^e_i. Based on the energy conservation law, the potential strain energy stored in the flexible beam element is equal to the potential strain energy stored in the cantilever beam. Therefore, the potential strain energy stored in the flexible beam element (i) can be derived as below:

$$\begin{aligned} U_i &= \tfrac{1}{2}\Delta x_i^{e\mathrm{T}} K_i \Delta x_i^e = \tfrac{1}{2}(x_{i,k}^e - [Ad]x_{i,j}^e)^{\mathrm{T}} K_i (x_{i,k}^e - [Ad]x_{i,j}^e) \\ &= \tfrac{1}{2}\begin{bmatrix} x_{i,j}^{e\mathrm{T}} & x_{i,k}^{e\mathrm{T}} \end{bmatrix} \begin{bmatrix} [Ad]^{\mathrm{T}} K_i [Ad] & -[Ad]^{\mathrm{T}} K_i \\ -K_i[Ad] & K_i \end{bmatrix} \begin{bmatrix} x_{i,j}^e \\ x_{i,k}^e \end{bmatrix} \\ &= \tfrac{1}{2} x_i^{e\mathrm{T}} K_i^e x_i^e \end{aligned} \tag{9}$$

The element stiffness matrix, K_i^e, of the flexible beam element (i) in the local coordinate system can then be obtained as below:

$$K_i^e = \begin{bmatrix} [Ad]^{\mathrm{T}} K_i [Ad] & -[Ad]^{\mathrm{T}} K_i \\ -K_i[Ad] & K_i \end{bmatrix} \tag{10}$$

According to the above equation, Equation (1), which describes the force-displacement relationship of node j and node k in the local coordinate system, can be further given as follows:

$$\begin{Bmatrix} F_{i,j}^e \\ F_{i,k}^e \end{Bmatrix} = \begin{bmatrix} [Ad]^{\mathrm{T}} K_i [Ad] & -[Ad]^{\mathrm{T}} K_i \\ -K_i[Ad] & K_i \end{bmatrix} \begin{Bmatrix} x_{i,j}^e \\ x_{i,k}^e \end{Bmatrix} \tag{11}$$

Rewriting Equation (11), the following relationship can be obtained:

$$\begin{Bmatrix} x_{i,k}^e \\ F_{i,k}^e \end{Bmatrix} = T_i^e \begin{Bmatrix} x_{i,j}^e \\ F_{i,j}^e \end{Bmatrix} = \begin{bmatrix} [Ad] & -K_i^{-1}[Ad]^{-\mathrm{T}} \\ \mathbf{0} & -[Ad]^{-\mathrm{T}} \end{bmatrix} \begin{Bmatrix} x_{i,j}^e \\ F_{i,j}^e \end{Bmatrix} \tag{12}$$

where T_i^e is the element transfer matrix of the flexible beam element (i) in the local coordinate system. The element transfer matrix can also describe the force-displacement relationship for two nodes at the free end in the local coordinate system.

Equations (11) and (12) establish the explicit relationships between theoretical stiffness matrix, element stiffness matrix, and element transfer matrix for the common flexible beam element. It can be observed that the element transfer matrix only depends on the coordinate transformation matrix $[Ad]$ and the theoretical stiffness matrix K_i. In order to reduce the degrees of freedom of the kinetostatic model, Ling [24] developed the kinetostatic model of serial–parallel compliant mechanisms by employing the transfer matrix approach. However, the author did not establish the explicit relationship between the theoretical stiffness matrix and the element transfer matrix. Compared with that approach, the general expression of the element transfer matrix with respect to the theoretical stiffness matrix for the common flexible element is established based on the energy conservation law. The explicit relationships between the theoretical stiffness matrix, the element stiffness matrix, and the element transfer matrix for the common flexible beam element are shown in Figure 2.

Due to this element stiffness matrix being developed in the local coordinate system, it needs to be converted into the reference coordinate system. This force-displacement relationship of node j and node k in the reference coordinate system is obtained as below:

$$\begin{Bmatrix} x_{i,k} \\ F_{i,k} \end{Bmatrix} = T_i \begin{Bmatrix} x_{i,j} \\ F_{i,j} \end{Bmatrix} \tag{13}$$

where $x_{i,j}$, $x_{i,k}$, and $F_{i,j}$, $F_{i,k}$ are the nodal displacements and nodal forces at two free ends in the reference coordinate system, respectively. T_i is a 12 × 12 element transfer matrix in the reference coordinate system and can be calculated as:

$$[T_i] = [Tr_i]^{\mathrm{T}}[T_i^e][Tr_i] \tag{14}$$

where $\boldsymbol{Tr}_i$ is a 12×12 rotation transformation matrix, which can be expressed as below:

$$\boldsymbol{Tr}_i = \begin{bmatrix} \boldsymbol{R}_i & 0 & 0 & 0 \\ 0 & \boldsymbol{R}_i & 0 & 0 \\ 0 & 0 & \boldsymbol{R}_i & 0 \\ 0 & 0 & 0 & \boldsymbol{R}_i \end{bmatrix} \tag{15}$$

where $\boldsymbol{R}_i$ is a 3×3 coordinate rotation matrix of the local coordinate system to the reference coordinate system.

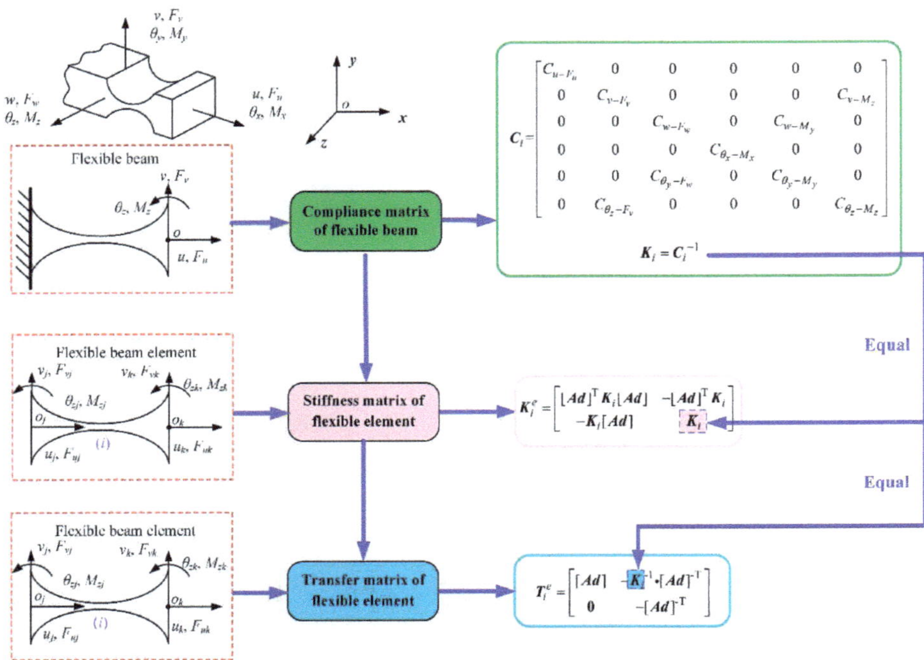

Figure 2. Transfer matrix of a flexible beam element.

3. Kinetostatic Modeling Based on the Substructure Condensed Approach

3.1. Transfer Matrices for Three Types of Substructures

In most cases, compliant mechanisms usually consist of flexible beams (equal-section flexible beam), flexure hinges (variable-section flexible beam), and lumped masses with series and/or parallel connections. For example, Figure 3 provides several typical compliant mechanisms with complex topology. The common characteristics of these compliant mechanisms are that the connection relationships can be divided into three types, namely series substructure, closed-loop parallel substructure, and opened-loop parallel substructure. Figure 4 illustrates the topology of these three types of connections abstracted from Figure 3.

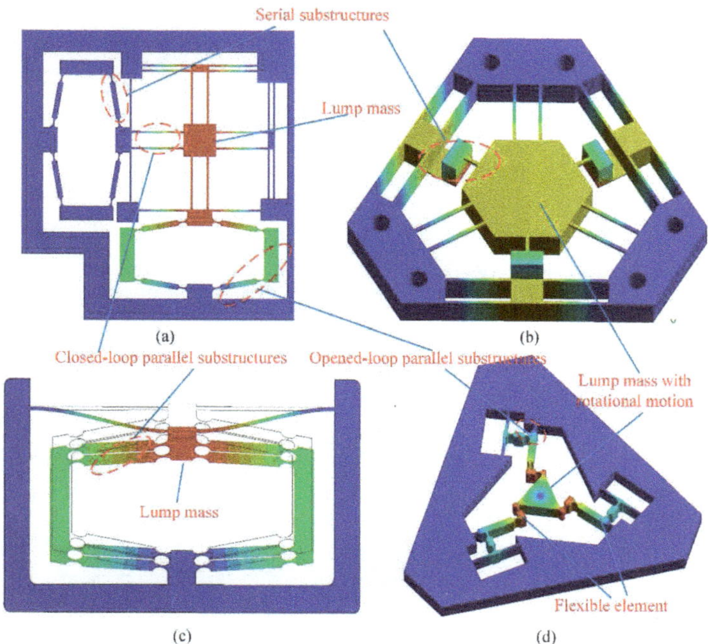

Figure 3. Typical compliant mechanisms: (**a**) a *XY* monolithic precision platform [35]; (**b**) a monolithic tip-tilt-piston spatial compliant manipulator [35]; (**c**) a one-dimensional precision positioning platform [38]; (**d**) a 3-DOF precision positioning platform [24,35].

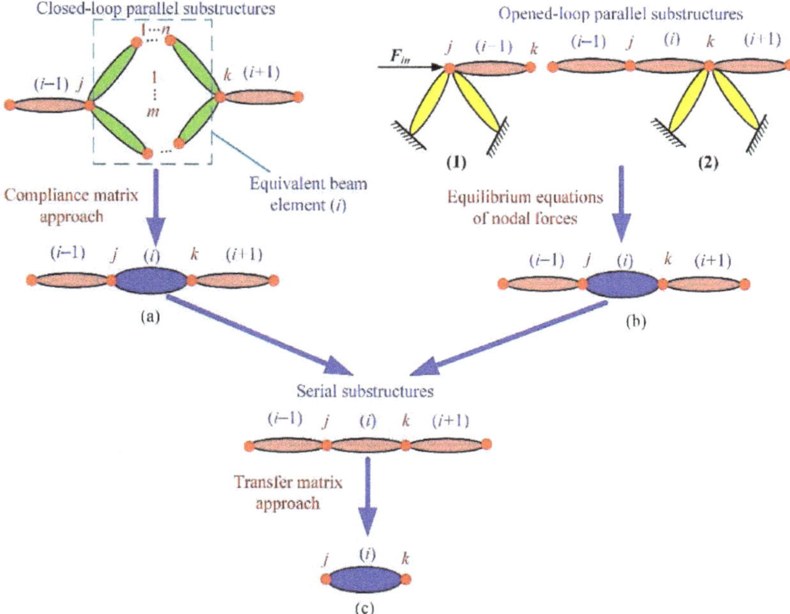

Figure 4. Illustration of serial and parallel substructures: (**a**) the closed-loop parallel substructure; (**b**) the opened-loop parallel substructure; (**c**) the series substructure.

For the closed-loop parallel substructure in Figure 4a, it can be condensed into an equivalent beam when the j-end of the equivalent beam element is clamped. This equivalent beam consists of m branches with a parallel connection where each branch consists of n flexible elements with a series connection. Based on the compliance matrix approach, the compliance matrix of the t-th parallel branch can be expressed as:

$$C_t = \sum_{g=1}^{n}\left([\boldsymbol{Ad}]_g \boldsymbol{K}_g{}^{-1}[\boldsymbol{Ad}]_g^{\mathrm{T}}\right) \tag{16}$$

where $\boldsymbol{K}_g$ is the theoretical stiffness matrix of the g-th flexible beam of the t-th parallel branch. $[\boldsymbol{Ad}]_g$ is a 6×6 coordinate transformation matrix, which can be expressed by Equation (7).

In the same way, the theoretical stiffness matrix of the equivalent beam is obtained as follows:

$$\boldsymbol{K}_i = \sum_{t=1}^{m}\left([\boldsymbol{Ad}]_t C_t [\boldsymbol{Ad}]_t{}^{\mathrm{T}}\right)^{-1} \tag{17}$$

According to Equation (12), the element transfer matrix of the equivalent beam element (i) in the local coordinate system can be expressed as:

$$\boldsymbol{T}_{eq}^{e} = \begin{bmatrix} [\boldsymbol{Ad}] & -\boldsymbol{K}_i{}^{-1}[\boldsymbol{Ad}]^{-\mathrm{T}} \\ \boldsymbol{0} & -[\boldsymbol{Ad}]^{-\mathrm{T}} \end{bmatrix} \tag{18}$$

By performing rotation transformation, the element transfer matrix $\boldsymbol{T}_{eq}$ of the equivalent beam element (i) in the reference coordinate system can be obtained. The force-displacement relation of node j and node k of equivalent beam element (i) in the reference coordinate system can then be expressed as:

$$\begin{Bmatrix} \boldsymbol{x}_{i,k} \\ \boldsymbol{F}_{i,k} \end{Bmatrix} = \boldsymbol{T}_{eq} \begin{Bmatrix} \boldsymbol{x}_{i,j} \\ \boldsymbol{F}_{i,j} \end{Bmatrix} \tag{19}$$

Therefore, the closed-loop parallel substructure can be condensed into an equivalent beam element by employing the compliance matrix approach.

For the opened-loop parallel substructure, the relationships between the nodal displacement/force at node j of flexible element (i + 1) and node k of flexible element (i) can be obtained based on the equilibrium equations of nodal forces.

$$\begin{cases} \text{Situations 1(input node)}: \begin{Bmatrix} \boldsymbol{x}_{i+1,j} \\ \boldsymbol{F}_{i+1,j} \end{Bmatrix} = \begin{bmatrix} \boldsymbol{E}_{6\times6} & \boldsymbol{0}_{6\times6} \\ \boldsymbol{0}_{6\times6} & -\boldsymbol{E}_{6\times6} \end{bmatrix} \begin{bmatrix} \boldsymbol{E}_{6\times6} & \boldsymbol{0}_{6\times6} \\ \sum\limits_{t=1}^{n} \boldsymbol{K}_i & -\boldsymbol{E}_{6\times6} \end{bmatrix} \begin{Bmatrix} \boldsymbol{x}_{in} \\ \boldsymbol{F}_{in} \end{Bmatrix} = \boldsymbol{S}\boldsymbol{Q}_{in,i} \begin{Bmatrix} \boldsymbol{x}_{in} \\ \boldsymbol{F}_{in} \end{Bmatrix} \\ \text{Situations 2(other node)}: \begin{Bmatrix} \boldsymbol{x}_{i+1,j} \\ \boldsymbol{F}_{i+1,j} \end{Bmatrix} = \begin{bmatrix} \boldsymbol{E}_{6\times6} & \boldsymbol{0}_{6\times6} \\ \boldsymbol{0}_{6\times6} & -\boldsymbol{E}_{6\times6} \end{bmatrix} \begin{bmatrix} \boldsymbol{E}_{6\times6} & \boldsymbol{0}_{6\times6} \\ \sum\limits_{t=1}^{n} \boldsymbol{K}_i & \boldsymbol{E}_{6\times6} \end{bmatrix} \begin{Bmatrix} \boldsymbol{x}_{i,k} \\ \boldsymbol{F}_{i,k} \end{Bmatrix} = \boldsymbol{S}\boldsymbol{Q}_i \begin{Bmatrix} \boldsymbol{x}_{i,k} \\ \boldsymbol{F}_{i,k} \end{Bmatrix} \end{cases} \tag{20}$$

where $\boldsymbol{E}_{6\times6}$ is a 6×6 identity matrix and $\boldsymbol{0}_{6\times6}$ is a 6×6 zero matrix. $\boldsymbol{x}_{in}$ and $\boldsymbol{F}_{in}$ are input displacement and input force, respectively. $\boldsymbol{Q}_{in,i}$ and $\boldsymbol{Q}_i$ indicate the force summation. $\boldsymbol{S}$ is a transfer matrix, which can transfer the force and displacement at node k of flexible element (i) to node j of the flexible element (i).

Therefore, the parallel substructure, such as that in Figure 4a,b, can be condensed into an equivalent beam element based on the compliance matrix approach and nodal forces equilibrium equations.

As to the series substructure, the relationship between the nodal displacement/force at node j of the flexible element (i + 1) and at node k of the flexible element (i) can be obtained in a matrix form based on Newton's third law.

$$\left\{ \begin{array}{c} \boldsymbol{u}_{i+1,j} \\ \boldsymbol{F}_{i+1,j} \end{array} \right\} = \begin{bmatrix} \boldsymbol{E}_{6\times6} & \boldsymbol{0}_{6\times6} \\ \boldsymbol{0}_{6\times6} & -\boldsymbol{E}_{6\times6} \end{bmatrix} \left\{ \begin{array}{c} \boldsymbol{u}_{i,k} \\ \boldsymbol{F}_{i,k} \end{array} \right\} = \boldsymbol{S} \left\{ \begin{array}{c} \boldsymbol{u}_{i,k} \\ \boldsymbol{F}_{i,k} \end{array} \right\} \tag{21}$$

Substituting Equation (19) into Equation (21) and substituting Equation (13) into Equations (20) and (21), the relationship between the nodal displacement/force at node j of flexible element (i + 1) and at node j of flexible element (i) can be calculated as follows:

$$\begin{cases} \text{Closed-loop parallel substructure :} & \left\{ \begin{array}{c} \boldsymbol{u}_{i+1,j} \\ \boldsymbol{F}_{i+1,j} \end{array} \right\} = (\boldsymbol{ST}_{eq}) \left\{ \begin{array}{c} \boldsymbol{u}_{i,j} \\ \boldsymbol{F}_{i,j} \end{array} \right\} \\ \text{Opened-loop parallel substructure 1 :} & \left\{ \begin{array}{c} \boldsymbol{u}_{i+1,j} \\ \boldsymbol{F}_{i+1,j} \end{array} \right\} = (\boldsymbol{SQ}_{in,i}) \left\{ \begin{array}{c} \boldsymbol{u}_{in} \\ \boldsymbol{F}_{in} \end{array} \right\} \\ \text{Opened-loop parallel substructure 2 :} & \left\{ \begin{array}{c} \boldsymbol{u}_{i+1,j} \\ \boldsymbol{F}_{i+1,j} \end{array} \right\} = (\boldsymbol{SQ}_i \boldsymbol{T}_i) \left\{ \begin{array}{c} \boldsymbol{u}_{i,j} \\ \boldsymbol{F}_{i,j} \end{array} \right\} \\ \text{Serise substructure :} & \left\{ \begin{array}{c} \boldsymbol{u}_{i+1,j} \\ \boldsymbol{F}_{i+1,j} \end{array} \right\} = (\boldsymbol{ST}_i) \left\{ \begin{array}{c} \boldsymbol{u}_{i,j} \\ \boldsymbol{F}_{i,j} \end{array} \right\} \end{cases} \tag{22}$$

3.2. *Establishing Kinetostatic Model*

To better describe the proposed modeling approach, a general topology for typical compliant mechanisms with complex connections abstracted from Figure 3 is employed, as shown in Figure 5, which can represent most cases in applications. The compliant mechanism consists of three branch chains, each of which comprises flexible elements and lumped masses with series and/or parallel connections. The lumped mass with rotational motion is simplified to one node at the rotary center and three sub-nodes at the connection nodes with three branch chains, which is shown in Figure 5. The proposed modeling procedure consists of five steps:

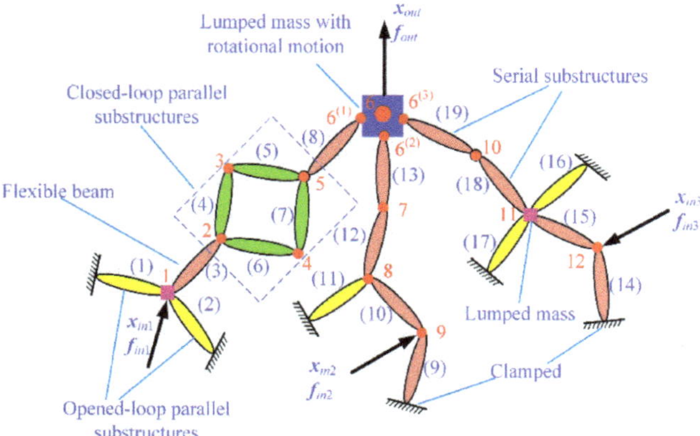

Figure 5. General topology of typical compliant mechanisms.

3.2.1. Discretizing and Numbering

The flexible beams of three branch chains are denoted from (1) to (19), and the connection nodes are numbered from 1 to 12, in which all clamped nodes are denoted as 0. The input forces are denoted as f_{in1}, f_{in2}, and f_{in3}. The input displacements are denoted as x_{in1}, x_{in2}, and x_{in3}. The output force (external force) and output displacement of the output platform are, respectively, numbered as f_{out} and x_{out}. These three sub-nodes are numbered as $6^{(1)}$, $6^{(2)}$, and $6^{(3)}$.

3.2.2. Calculating Transfer Matrices of Flexible Beams and Lumped Mass

For the common flexible element with nodes on the axis of symmetry, the element transfer matrix in the local coordinate system can be formulated using Equation (12). However, the flexible element whose nodes are not on the axis of symmetry, such as the flexible element in Figure 3d, are widely used in compliant mechanisms. Their element transfer matrix cannot be calculated directly based on Equation (12). As shown in Figure 6a, the nodal forces at node 1 and node 2 of the flexible element in regard to node j and node k can be obtained by performing the coordinate transformation matrix.

$$F^e_{i,1} = [Ad_1]^{-T} F^e_{i,j},\ \ F^e_{i,2} = [Ad_2]^{-T} F^e_{i,k} \tag{23}$$

where $[Ad_1]$ and $[Ad_2]$ are coordinate transformation matrices from the coordinate systems $o_j x_j y_j z_j$ and $o_k x_k y_k z_k$ into the coordinate systems $o_1 x_1 y_1 z_1$ and $o_2 x_2 y_2 z_2$, respectively.

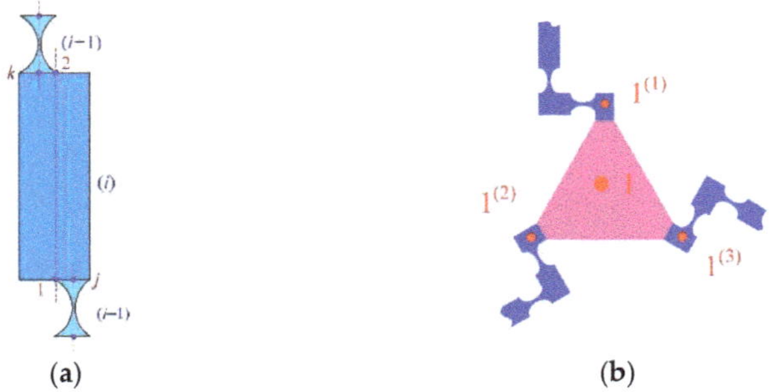

Figure 6. Nodal displacements and nodal forces: (**a**) the flexible beam; (**b**) the lumped mass.

Similarly, the nodal displacements at node 1 and node 2 of the flexible element (i) in regard to node j and node k can be expressed as:

$$x^e_{i,1} = [Ad_1] x^e_{i,j},\ \ x^e_{i,2} = [Ad_2] x^e_{i,k} \tag{24}$$

The force-displacement relationship of node 1 and node 2 of this element in the local coordinate system can then be obtained based on the elastic beam theory.

$$\begin{Bmatrix} F^e_{i,1} \\ F^e_{i,2} \end{Bmatrix} = \begin{bmatrix} [Ad]^T K_i [Ad] & -[Ad]^T K_i \\ -K_i [Ad] & K_i \end{bmatrix} \begin{Bmatrix} x^e_{i,1} \\ x^e_{i,2} \end{Bmatrix} \tag{25}$$

By substituting Equations (23) and (24) into Equation (12), the force-displacement relationship of the flexible element (i) in regard to node j and node k can be transferred to:

$$\begin{Bmatrix} x^e_{i,k} \\ F^e_{i,k} \end{Bmatrix} = \begin{bmatrix} [Ad_2]^{-1}[Ad][Ad_1] & -[Ad_2]^{-1} K_i^{-1} [Ad]^{-T} [Ad_1]^{-1} \\ 0 & -[Ad_2]^T [Ad]^{-T} [Ad_1]^{-T} \end{bmatrix} \begin{Bmatrix} x^e_{i,j} \\ F^e_{i,j} \end{Bmatrix} \tag{26}$$

For the lumped mass with only translational motion, such as the output platforms in Figure 3a,c, it can be regarded as a node. However, the lumped mass with rotational motion, such as the output platforms in Figure 3b,d, cannot be simply regarded as a node. As shown in Figure 6b, it is equivalent to one node at its rotary center and three sub-nodes at the connection nodes with three branch chains. It should be noted that the displacement of sub-node $1^{(1)}$, sub-node $1^{(2)}$, and sub-node $1^{(3)}$ is not equal to the displacement of node 1 due to the rotational motion. The nodal displacement and nodal force at the rotary center can be transferred to three sub-nodes at the connection nodes with three branch chains by performing a coordinate transformation matrix. The nodal displacement and nodal force of the lumped mass are then contained in the element transfer matrix of the adjacent flexible element. Therefore, the issue of the lumped mass with rotational motion in compliant mechanisms is solved.

3.2.3. Calculating the Transfer Matrix of Each Branch Chain

The compliant mechanism consists of three branch chains, and each branch chain includes one input node and one output node. According to the three connection types, the transfer matrix of each branch chain from input node to output node can be calculated as:

$$\begin{cases} T_{ch1} = (T_8)(ST_{eq})(ST_2)(SQ_{in,1}) \\ T_{ch2} = (T_{13})(ST_{12})(SQ_{10}T_{10})(SQ_{in,2}) \\ T_{ch3} = (T_{19})(ST_{18})(SQ_{15}T_{15})(SQ_{in,3}) \end{cases} \tag{27}$$

where T_i and S are transfer matrices and given by Equations (12), (26), and (21). Q_i, $Q_{in,1}$, $Q_{in,2}$, and $Q_{in,3}$ are summation matrices, which can be expressed as:

$$\begin{cases} Q_{in,1} = \begin{bmatrix} E_{6\times6} & 0_{6\times6} \\ K_1 + K_2 & -E_{6\times6} \end{bmatrix}, Q_{in,2} = \begin{bmatrix} E_{6\times6} & 0_{6\times6} \\ K_9 & -E_{6\times6} \end{bmatrix}, Q_{in,3} = \begin{bmatrix} E_{6\times6} & 0_{6\times6} \\ K_{14} & -E_{6\times6} \end{bmatrix} \\ Q_{10} = \begin{bmatrix} E_{6\times6} & 0_{6\times6} \\ K_{11} & E_{6\times6} \end{bmatrix}, Q_{15} = \begin{bmatrix} E_{6\times6} & 0_{6\times6} \\ K_{16} + K_{17} & E_{6\times6} \end{bmatrix} \end{cases} \tag{28}$$

3.2.4. Establishing the Kinetostatic Model of the Compliant Mechanism

Through the above analysis, the general topology of the typical compliant mechanisms in Figure 5 can be condensed into the equivalent topology including only the input node and the output node, as shown in Figure 7. Based on the transfer matrices of three branch chains, the force-displacement relationships of three equivalent beam elements can be obtained as:

$$\begin{Bmatrix} f_{i,j} \\ f_{i,k} \end{Bmatrix} = \begin{bmatrix} k_{i,1} & k_{i,2} \\ k_{i,3} & k_{i,4} \end{bmatrix} \begin{Bmatrix} x_{in,j} \\ x_{i,k} \end{Bmatrix} \tag{29}$$

where $k_{i,1}$, $k_{i,2}$, $k_{i,3}$, and $k_{i,4}$ (i = 1,2,3) are block sub-matrices and can be obtained from the transfer matrix of each branch chain, T_{chi}.

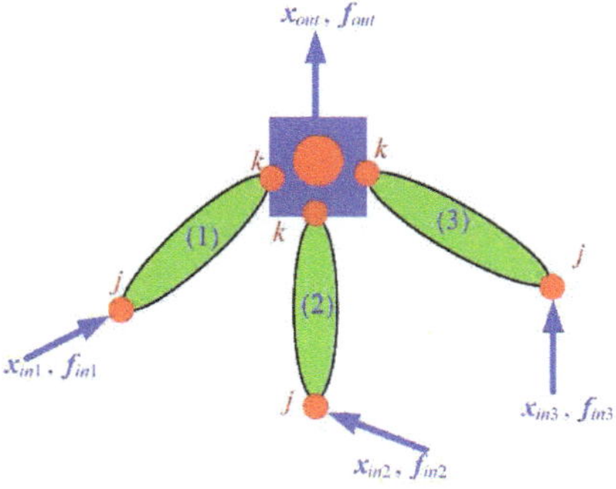

Figure 7. The equivalent topology of typical compliant mechanisms.

Taking the input/output nodes as research objects, the input/output force-displacement relations of the compliant mechanism can be calculated based on the equilibrium equations of nodal forces.

$$\begin{cases} f_{in1} = k_{1,1}x_{in1} + k_{1,2}x_{out} \\ f_{in2} = k_{2,1}x_{in2} + k_{2,2}x_{out} \\ f_{in3} = k_{3,1}x_{in3} + k_{3,2}x_{out} \\ f_{out} = k_{1,3}x_1 + k_{2,3}x_{in2} + k_{3,3}x_{in3} + (k_{1,4} + k_{2,4} + k_{3,4})x_{out} \end{cases} \tag{30}$$

Rewriting the above equations into a matrix form, the following kinetostatic model of compliant mechanisms with complex topology can be expressed as:

$$\begin{Bmatrix} f_{in1} \\ f_{in2} \\ f_{in3} \\ f_{out} \end{Bmatrix} = \begin{bmatrix} k_{1,1} & 0 & 0 & k_{1,2} \\ 0 & k_{2,1} & 0 & k_{2,2} \\ 0 & 0 & k_{3,1} & k_{3,2} \\ k_{1,3} & k_{2,3} & k_{3,3} & k_{1,4}+k_{2,4}+k_{3,4} \end{bmatrix} \begin{Bmatrix} x_{in1} \\ x_{in2} \\ x_{in3} \\ x_{out} \end{Bmatrix} \tag{31}$$

It can be seen from the modeling process that a kinetostatic model of compliant mechanisms with complex topology can be easily established with a low number of elements from the point of view of input/output nodes while using simple steps and concise equations.

3.2.5. Calculating the Kinetostatic Performances of the Compliant Mechanism

Based on this kinetostatic model, the kinetostatic performances, such as input/output stiffness, coupling stiffness, and input/output displacements relations can be derived in concise and explicit matrix forms. Depending on the applied input and output forces, the following two cases will be explained in detail, as follows:

(1) When external loads are applied to input nodes, i.e., $f_{ini} \neq 0$ (i = 1,2,3) and $f_{out} = 0$

The relationship between the output displacement and input forces can be obtained from Equation (31) in the form of an explicit matrix.

$$x_{out} = \begin{bmatrix} C_1 & C_2 & C_3 \end{bmatrix} \begin{Bmatrix} f_{in1} \\ f_{in2} \\ f_{in3} \end{Bmatrix} \tag{32}$$

where C_1, C_2, and C_3 are output coupling compliances, which are used to describe the output displacement under the input forces. Moreover, C_1, C_2, and C_3 can also be utilized to analyze the output coupling of the compliant mechanism.

Similarly, the force/displacement relations of input nodes can be expressed as follows:

$$\begin{Bmatrix} x_{in1} \\ x_{in2} \\ x_{in3} \end{Bmatrix} = \begin{bmatrix} C_{1,1} & C_{1,2} & C_{1,3} \\ C_{2,1} & C_{2,2} & C_{2,3} \\ C_{3,1} & C_{3,2} & C_{3,3} \end{bmatrix} \begin{Bmatrix} f_{in1} \\ f_{in2} \\ f_{in3} \end{Bmatrix} \tag{33}$$

where C_{ij} ($i = j$) is the input compliance, which can be utilized to obtain the required input force of the compliant mechanism. This is crucial for the design and selection of actuators. C_{ij} ($i \neq j$) is the input coupling compliance, which describes the displacement coupling of the input forces of the compliant mechanism. Hence, it can also be utilized to analyze the input coupling of the compliant mechanism.

By combining Equations (32) and (33), the input/output displacements relationship of the compliant mechanism can be expressed as:

$$x_{out} = T \begin{Bmatrix} x_{in1} \\ x_{in2} \\ x_{in3} \end{Bmatrix} = \begin{bmatrix} C_1 & C_2 & C_3 \end{bmatrix} \begin{bmatrix} C_{1,1} & C_{1,2} & C_{1,3} \\ C_{2,1} & C_{2,2} & C_{2,3} \\ C_{3,1} & C_{3,2} & C_{3,3} \end{bmatrix}^{-1} \begin{Bmatrix} x_{in1} \\ x_{in2} \\ x_{in3} \end{Bmatrix} \tag{34}$$

where T is a kinematics matrix, which describes the kinematic relationship of the input/output displacements.

Differing from conventional rigid-body mechanisms, the forward kinematic of the compliant mechanism can be easily and accurately derived based on Equation (34) with an explicit matrix form. The motion control and workspace analysis of the compliant mechanism can then be conducted.

(2) When the external load is applied to the output node, i.e., $f_{out} \neq 0$ and $f_{ini} = 0$ (i = 1,2,3)

The force/displacement relationship of the output node can be derived from the kinetostatic model.

$$x_{out} = K^{-1} f_{out} = C f_{out} \tag{35}$$

where $C = K^{-1}$ is the output compliance matrix, which represents the accuracy of this compliant mechanism under the external force applied on the output node.

Similarly, the relationship between the output force and the input displacements can be calculated as follows:

$$\begin{Bmatrix} x_{in1} \\ x_{in2} \\ x_{in3} \end{Bmatrix} = \begin{bmatrix} K_1 \\ K_2 \\ K_3 \end{bmatrix} f_{out} \tag{36}$$

where K_i (i = 1,2,3) can be utilized to analyze the influences of the output force on the input displacements.

4. Verification and Discussion

To validate the conciseness and accuracy of the derived modeling method, two typical precision positioning platforms with complex serial–parallel topology are exemplarily conducted. The first case is employed to compare the results of the derived modeling method with those of commercial FEA software Workbench 15.0 (Ansys, Pittsburgh, Pennsylvania, USA). The second case is employed to compare the derived modeling method with an energy-based approach and the matrix displacement approach in previous literature.

4.1. First Example

In the first example, a 2-DOF precision positioning platform is considered to help illustrate and validate the conciseness and accuracy of this modeling method, which is shown in Figure 8. This platform consists of two compliant amplifiers and flexible guiding beams. The motions in the x-and y-directions are decoupled by flexible guiding beams, and it has similar characteristics in the two directions. The connection relationships of each branch chain can be divided into series substructure, closed-loop parallel substructure, and opened-loop parallel substructure.

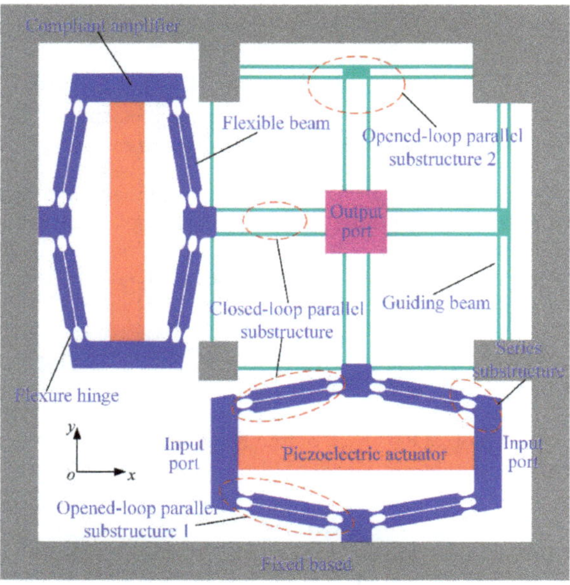

Figure 8. The 2-DOF precision positioning platform.

In order to establish the kinetostatic model, the compliant amplifier is first discretized into flexure hinges, flexible beams, and lumped masses. After discretization, it is simplified as a topology including a finite number of elements, which is shown in Figure 9a. The flexible elements are denoted from (1) to (30) and are connected with nodes from 1 to 23. The input forces and output force are numbered as f_{in1}, f_{in2}, and f_{out1}, respectively. According to the connection types of the flexible elements, the transfer matrices of two branch chains from input nodes to the output node is obtained as follows:

$$\begin{cases} T_{ch1} = (Q_{eq1}T_{eq1})(ST_1)(SQ_{in,1}) \\ T_{ch2} = (Q_{eq2}T_{eq2})(ST_9)(SQ_{in,2}) \end{cases} \tag{37}$$

where T_i, T_{eqi}, and S are transfer matrices and can be obtained by Equations (12) and (21). $Q_{in,1}$, Q_{eq1}, $Q_{in,2}$, and Q_{eq2} are summation matrices, which can be expressed as:

$$\begin{cases} Q_{in,1} = \begin{bmatrix} E_{6\times6} & 0_{6\times6} \\ K_{eq1} & -E_{6\times6} \end{bmatrix}, Q_{eq1} = \begin{bmatrix} E_{6\times6} & 0_{6\times6} \\ K_8 & E_{6\times6} \end{bmatrix} \\ Q_{in,2} = \begin{bmatrix} E_{6\times6} & 0_{6\times6} \\ K_{eq2} & -E_{6\times6} \end{bmatrix}, Q_{eq2} = \begin{bmatrix} E_{6\times6} & 0_{6\times6} \\ K_{17} & E_{6\times6} \end{bmatrix} \end{cases} \tag{38}$$

where K_{eq1} and K_{eq2} are equivalent stiffness matrices of elements (24)–(30) and elements (17)–(23).

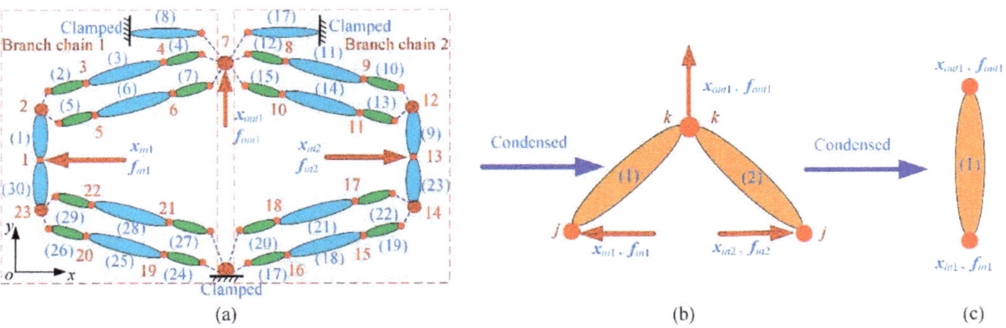

Figure 9. Condensed topology of the compliant amplifier: (**a**) numbering and discretization; (**b**) the equivalent topology; (**c**) the equivalent beam element.

Based on the above analysis, the compliant amplifier with complex serial–parallel topology is condensed into an equivalent topology including only the input nodes and the output node, which is shown in Figure 9b. The force-displacement relationship of node j and node k of equivalent beam elements can be obtained from Equation (29).

$$\begin{Bmatrix} f_{i,j} \\ f_{i,k} \end{Bmatrix} = \begin{bmatrix} k_{i,1} & k_{i,2} \\ k_{i,3} & k_{i,4} \end{bmatrix} \begin{Bmatrix} x_{i,j} \\ x_{i,k} \end{Bmatrix} \tag{39}$$

This research takes the input/output nodes as research objects; the force-displacement relationship of the input/output nodes of the compliant amplifier can be calculated based on the equilibrium equations of nodal forces in the form of a matrix.

$$\begin{Bmatrix} f_{in1} \\ f_{in2} \\ f_{out1} \end{Bmatrix} = \begin{bmatrix} k_{1,1} & 0 & k_{1,2} \\ 0 & k_{2,1} & k_{2,2} \\ k_{1,3} & k_{2,3} & k_{1,4}+k_{2,4} \end{bmatrix} \begin{Bmatrix} x_{in1} \\ x_{in2} \\ x_{out1} \end{Bmatrix} \tag{40}$$

Due to the symmetrical configuration, the input force f_{in2} can be calculated by the input force f_{in1}, f_{in2} = $(-)f_{in1}$. The force-displacement relationship of the input/output nodes of the compliant amplifier can be further expressed as:

$$\begin{Bmatrix} f_{in1} \\ f_{out1} \end{Bmatrix} = \begin{bmatrix} kk_{1,1} & kk_{1,2} \\ kk_{1,3} & kk_{1,4} \end{bmatrix} \begin{Bmatrix} x_{in1} \\ x_{out1} \end{Bmatrix} = K_{e1} \begin{Bmatrix} x_{in1} \\ x_{out1} \end{Bmatrix} \tag{41}$$

where $kk_{1,1}$, $kk_{1,2}$, $kk_{1,3}$, and $kk_{1,4}$ are block sub-matrices of the equivalent beam element, which can be expressed as:

$$\begin{cases} kk_{1,1} = k_{1,1} + k_{4,4}, kk_{1,2} = k_{1,2} \\ kk_{1,3} = k_{1,3} - k_{2,3}(k_{2,1} + k_{3,4})^{-1} kk_{1,1} \\ kk_{1,4} = (k_{1,4} + k_{2,4}) - k_{2,3}(k_{2,1} + k_{3,4})^{-1}(k_{1,2} - k_{2,2}) \end{cases} \tag{42}$$

Based on the above analysis, the compliant amplifier is condensed into an equivalent beam element, which is shown in Figure 9c. Equation (41) establishes the two-port kinetostatic model of bridge-type compliant amplifiers and can be further utilized to condense the general precision positioning platform with bridge-type compliant amplifiers.

The 2-DOF precision positioning platform is then simplified as an equivalent topology, as shown in Figure 10a. According to the connection types of the flexible elements, the transfer matrices of branch chains 1 and 2 can be expressed as follows:

$$\begin{cases} T_{E1} = T_b(ST_{e1}) \\ T_{E2} = [Tr_1]^{\mathrm{T}}[T_{E1}][Tr_1] \end{cases} \tag{43}$$

where T_b and T_{e1} are transfer matrices and can be obtained by Equations (12) and (41). Tr_1 is the rotation transformation matrix, and the rotary angle is 90°.

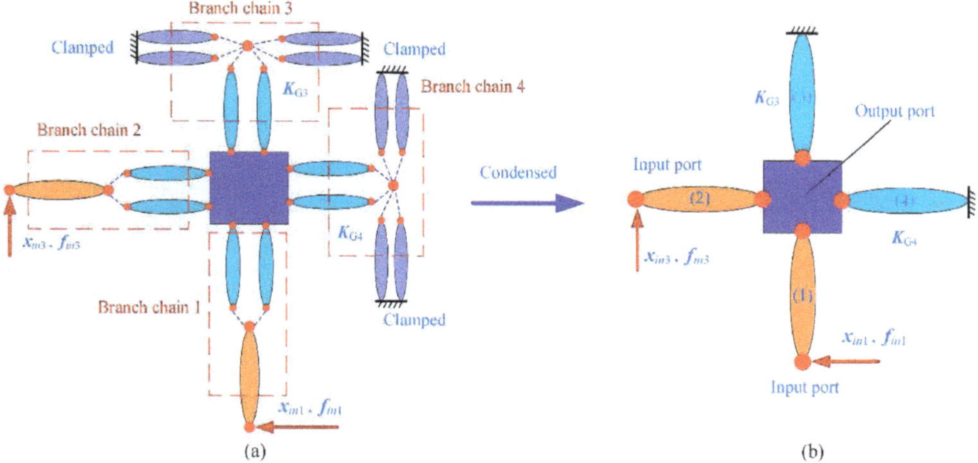

Figure 10. The 2-DOF precision positioning platform: (**a**) numbering and discretization; (**b**) the equivalent topology.

For the non-driven branch chains, it can be simplified as an equivalent flexible beam based on the compliance matrix approach. The theoretical stiffness matrices of branch chains 3 and 4 are numbered as K_{G3} and K_{G4}.

Through the above condensation, the 2-DOF precision positioning platform is further simplified as an equivalent topology, as shown in Figure 10b. The force-displacement relationship of the input/output nodes of the precision positioning platform can then be calculated based on the equilibrium equations of nodal forces in the form of a matrix.

$$\begin{Bmatrix} f_{in1} \\ f_{in3} \\ f_{out} \end{Bmatrix} = \begin{bmatrix} k_{E1,1} & 0 & k_{E1,2} \\ 0 & k_{E2,1} & k_{E2,2} \\ k_{E1,3} & k_{E2,3} & k_{E1,4} + k_{E2,4} + K_{G3} + K_{G4} \end{bmatrix} \begin{Bmatrix} x_{in1} \\ x_{in3} \\ x_{out} \end{Bmatrix} \tag{44}$$

The relationships between the output displacement, input displacements, and input forces of the precision positioning platform can be obtained from Equation (40) under the output force $f_{out} = \mathbf{0}$.

$$\begin{cases} x_{out} = \begin{bmatrix} C_1 & C_2 \end{bmatrix} \begin{Bmatrix} f_{in1} \\ f_{in3} \end{Bmatrix} \\ \begin{Bmatrix} x_{in1} \\ x_{in3} \end{Bmatrix} = \begin{bmatrix} C_{1,1} & C_{1,2} \\ C_{2,1} & C_{2,2} \end{bmatrix} \begin{Bmatrix} f_{in1} \\ f_{in3} \end{Bmatrix} \\ x_{out} = \begin{bmatrix} C_1 & C_2 \end{bmatrix} \begin{bmatrix} C_{1,1} & C_{1,2} \\ C_{2,1} & C_{2,2} \end{bmatrix}^{-1} \begin{Bmatrix} f_{in1} \\ f_{in3} \end{Bmatrix} \end{cases} \tag{45}$$

where C_1 and C_2 are the output coupling compliances. C_{ij} ($i = j$) is the input compliance and C_{ij} ($i \neq j$) is the input coupling compliance. $f_{in1} = [-f_{\text{pzt}}, 0, 0, 0, 0, 0]^{\text{T}}$ and $f_{in3} = [f_{\text{pzt}}, 0, 0, 0, 0, 0]^{\text{T}}$ denote the input forces from the piezoelectric actuators.

The static performances of the precision positioning platform, such as input stiffness K_{in} and displacement amplification ratio R, can be expressed as:

$$K_{in} = -\frac{f_{\text{pzt}}}{x_{in1(x)}} = \frac{f_{\text{pzt}}}{x_{in3(y)}}, \quad R = \frac{x_{out(y)}}{2 \cdot x_{in1(x)}} \tag{46}$$

where $x_{out(y)}$, $x_{in3(y)}$, and $x_{in1(x)}$ are the nodal displacements in the y-direction and the x-direction.

For comparison, the precision positioning platform is analyzed by employing Workbench 15.0. The geometric parameters of this platform are defined in Figure 11, and the specific values are listed in Table 1. When an actuating force of f_{pzt} = 10 N along x-direction is applied to the input nodes, the output displacement in the y-direction and input displacements in the x-direction can be obtained from the Workbench 15.0. The finite element analysis results of input stiffness and displacement amplification ratio can then be calculated based on Equation (46). As shown in Figure 12, four sets of simulation results for input stiffness and displacement amplification ratio are carried out by selecting a different angle θ of the bridge-type compliant amplifier, while keeping other parameters unchanged. Quantitative comparison between the derived kinetostatic model and commercial FEA software Workbench 15.0 is shown in Table 2.

Table 1. Geometric and material parameters of the platform.

Parameters	Values	Parameters	Values	Parameters	Values
l_1 (mm)	14.0	l_4 (mm)	16	b (mm)	0.6
h_1 (mm)	4.0	h_4 (mm)	0.5	h_e (mm)	0.3
l_2 (mm)	11.0	l_5 (mm)	15	w (mm)	10.0
h_2 (mm)	1.5	h_5 (mm)	0.5	θ (deg)	10.0
l_3 (mm)	15.0	l_6 (mm)	3.0	E (GPa)	200
h_3 (mm)	0.5	a (mm)	1.2	G (GPa)	77.64

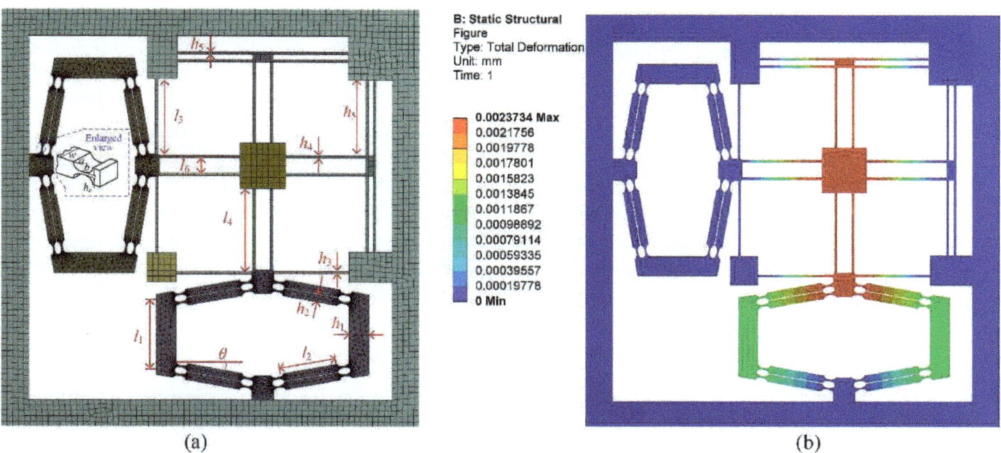

Figure 11. Finite element analysis. (**a**) geometric parameters; (**b**) result of one simulation set.

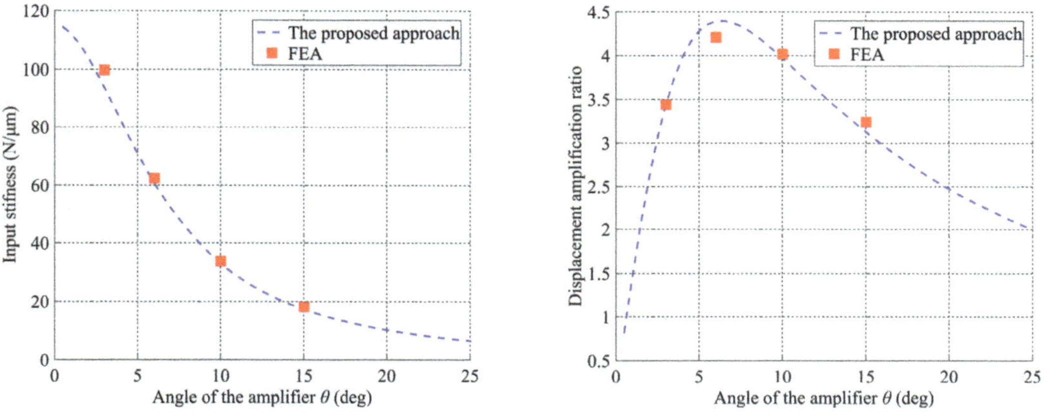

Figure 12. Simulation results of static performances.

Table 2. Comparison results of static performances.

Angle θ (deg)	K_{in} (N/µm)			R		
	The Proposed Approach	FEA	Error (%)	The Proposed Approach	FEA	Error (%)
3	93.61	99.80	6.20	3.44	3.24	6.17
6	60.72	62.50	2.85	4.39	4.21	4.28
10	33.07	33.90	2.45	3.97	4.02	1.24
15	17.30	18.18	4.84	3.12	3.05	2.30

From the calculated results shown in Figure 12, it can be observed that the theoretical calculation results of the derived kinetostatic model can match well with those of the commercial software Workbench 15.0. At the same time, the comparison results listed in Table 2 show that relative errors of the theoretical model in regard to the commercial software Workbench 15.0 are less than 6.2% regarding both the input stiffness and displacement amplification ratio. These results also indicate that the derived modeling approach can predict the kinetostatic performances of compliant mechanisms with satisfactory accuracy.

4.2. Second Example

In the second example, the 3-DOF precision positioning platform in references [24,35], which is shown in Figure 13, is employed to validate the proposed modeling method by comparing the kinematic and static performances of this positioning platform with four approaches: (a) the proposed modeling approach; (b) the matrix displacement approach; (c) the energy-based approach; (d) the finite element analysis by the commercial software Workbench 15.0. For comparability, the material and geometric parameters of this platform are consistent with those in references [24,35]. The 3-DOF precision positioning platform is then discretized and numbered. It can be observed that this precision positioning platform is composed of 33 elements and 25 nodes. The input forces and output force are, respectively, numbered as f_{in1}, f_{in2}, f_{in3}, and f_{out}. Considering the symmetry configuration of the platform, only one branch chain is employed to illuminate the proposed modeling approach. According to the connection types of flexible elements, the transfer matrix of branch chain (1) can be calculated as:

$$T_{ch1} = (T_{11})(ST_{10})(ST_9)(ST_8)(SQ_5T_5)(ST_4)(ST_3)(SQ_{in,1}) \tag{47}$$

where T_i and S are transfer matrices and given by Equations (12) and (21). $Q_{in,1}$ and Q_5 are summation matrices, which can be expressed as:

$$Q_{in,1} = \begin{bmatrix} E_{6\times6} & 0_{6\times6} \\ K_1 + K_2 & -E_{6\times6} \end{bmatrix}, Q_5 = \begin{bmatrix} E_{6\times6} & 0_{6\times6} \\ K_{eq} & E_{6\times6} \end{bmatrix} \tag{48}$$

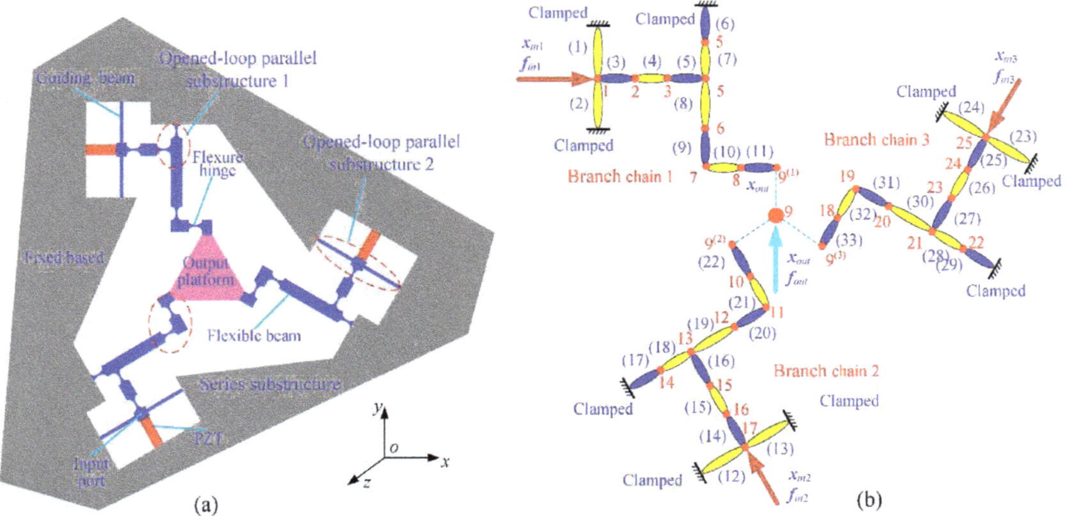

Figure 13. The 3-DOF precision positioning platform: (**a**) schematic; (**b**) discretization and numbering.

By performing rotation transformation, the transfer matrices of branch chains (2) and (3) can be expressed as:

$$\begin{cases} [T_{ch2}] = [Tr_2]^{\mathrm{T}}[T_{ch1}][Tr_2] \\ [T_{ch3}] = [Tr_3]^{\mathrm{T}}[T_{ch1}][Tr_3] \end{cases} \tag{49}$$

where Tr_2 and Tr_3 are rotation transformation matrices. The rotary angles of Tr_2 and Tr_3 are 120° and 240°, respectively.

Through the above analysis, the 3-DOF precision positioning platform in Figure 13 can be simplified as an equivalent topology only including input nodes and output node, which is shown in Figure 14. This research takes input/output nodes as the research objects and calculates the force-displacement relationship of the input/output node of the precision positioning platform based on the principle of equilibrium of nodal forces in the form of a matrix.

$$\begin{Bmatrix} f_{in1} \\ f_{in2} \\ f_{in3} \\ f_{out} \end{Bmatrix} = \begin{bmatrix} k_{in1,1} & 0 & 0 & k_{in1,2} \\ 0 & k_{in2,1} & 0 & k_{in2,2} \\ 0 & 0 & k_{in3,1} & k_{in3,2} \\ k_{in1,3} & k_{in2,3} & k_{in3,3} & k_{in1,4} + k_{in2,4} + k_{in3,4} \end{bmatrix} \begin{Bmatrix} x_{in1} \\ x_{in2} \\ x_{in3} \\ x_{out} \end{Bmatrix} \tag{50}$$

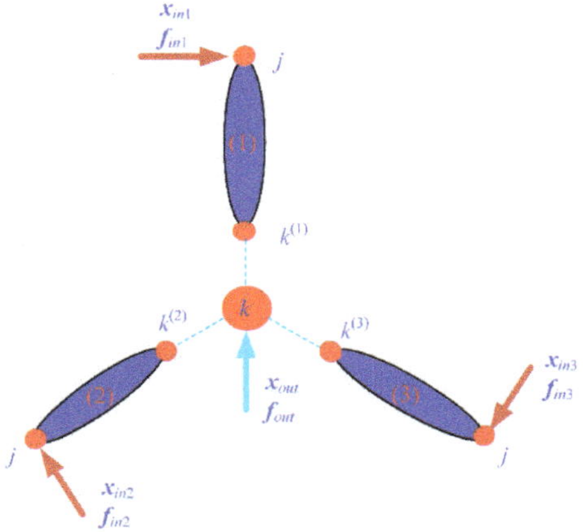

Figure 14. The equivalent topology of this platform.

According to the force-displacement relationship of the 3-DOF precision positioning platform, the kinetostatic performances, such as input/output stiffness, coupling stiffness, and input/output displacements relations can be derived by Equations (32)–(36).

For finite element analysis, the 3-DOF precision positioning platform is modeled through SolidWorks, and then the kinetostatic performances can be calculated with Workbench 15.0. The material and geometric parameters of the platform are obtained from references [24,35]. In order to calculate the kinetostatic performances in Workbench 15.0, two sets of load conditions of $f_{in1} = f_{in2} = f_{in3}$ = 100 N and f_{out} = [0 N, 50 N, 0 N, 0 N·m, 0 N·m, 1 N·m] are, respectively, applied on the input nodes and the output node of the precision positioning platform. The input displacements and output displacement results obtained by commercial FEA software Workbench 15.0 under two conditions are shown in Figure 15. Meanwhile, the detailed comparative results with four different approaches are listed in Tables 3 and 4.

Table 3. The comparative results of the four different methods under the input force of $f_{in1} = f_{in2} = f_{in3}$ = 100 N.

Approaches	$f_{in1} = f_{in2} = f_{in3}$ = 100 N								Elements
	x_{in1} (μm)	x_{in2} (μm)		x_{in3} (μm)		x_{out} (μm/rad)			
	u	u	v	u	v	u	v	θ	
FEA	42.5	−21.4	36.9	−21.4	−36.9	0	0	0.0116	120,000
Reference [24]	46.3	−23.1	40.1	−23.2	−40.1	0	0	0.0130	9
Reference [35]	41.1	−20.5	35.6	−20.5	35.6	0	0	0.0114	33
The proposed approach	41.1	−20.5	35.6	−20.6	35.5	0	0	0.0114	3
Error	3.1%	4.2%	3.5%	4.2%	3.5%	0	0	1.7%	

Table 4. The input/output displacements under the output force of f_{out} = [0 N, 50 N, 0 N, 0 N·m, 0 N·m, 1 N·m].

Approaches	f_{out} = [0 N, 50 N, 0 N, 0 N·m, 0 N·m, 1 N·m]								Elements
	x_{in1} (μm)	x_{in2} (μm)		x_{in3} (μm)		x_{out} (μm/rad)			
	u	u	v	u	v	u	v	θ	
FEA	−47.5	10.7	−18.1	22.1	39.4	0	68.4	0.0149	120,000
The proposed approach	−47.8	10.5	−18.3	22.7	39.2	0	68.7	0.0150	3
Error	0.6%	1.9%	1.1%	2.7%	0.5%	0	0.4%	0.7%	

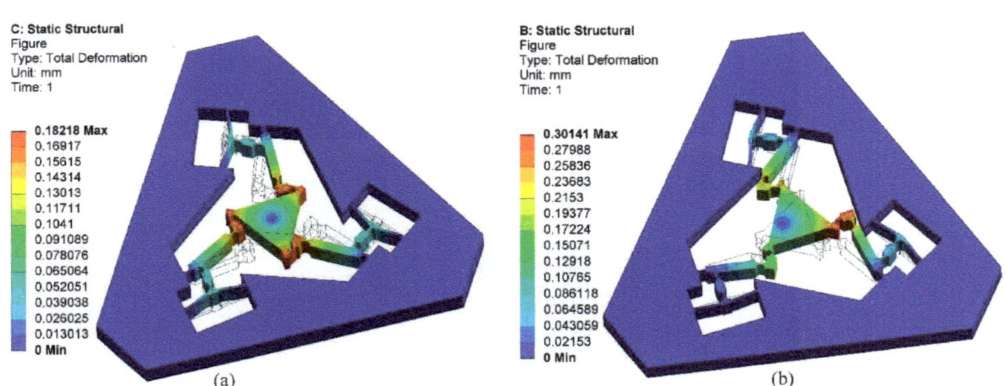

Figure 15. Finite element results. (**a**) $f_{in1} = f_{in2} = f_{in3}$ = 100 N; (**b**) f_{out} = [0 N, 50 N, 0 N, 0 N·m, 0 N·m, 1 N·m].

It can be observed that the theoretical calculation results of the proposed modeling approach and the energy-based approach in reference [35] are closer to those of the commercial FEA software Workbench 15.0 than the matrix displacement approach in reference [24]. However, the proposed modeling approach can describe the kinetostatic performances of the precision positioning platform with a lower number of elements than the energy-based approach in reference [35]. The maximum deviation between the proposed modeling approach and commercial FEA software is less than 4.3% under the input force of $f_{in1} = f_{in2} = f_{in3}$ = 100 N and less than 2.8% under the output force of f_{out} = [0 N, 50 N, 0 N, 0 N·m, 0 N·m, 1 N·m] for the input/output displacements. However, the number of elements in the finite element analysis is one hundred and twenty thousand, while the proposed modeling approach only needs three elements. To sum up, the proposed modeling approach can accurately calculate the kinetostatic performances of compliant mechanisms with a low number of elements by using simple steps and concise equations.

5. Conclusions

This paper develops a substructure condensed approach for the design and analysis of compliant mechanisms with complex serial–parallel topology. The element transfer matrix of the common flexible element is established by employing the energy conservation law. Based on the equilibrium equations of nodal forces and the transfer matrix approach, this paper calculates the general kinetostatic model of compliant mechanisms with a low number of elements. The kinematic and static performances, for instance, input stiffness, output stiffness, coupling stiffness, and input/output displacement relations with concise and explicit forms, can be easily obtained from the proposed kinetostatic model. The proposed modeling approach is verified by comparing this approach with existing modeling approaches and finite element analysis. The comparison results validate the conciseness and efficiency of this approach. The substructure condensed approach proposed in this paper can be further utilized in the optimization design and motion control for complex compliant mechanisms.

Author Contributions: Conceptualization, S.W. and Z.S.; methodology, S.W.; software, S.W.; validation, S.W.; formal analysis, S.W.; writing—original draft preparation, S.W.; writing—review and editing, Z.S. and H.F.; visualization, S.W.; supervision, Z.S. and H.F.; project administration, Z.S. and H.F.; funding acquisition, Z.S. and H.F. All authors have read and agreed to the published version of the manuscript.

Funding: This work is supported by the National Natural Science Foundation of China (grant no. 51475114) and the Opening Project of Key Laboratory of Operation Safety Technology on Transport Vehicles, Ministry of Transport, PRC (grant no. 2020-8409).

Institutional Review Board Statement: Not applicable.

Informed Consent Statement: Not applicable.

Data Availability Statement: Not applicable.

Conflicts of Interest: The authors declare no conflict of interest.

References

1. Howell, L.L. *Compliant Mechanisms;* John Wiley & Sons: New York, NY, USA, 2001.
2. Chen, G.; Han, Q.; Jin, K. A fully compliant tristable mechanism employing both tensural and compresural segments. *J. Mech. Robot.* **2019**, *12*, 011003. [CrossRef]
3. Valentini, P.P.; Pennestrì, E. Compliant four-bar linkage synthesis with second-order flexure hinge approximation. *Mech. Mach. Theory* **2018**, *128*, 225–233. [CrossRef]
4. Valentini, P.P.; Cirelli, M.; Pennestrì, E. Second-order approximation pseudo-rigid model of flexure hinge with parabolic variable thickness. *Mech. Mach. Theory* **2019**, *136*, 178–189. [CrossRef]
5. Liao, S.; Ding, B.; Li, Y. Design, assembly, and simulation of flexure-based modular micro-positioning stages. *Machines* **2022**, *10*, 421. [CrossRef]
6. Xiao, X.; Xi, R.; Li, Y.; Tang, Y.; Ding, B.; Ren, H.; Meng, M. Design and control of a novel electromagnetic actuated 3-DoFs micropositioner. *Microsyst. Technol.* **2021**, *27*, 3763–3772. [CrossRef]
7. Yang, M.; Du, Z.; Chen, F.; Dong, W.; Zhang, D. Kinetostatic modelling of a 3-PRR planar compliant parallel manipulator with flexure pivots. *Precis. Eng.* **2017**, *48*, 323–330. [CrossRef]
8. Shao, Z.; Wu, S.; Wu, J.; Fu, H. A novel 5-DOF high-precision compliant parallel mechanism for large-aperture grating tiling. *Mech. Sci.* **2017**, *8*, 349–358. [CrossRef]
9. Zhu, Z.; To, S.; Zhang, S. Theoretical and experimental investigation on the novel end-fly-cutting-servo diamond machining of hierarchical micro-nanostructures. *Int. J. Mach. Tools Manuf.* **2015**, *94*, 15–25. [CrossRef]
10. Wei, Y.; Xu, Q. Design of a new passive end-effector based on constant-force mechanism for robotic polishing. *Robot. Comput.-Integr. Manuf.* **2022**, *74*, 102278. [CrossRef]
11. Ding, B.; Zhao, J.; Li, Y. Design of a spatial constant-force end-effector for polishing/deburring operations. *Int. J. Adv. Manuf. Technol.* **2021**, *116*, 3507–3515. [CrossRef]
12. Ding, B.; Li, X.; Li, Y. Configuration design and experimental verification of a variable constant-force compliant mechanism. *Robotica* **2022**, *40*, 3463–3475. [CrossRef]
13. Ursi, P.; Rossi, A.; Botta, F.; Belfiore, N.P. Analytical Modeling of a New Compliant Microsystem for Atherectomy Operations. *Micromachines* **2022**, *13*, 1094. [PubMed]
14. Shi, H.; Su, H.J.; Dagalakis, N. A stiffness model for control and analysis of a MEMS hexapod nanopositioner. *Mech. Mach. Theory* **2014**, *80*, 246–264. [CrossRef]
15. Su, H.J.; Shi, H.; Yu, J. A symbolic formulation for analytical compliance analysis and synthesis of flexure mechanisms. *J. Mech. Des. Trans. ASME* **2012**, *134*, 051009. [CrossRef]
16. Ling, M.; Howell, L.L.; Cao, J.; Chen, G. Kinetostatic and dynamic modeling of flexure-based compliant mechanisms: A survey. *Appl. Mech. Rev.* **2019**, *72*, 030802. [CrossRef]
17. Ling, M.; Cao, J.; Jiang, Z.; Lin, J. Modular kinematics and statics modeling for precision positioning stage. *Mech. Mach. Theory* **2017**, *107*, 274–282. [CrossRef]
18. Ling, M.; Cao, J.; Zeng, M.; Lin, J.; Inman, D.J. Enhanced mathematical modeling of the displacement amplification ratio for piezoelectric compliant mechanisms. *Smart Mater. Struct.* **2016**, *25*, 075022. [CrossRef]
19. Lobontiu, N.; Garcia, E. Analytical model of displacement amplification and stiffness optimization for a class of flexure-based compliant mechanisms. *Comput. Struct.* **2003**, *81*, 2797–2810. [CrossRef]
20. Chen, W.; Zhang, X.; Li, H.; Wei, J.; Fatikow, S. Nonlinear analysis and optimal design of a novel piezoelectric-driven compliant microgripper. *Mech. Mach. Theory* **2017**, *118*, 32–52. [CrossRef]

21. Zhu, Z.; To, S.; le Zhu, W.; Li, Y.; Huang, P. Optimum Design of a Piezo-Actuated Triaxial Compliant Mechanism for Nanocutting. *IEEE Trans. Ind. Electron.* **2018**, *65*, 6362–6371. [CrossRef]
22. Li, Y.; Huang, J.; Tang, H. A compliant parallel XY micromotion stage with complete kinematic decoupling. *IEEE Trans. Autom. Sci. Eng.* **2012**, *9*, 538–553. [CrossRef]
23. Jiang, Y.; Li, T.; Wang, L. Stiffness modeling of compliant parallel mechanisms and applications in the performance analysis of a decoupled parallel compliant stage. *Rev. Sci. Instrum.* **2015**, *86*, 095109. [CrossRef] [PubMed]
24. Ling, M.; Cao, J.; Howell, L.L.; Zeng, M. Kinetostatic modeling of complex compliant mechanisms with serial-parallel substructures: A semi-analytical matrix displacement method. *Mech. Mach. Theory* **2018**, *125*, 169–184. [CrossRef]
25. Ling, M.; Cao, J.; Pehrson, N. Kinetostatic and dynamic analyses of planar compliant mechanisms via a two-port dynamic stiffness model. *Precis. Eng.* **2019**, *57*, 149–161. [CrossRef]
26. Zhu, W.; Rui, X.T. Modeling of a three degrees of freedom piezo-actuated mechanism. *Smart Mater. Struct.* **2017**, *26*, 015006. [CrossRef]
27. Ma, H.W.; Yao, S.M.; Wang, L.Q.; Zhong, Z. Analysis of the displacement amplification ratio of bridge-type flexure hinge. *Sens. Actuators A Phys.* **2006**, *132*, 730–736. [CrossRef]
28. Lobontiu, N.; Paine, J.S.N.; Garcia, E.; Goldfarb, M. Corner-filleted flexure hinges. *J. Mech. Des. Trans. ASME* **2001**, *123*, 346–352. [CrossRef]
29. Shi, R.C.; Dong, W.; Du, Z.J. Design methodology and performance analysis of application-oriented flexure hinges. *Rev. Sci. Instrum.* **2013**, *84*, 075005. [CrossRef]
30. Pham, H.H.; Chen, I.M. Stiffness modeling of flexure parallel mechanism. *Precis. Eng.* **2005**, *29*, 467–478. [CrossRef]
31. Li, Y.; Xu, Q. Design and robust repetitive control of a new parallel-kinematic XY piezostage for micro-nano manipulation. *IEEE/ASME Trans. Mechatron.* **2012**, *17*, 1120–1132. [CrossRef]
32. Li, Y.; Xu, Q. Development and assessment of a novel decoupled XY parallel micropositioning platform. *IEEE/ASME Trans. Mechatron.* **2010**, *15*, 125–135.
33. Noveanu, S.; Lobontiu, N.; Lazaro, J.; Mandru, D. Substructure compliance matrix model of planar branched flexure-hinge mechanisms: Design, testing and characterization of a gripper. *Mech. Mach. Theory* **2015**, *91*, 1–20. [CrossRef]
34. Lobontiu, N. Compliance-based matrix method for modeling the quasi-static response of planar serial flexure-hinge mechanisms. *Precis. Eng.* **2014**, *38*, 639–650. [CrossRef]
35. Wu, S.; Shao, Z.; Su, H.; Fu, H. An energy-based approach for kinetostatic modeling of general compliant mechanisms. *Mech. Mach. Theory* **2019**, *142*, 125–135. [CrossRef]
36. Chen, G.; Liu, X.; Gao, H.; Jia, J. A generalized model for conic flexure hinges. *Rev. Sci. Instrum.* **2009**, *80*, 055106. [CrossRef]
37. Chen, G.; Shao, X.; Huang, X. A new generalized model for elliptical arc flexure hinges. *Rev. Sci. Instrum.* **2008**, *79*, 095103. [CrossRef]
38. Ling, M.; Cao, J.; Jiang, Z.; Lin, J. A semi-analytical modeling method for the static and dynamic analysis of complex compliant mechanism. *Precis. Eng.* **2018**, *52*, 64–72. [CrossRef]

Article

Optimal Design for 3-PSS Flexible Parallel Micromanipulator Based on Kinematic and Dynamic Characteristics

Jun Ren *, Qiliang Li, Hanhai Wu and Qiuyu Cao

School of Mechanical Engineering, Hubei University of Technology, Wuhan 430068, China
* Correspondence: renjun@mail.hbut.edu.cn

Abstract: This paper proposes two optimal design schemes for improving the kinematic and dynamic performance of the 3-PSS flexible parallel micromanipulator according to different application requirements and conditions. Firstly, the workspace, dexterity, frequencies, and driving forces of the mechanism are successively analyzed. Then, a progressive optimization design is carried out, in which the scale parameters of this mechanism are firstly optimized to maximize the workspace, combining the constraints of the minimum global dexterity of the mechanism. Based on the optimized scale parameters, the minimum thickness and the cutting radius of the flexure spherical hinge are further optimized for minimizing the required driving forces, combined with constraints of the minimum first-order natural frequency of the mechanism and the maximum stress of the flexure spherical hinge during the movement of the mechanism. Afterward, a synchronous optimization design is proposed, in which the scale parameters are optimized to maximize the first-order natural frequency of the mechanism, combined with the constraints of a certain inscribed circle of the maximum cross-section of the workspace, the maximum stroke of the selected piezoelectric stages, and the maximum ultimate angular displacement of the flexure spherical hinge. The effectiveness of both optimization methods is verified by the comparison of the kinematic and dynamic characteristics of the original and optimized mechanism. The advantage of the progressive optimization method is that both the workspace and the driving forces are optimized and the minimum requirements for global dexterity and first-order natural frequency are ensured. The merit of the synchronous optimization method is that only the scale parameters of the mechanism need to be optimized without changing the structural parameters of the flexible spherical hinge.

Keywords: flexible parallel mechanism; 3-PSS; optimal design; kinematics; dynamics

Citation: Ren, J.; Li, Q.; Wu, H.; Cao, Q. Optimal Design for 3-PSS Flexible Parallel Micromanipulator Based on Kinematic and Dynamic Characteristics. *Micromachines* **2022**, *13*, 1457. https://doi.org/10.3390/mi13091457

Academic Editor: Alessandro Cammarata

Received: 9 August 2022
Accepted: 31 August 2022
Published: 2 September 2022

1. Introduction

The flexible parallel micromanipulator transmits force and motion through the deformation of the flexure hinge [1–3] and combines a series of advantages of the parallel mechanism and flexible mechanism such as high load carrying capacity, high accuracy, no friction, no gap, easy assembly, and so on [4–8]. At present, it has a wide range of application prospects in the fields of microelectronics, microbial experiments, precision measurement, aerospace, and other fields [9–12].

In theoretical analysis and practical application, the optimal design is an effective means to improve the performance of a flexible parallel mechanism and expand its application fields [13,14]. The performance of a flexible parallel mechanism mainly refers to the kinematic characteristics (e.g., workspace and dexterity) and dynamic characteristics (e.g., natural frequency and driving force). Ding et al. [15] proposed a novel planar micromanipulation stage with large rotational displacement and obtained the kinematic model and reachable workspace of the platform analytically. Lu et al. [16] analyzed the kinematic characteristics of a 1-translational-3-rotational (1T3R) parallel manipulator, including a reachable position workspace and orientation workspace. The size and shape of the mechanism can be determined according to the given design index to meet the design requirements [17]. The optimal design of flexible parallel micromanipulators is mostly based on kinematic performance, such as workspace, dexterity, output/input displacement amplification ratio, etc. He et al. [18] optimized the dimensions of the TCP-actuated finger mechanism with the local and global performance concerning the dexterity and the extreme value of the velocity as the evaluation indices. Ding et al. [19] proposed an FEA-based optimization method based on structural parameters to solve the problem of insufficient constant-force stroke of the compliant constant force mechanism (CFM) based on a flexible Z-shaped beam and a bistable beam. Yang et al. [20] proposed a 3-PRR-compliant parallel robot and optimized the geometric parameters of the mechanism and the flexure hinge by a genetic algorithm to obtain the desired motion performance. Xu et al. [21] established a kinematics model of a compliant mechanism with one flexible joint designed from a rigid four-bar linkage. They optimized the structural parameters of the mechanism by taking the path deviation and strain energy as two objectives. Li et al. [22] proposed a new 2-DOF-compliant micromanipulator and established a kinematics model. Then, kinematic optimization of the design parameters was carried out. Li et al. [23] also performed performance improvements and dimension optimizations on the 3-PRC-compliant parallel micromanipulator to improve several disadvantages of the mechanism in the aspects of stiffening, buckling, and parasitic motions. Xu et al. [24] carried out the optimal design of a 3-PUU flexible parallel micromanipulator by taking the maximum value of the mechanism's workspace and the weighted combination of the mechanism's dexterity as the optimization objective. Jia et al. [25] optimized the 3-PRR flexible parallel mechanisms with the workspace as the optimization objective and dexterity as the constraint condition. In addition to kinematic performance, a few scholars have also carried out research on the optimal design of flexible parallel mechanisms for dynamic performance. Wang et al. [26] proposed a compliant mechanisms optimization method based on dynamic characteristics and verified the feasibility of the optimization method based on a specific configuration. Li et al. [27] proposed a dynamics modeling and optimization method for a 2-DOF translational parallel robot with flexible links for a high-speed pick-and-place operation. The dimension of the mechanism can be optimized according to sensitivity and dynamic stress to improve the dynamic accuracy of the end-effector at high speed. Du et al. [28] carried out the optimal design of the 3-RRR flexible parallel mechanism based on dynamic performance and obtained the optimal mechanism with lightweight and small deformation. Li et al. [29] proposed a class of an XY totally decoupled parallel stage and established the kinematics and dynamics model of the mechanisms. The stage structure optimization was then carried out to achieve a maximal natural frequency under the performance constraints such as workspace. After that, Li et al. [30] presented a novel compliant parallel XY micro-motion stage, and the dimensions of the mechanism were optimized for maximizing the natural frequencies. To sum up, current optimal designs of flexible parallel micromanipulators are mostly based on kinematic performance or dynamic performance, but few of them take into account both. In practical applications, however, it is sometimes necessary to consider both kinematics and dynamics characteristics to meet specific requirements.

In our previous studies [31], a novel 3-PSS flexible parallel micromanipulator was proposed, its dynamics model was established, and its frequency characteristics were analyzed. In order to optimize the performance of the mechanism to meet both the kinematic characteristics and dynamic characteristics, this paper proposes two optimization schemes for designing the 3-PSS flexible parallel micromanipulator. One of the schemes is called the progressive optimization strategy, in which the scale parameters of the mechanism are firstly optimized according to the kinematic performance requirement and then the structural parameters of the flexure spherical hinge are further optimized according to the dynamic characteristic requirements based on the optimized scale parameters. Another scheme is a synchronized optimization strategy, in which the scale parameters of the flexible parallel micromanipulator are optimized by taking the kinematic performance and the dynamic performance as the constraint condition and the optimization object, respectively. Since the range of motion of the flexible parallel mechanism is usually small, the workspace and motion accuracy (directly related to the dexterity of the mechanism) are usually the major consideration in the kinematic performance. In dynamic performance, the natural frequency of the structure and driving forces required are often concerned more. Therefore, the optimization of kinematic performance and dynamic performance mentioned in this paper mainly refers to the workspace, dexterity, natural frequencies, and driving forces.

2. Kinematics and Dynamics Performance Analysis

The 3-PSS (P—prismatic pair; S—spherical joint) flexible parallel micromanipulator has three translational degrees of freedom in space, and its 3D structure is shown in Figure 1a. This mechanism is composed of a fixed platform, a moving platform, and three identical branches. Three branches connecting to the moving platform and the fixed platform of the mechanism are equally distributed around the moving platform by 120°. Each branch consists of a piezo stage, four flexure spherical hinges, and two rigid rods. Three translational DOFs of the moving platform can be achieved by coordinating up-and-down movements of the three piezo stages.

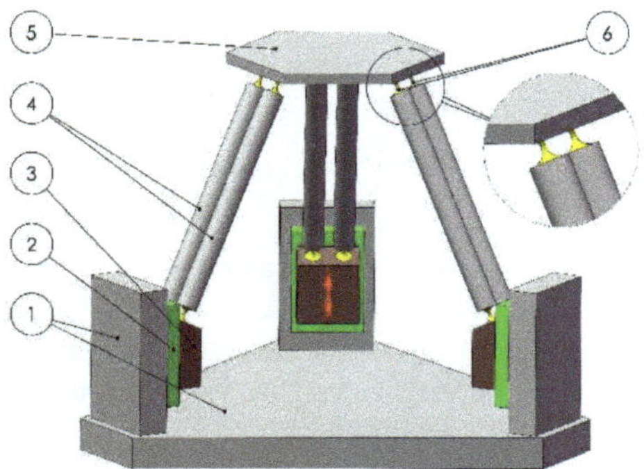

1. Fixed platform; 2. Piezo stages; 3. Slider;
4. Rod; 5. Moving platform; 6. Flexure spherical hinge.

(a)

(b)

Figure 1. (**a**) 3D structure of 3-PSS flexible parallel micromanipulator; (**b**) simplified pseudo-rigid body model and coordinate frame settings.

2.1. Kinematics Analysis of 3-PSS Flexible Parallel Micromanipulator

In order to simplify the analysis, the "simplified pseudo-rigid body model" of the 3-PSS flexible parallel micromanipulator is established, as shown in Figure 1b. The reference coordinate frame O$\{x, y, z\}$ is set at the circumcircle center of the equilateral triangle formed by centroid A_i (i = 1, 2, 3) of three sliders in the initial state, and the radius r_a of the circle is defined as the radius of the fixed platform. Similarly, the moving coordinate system P $\{x_p, y_p, z_p\}$ is established at the center P of the moving platform, and the radius of the moving platform is defined as r_p. The x-axis and y-axis of the two coordinate frames are parallel and their z-axis overlaps. In the initial state, the vertical distance between the moving platform and the fixed platform is h. The length of each link B_iP_i is l, where points B_i and P_i, respectively, represent the center points of the flexure spherical hinges at both ends of the link. For the convenience of analysis, it is assumed that the center of mass of the slider coincides with the center of mass of the flexible spherical hinge at the lower end of the connecting rod. Thus, A_iB_i represents the input displacement of the i-th slider. OA_i is the position vector from the center point O of the stationary platform to the center point A_i of the slider in the initial state, and φ_i ($\varphi_1 = 0$; $\varphi_{i+1} = \varphi_i + 2/3\pi$) is the angle between OA_i and the x-axis of the reference coordinate frame. The angle between the B_iP_i and the z-axis of the reference coordinate frame is defined as θ_l, as shown in Figure 1b.

Workspace and dexterity are important indicators for evaluating the kinematic performance of a mechanism. Therefore, the kinematic model of the mechanism is established first and the workspace and dexterity characteristics of the micromanipulator are further analyzed. The position vector of the center point P of the moving platform in the reference coordinate frame $O\{x, y, z\}$ is expressed as $P = (x, y, z)^T$, as shown in Figure 1b. The closed vector equation of the i-th branch can then be established according to the closed-loop vector method.

$$\boldsymbol{OA}_i + \boldsymbol{A}_i\boldsymbol{B}_i = \boldsymbol{OP} + \boldsymbol{PP}_i + \boldsymbol{P}_i\boldsymbol{B}_i \tag{1}$$

The inverse kinematics equation of the micromanipulator can be obtained by arranging Equation (1) as follows:

$$d_i = z - \sqrt{l^2 - (x - E_r \cos \varphi_i)^2 - (y - E_r \sin \varphi_i)^2} \tag{2}$$

where d_i is the displacement of the i-th slider.

The relationship between the input and output of the micromanipulator can be obtained by taking the derivative of Equation (2) with respect to time.

$$\dot{\boldsymbol{B}} = \boldsymbol{J}\dot{\boldsymbol{P}} \tag{3}$$

where $\boldsymbol{J}$ is the velocity Jacobian matrix of the micromanipulator. $\dot{\boldsymbol{B}}$ and $\dot{\boldsymbol{P}}$ are the velocities of the sliders and the moving platform, respectively.

2.1.1. Workspace Analysis of 3-PSS Flexible Parallel Micromanipulator

The workspace of the flexible parallel micromanipulator refers to the working area that can be achieved by the center P of the moving platform, which is an important indicator for evaluating the motion performance of the micromanipulator. Here we choose the same set of parameters as in the literature [31], as listed in Table 1.

Table 1. Dimension parameters and material characteristics of the micromanipulator.

Parameter	Value	Parameter	Value
r_p (mm)	25	θ_l (°)	60
r_a (mm)	45	E (Gpa)	200
l (mm)	65	ν	0.3
t_s (mm)	1	ρ (g/cm^3)	7.85
R_s (mm)	2.5		

During the working process of the micromanipulator, the angular displacement of the flexure spherical hinge will change with the change of the position of the moving platform. However, the angular displacements of the flexible spherical hinges on the same branch are always the same. Hence, the angular displacement of the flexure spherical hinge on the i-th branch can be expressed as:

$$\psi_i = \frac{\boldsymbol{n}_i^0 \cdot \boldsymbol{n}_i}{|\boldsymbol{n}_i^0||\boldsymbol{n}_i|} \tag{4}$$

where, $i = 1, 2, 3$ represents the i-th branch, $\boldsymbol{n}_i^0$ represents the axial direction vector of the rod in the initial state, and $\boldsymbol{n}_i$ represents the axial direction vector of the rod at any time.

In order to avoid the fracture of the flexure spherical hinge due to excessive angular displacement, the following constraint is established:

$$0 \le \psi_i \le \psi_{\max} \tag{5}$$

where $\psi_{\max}$ is the maximum ultimate angular displacement of the flexure spherical hinge, which is assumed to be $1°$.

The maximum displacement $d_{\max}$ of the selected piezoelectric stages is 200 μm, and the following constraint is established for the slider displacement d_i.

$$0 \leq d_i \leq d_{\max} \tag{6}$$

Based on the above constraints, the workspace of the micromanipulator can be calculated by employing the cylindrical limit search method [25] with the parameters given in Table 1, as shown in Figure 2a. It is shown that the workspace of the mechanism is a closed symmetrical shape, the total height z_{max} of the workspace is 200 μm, which is equal to the stroke of the slider d_{max}. Sectional shapes of the workspace at different heights are also provided in Figure 2c–f. It can be seen that the maximum cross-section of the workspace is on the plane at the height $z_{max}/2$, that is 100 μm.

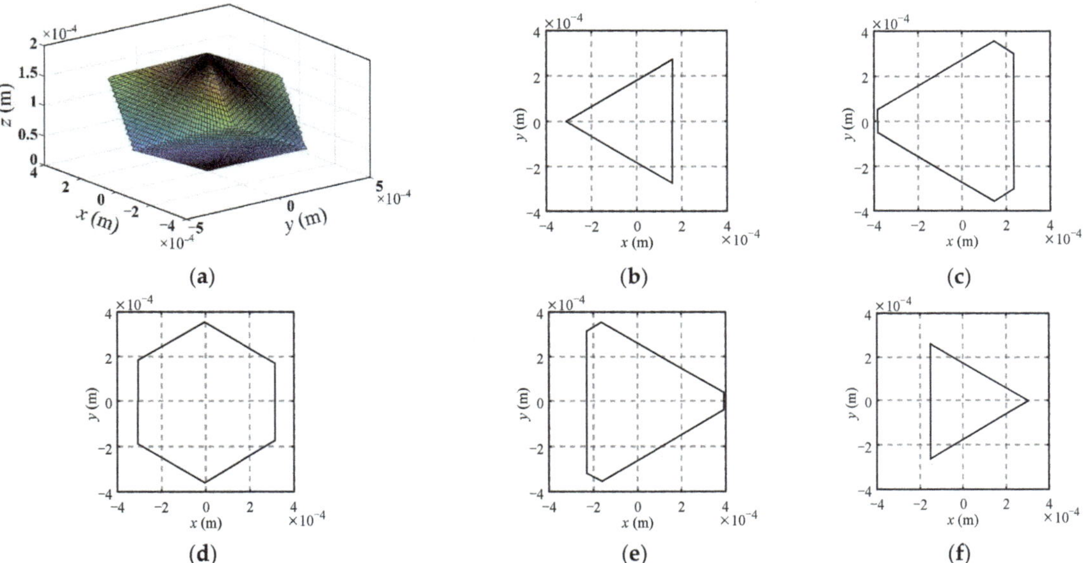

Figure 2. The workspace of the 3-PSS flexible parallel micromanipulator and sectional shapes of workspace at different heights. (**a**) The workspace; (**b**) z = 50 μm; (**c**) z = 75 μm; (**d**) z = 100 μm; (**e**) z = 125 μm; (**f**) z = 150 μm.

In order to quantify the size of the workspace, a cube covering the entire workspace of the micromanipulator is selected with a volume V which is evenly divided into N units. Then we calculate the angular displacements of the flexible spherical hinge and the displacements of the slider when the center of the mobile platform is located at the center of each unit. If the angular displacements of the flexure spherical hinge and the stroke of the driver satisfy the constraints, the unit is reserved, and the total number of reserved units is denoted as n. Thus, the volume V_w of the workspace can be expressed as:

$$V_w = \frac{n}{N} V \tag{7}$$

To clarify the influence of the scale parameters on the workspace volume of the mechanism, the variation of the workspace volume with the scale parameters is obtained by changing the rod length l of the mechanism and the difference between the moving and fixed platform radius E_r, as shown in Figure 3. It can be seen that the workspace volume of the mechanism increases with the increase in the rod length l and decreases with the increase in the difference in radius E_r.

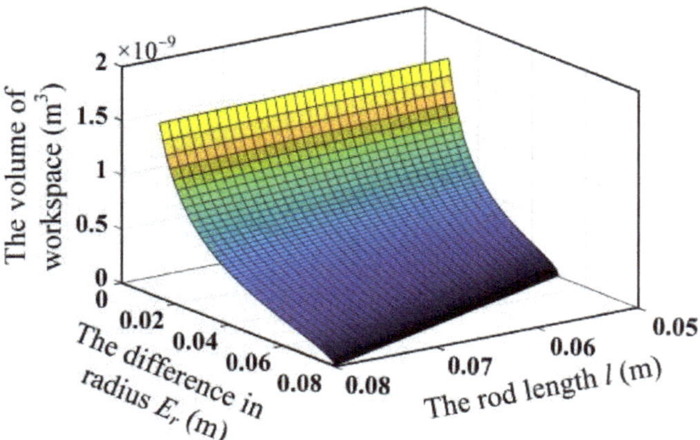

Figure 3. Relationship between workspace volume and scale parameters.

2.1.2. Dexterity Analysis of 3-PSS Flexible Parallel Micromanipulator

Dexterity is an important kinematic property of the flexible parallel micromanipulator. The general Jacobian matrix condition number k is used as a measure of the dexterity of the mechanism, where $k = \|J\|\bullet\|J^{-1}\|$, $\|\bullet\|$ represents the two-norm of the matrix. Furthermore, the dexterity of the mechanism is usually represented by the inverse of the Jacobian matrix condition number, that is $u = 1/k$. When $u = 0$, the mechanism is in odd isotropy. When $u = 1$, the mechanism is in isotropy. According to the mechanism model parameters given in Table 1, the dexterity distribution on the maximum cross-section ($z = z_{max}/2$) in a micromanipulator's workspace can be obtained, as shown in Figure 4.

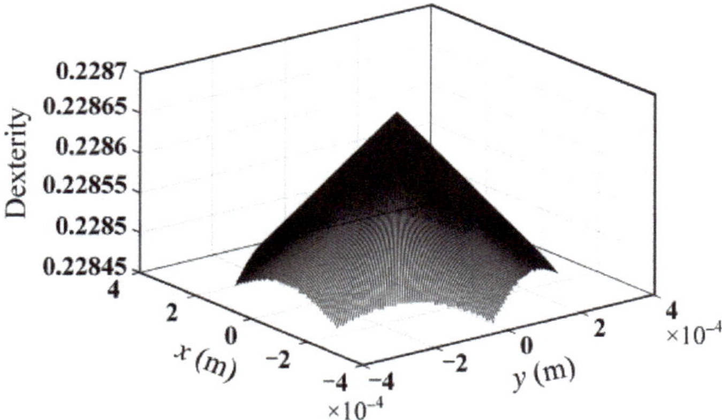

Figure 4. Dexterity within the maximum cross-section of the workspace of the mechanism.

According to the literature [24], the global dexterity in the workspace of the mechanism can be expressed as:

$$GDI \approx \frac{1}{N_w} \sum_{w \in N_w} \frac{1}{k} \tag{8}$$

where w is one of the N_w points uniformly distributed in the workspace. With the dimension parameters given in Table 1, the calculation result of the global dexterity in the workspace of the mechanism is 0.2287.

To clarify, the effect of the scale parameters on the global dexterity of the mechanism, the radius difference E_r, and the rod length l were varied, respectively. Then the variation of the global dexterity of the mechanism with the scale parameters can be obtained, as shown in Figure 5a. It can be seen from Figure 5b that the global dexterity of the mechanism decreases with the increase in rod length. Furthermore, it first increases to 1 and then decreases during the increase in the radius difference E_r, as shown in Figure 5c.

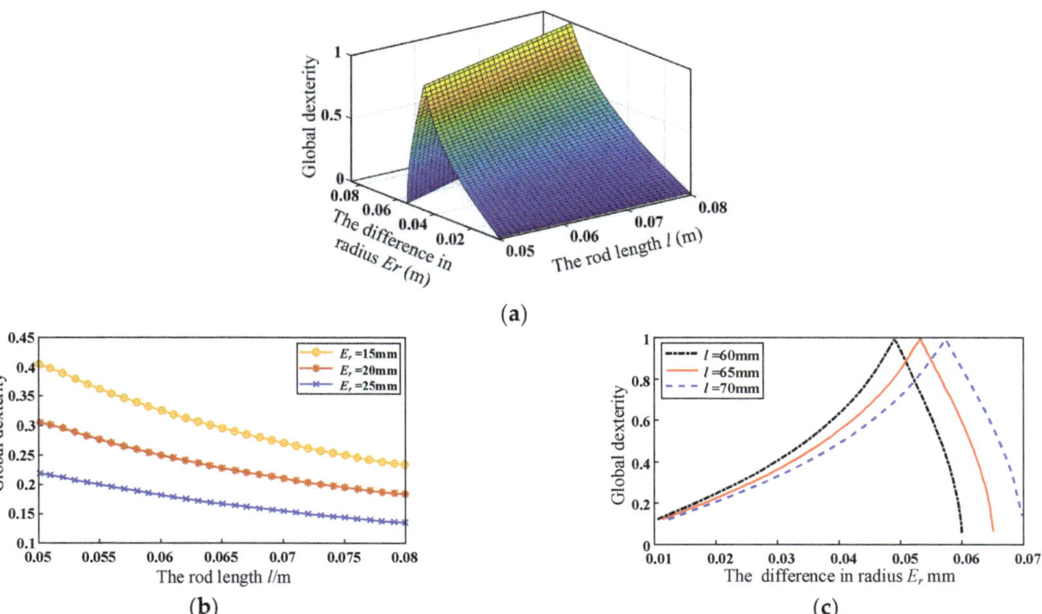

Figure 5. Relationship between scale parameters and global dexterity: (**a**) the effect of rod length l and radius difference E_r on global dexterity; (**b**) the effect of rod length l on global dexterity; (**c**) the effect of radius difference E_r on global dexterity.

2.2. Dynamics Analysis of 3-PSS Flexible Parallel Micromanipulator

Dynamic analysis is a necessary way to acquire the dynamic performance of a micromanipulator, including the relationship between the motion trajectory of the moving platform and the driving forces and natural frequency characteristics. The dynamic model of the mechanism is established by utilizing the Lagrange equation method. According to the dynamic model of the 3-PSS flexible parallel micromanipulation mechanism [31], the dynamic equation of the mechanism is expressed as:

$$\boldsymbol{M}\ddot{\boldsymbol{s}}+\boldsymbol{K}\boldsymbol{s}+\boldsymbol{G}=\boldsymbol{F} \tag{9}$$

where $\boldsymbol{M}$ is the mass matrix of the mechanism, $\boldsymbol{K}$ is the stiffness matrix of the mechanism, $\boldsymbol{G}$ is the gravity matrix of the mechanism, $\boldsymbol{s}$ is the displacement of the moving platform, and $\boldsymbol{F}$ is the generalized force exerted on the moving platform.

According to Equation (9), the undamped natural frequency of the mechanism can be determined by:

$$\left|\boldsymbol{K}-\omega^2\boldsymbol{M}\right|=0 \tag{10}$$

where ω represents the circular frequency of the mechanism, and the natural frequency of the mechanism is $f = \omega/(2\pi)$.

According to the virtual work principle, the driving force of the sliders is computed as:

$$\boldsymbol{F}_b=\boldsymbol{J}^{-\mathrm{T}}\boldsymbol{F} \tag{11}$$

where J is the velocity Jacobian matrix of the mechanism.

The motion trajectory of the moving platform is selected within the maximum cross-section of the workspace. The trajectory equation is given in Equation (12), and the driving displacement of the sliders can be obtained according to Equation (2). The motion trajectory of the moving platform and the corresponding driving displacements of the sliders are shown in Figure 6a,b, respectively.

$$\begin{cases} x = 1 \times 10^{-5}\sin(\omega t) \\ y = 1 \times 10^{-5}\cos(\omega t) \\ z = 1 \times 10^{-4} \end{cases} \tag{12}$$

where $\omega = \pi/4$.

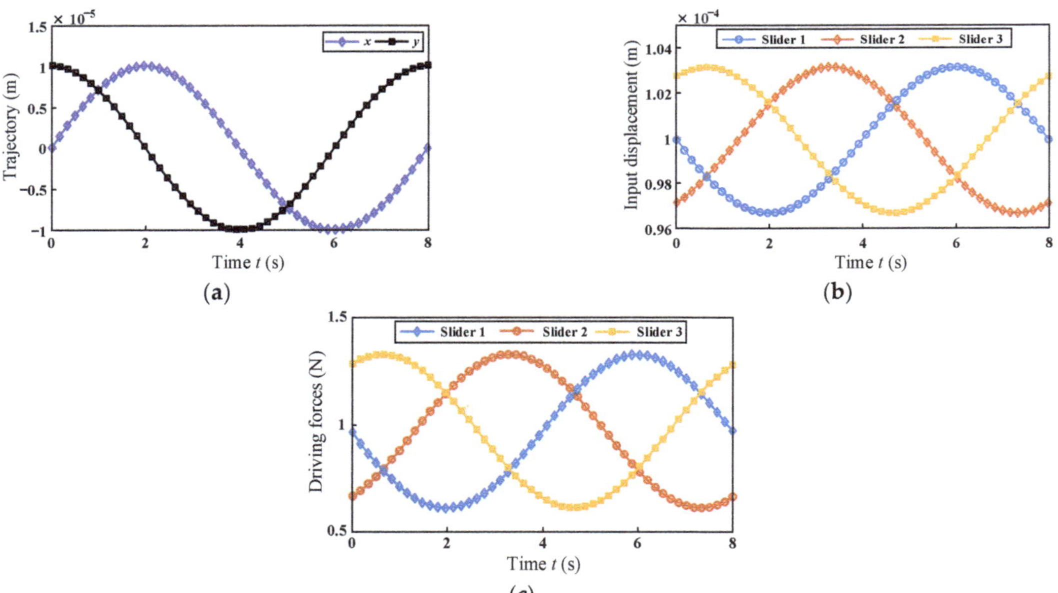

Figure 6. (**a**) Movement trajectory of the moving platform; (**b**) displacements of the sliders; (**c**) the driving forces of the micromanipulator.

With the given parameters listed in Table 1, the driving force required by each branch can be obtained. As shown in Figure 6c, the maximum absolute value F_{bM} of the driving force on each branch is identical, and the driving force on each branch of the mechanism changes regularly. By comparing Figure 6b,c, it can be seen that the variations of the driving forces are consistent with the variations of the input displacements of the branches.

According to Equations (9) and (11), it can be seen that the driving forces $\boldsymbol{F}_b$ of the sliders are related to the mass, stiffness of the mechanism, and the motion trajectory of the moving platform. For simplicity, the maximum absolute value of the driving force is used as the evaluation index to analyze the influence of the mass and stiffness of the mechanism on the driving forces under the condition of a given platform motion trajectory. According to the configuration and motion characteristics of the mechanism, the stiffness that affects the driving force $\boldsymbol{F}_b$ is mainly derived from the bending stiffness of the flexible spherical hinge. Therefore, we analyzed the variation of the driving force by changing the bending stiffness k_{bm} of the flexure spherical hinge. As shown in Figure 7, with the increase in the bending stiffness k_{bm} of the flexure spherical hinge, the maximum absolute value F_{bM} of the driving force increases linearly.

On the other hand, the influence of the mass change of different components of the mechanism on the driving force can be obtained by individually increasing the mass of the main components according to Equations (9) and (11). As shown in Figure 8, the maximum absolute value of the driving force of the mechanism also increases as the mass of the components increases. When increasing the same absolute mass Δm, the variation of the rod mass m_c has the greatest influence on the driving force, followed by the slider mass m_b. The mass m_p of the moving platform has the least influence on it.

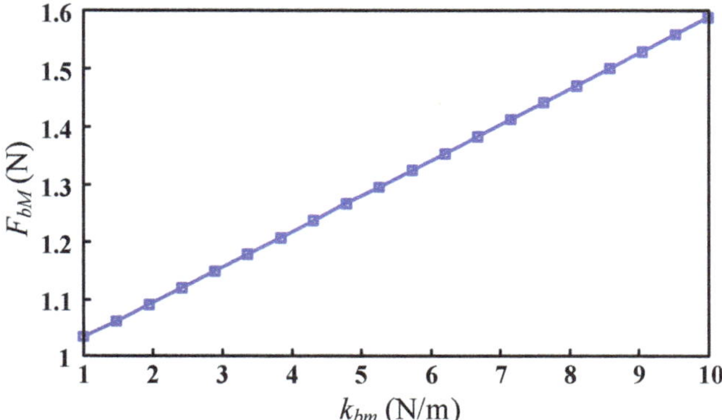

Figure 7. Influence of the bending stiffness of the flexure spherical hinge on the driving force of the mechanism.

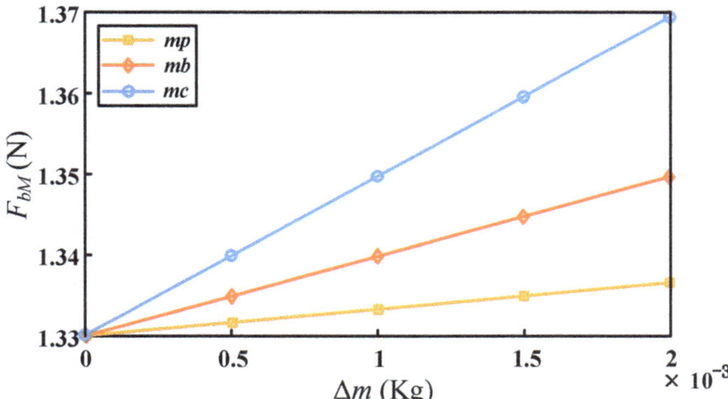

Figure 8. Influence of absolute mass change of different components on driving force.

3. Progressive Optimal Design Based on Kinematics and Dynamics

According to the analysis in the previous section, the kinematics and dynamic performance of the 3-PSS flexible parallel micromanipulator can be obtained. In order to make the mechanism meet the requirements involving certain kinematics and dynamic performance at the same time, a progressive optimization design is proposed. First, the scale parameters of the mechanism are optimized with the kinematic performance as the optimization objective. Based on the optimized scale parameters, the structural parameters of the flexure spherical hinge are further optimized by taking the dynamic performance as the optimization objective. The optimization process is shown in Figure 9.

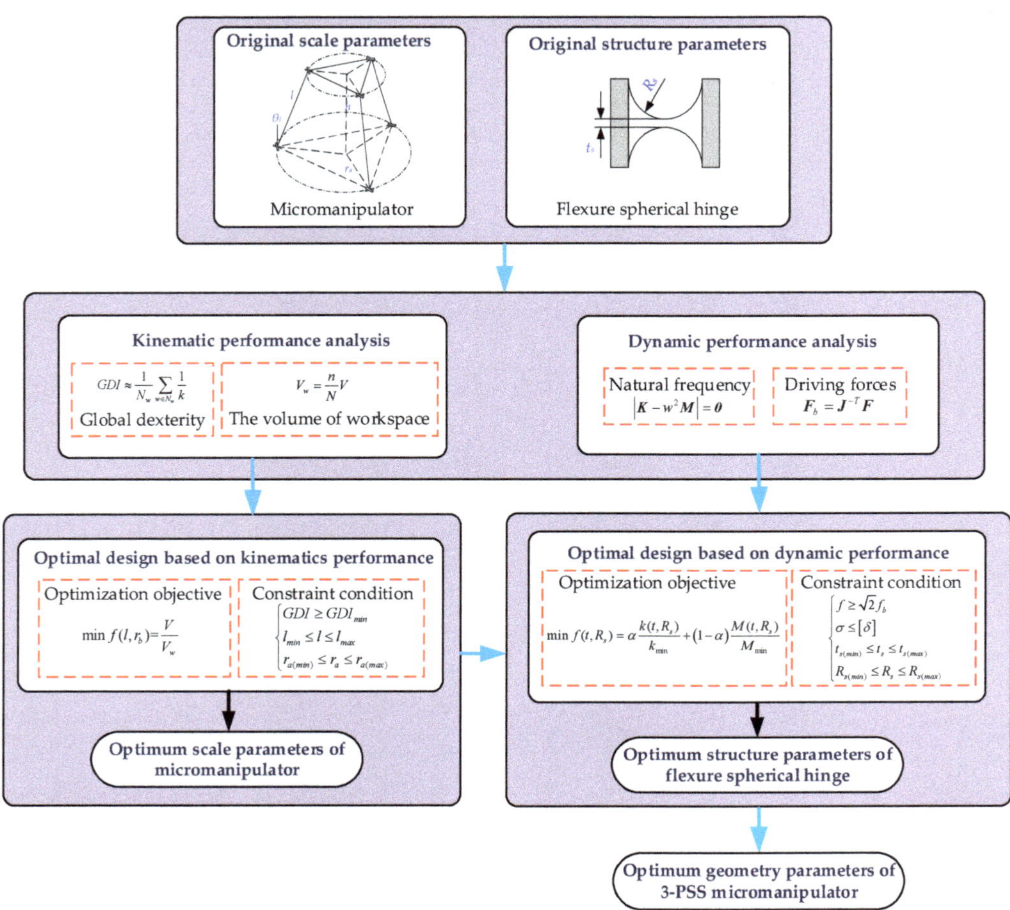

Figure 9. The progressive optimization procedure for analyzing and designing 3-PSS flexible parallel micromanipulator.

3.1. The Scale Parameters Optimization Based on Kinematic Performance

It can be known from the kinematic Equation (2) that the kinematic performance is mainly related to the rod length l and the difference between the moving and fixed platform radius E_r. In order to simplify the analysis, the radius of the moving platform of the mechanism is assumed to be a constant, and the fixed platform radius r_a and the rod length l are used as the optimization parameters.

For the convenience of optimization, the variation range of the rod length l is given as approximately $\pm 20\%$ of the original size. The corresponding variation range of the radius r_a of the fixed platform can be obtained through the variation range of the rod length l and the geometry relationship of the mechanism. It is assumed that the stroke d_{max} of the selected piezoelectric stages is 200 μm, and the ultimate angular displacement $\psi_{\max}$ of the flexure spherical hinge is 1°. When designing a flexible parallel mechanism, higher motion accuracy and a larger workspace are usually expected, but the two are contradictory. Since the motion accuracy is related to global dexterity, we take the minimum global dexterity (assumed to be 0.2 in this study) as one of the constraints, the minimum ratio of the mechanism's volume to the workspace volume as the objective function, and then combine the constraints of the scale parameters (l and r_a) to formulate the following optimization model.

$$\min f(l, r_a) = \frac{V}{V_w} \tag{13}$$

$$\begin{cases} GDI \geq 0.2 \\ 50 \text{ mm} \leq l \leq 80 \text{ mm} \\ 26 \text{ mm} \leq r_a \leq 105 \text{ mm} \end{cases} \tag{14}$$

where V_w represents the volume of the workspace of the micromanipulator and V represents the volume of the workspace of a cube covering the entire workspace of micromanipulator.

According to the above constraints and optimization objective, optimization is carried out by employing the genetic algorithm toolbox in MATLAB R2018b software (MathWorks, Natick, MA, USA). The optimized scale parameters are r_a = 38.66 mm and l = 50.13 mm, respectively. The global dexterity GDI of the optimized mechanism is 0.204, which meets the design requirements.

The dexterity distribution of the optimized mechanism on the maximum cross-section in the workspace is shown in Figure 10a. It can be seen that maximum dexterity occurs at the center of the maximum cross-section. When the position of the moving platform changes, the dexterity changes accordingly, and the dexterity gradually decreases along the direction away from the center of the maximum cross-section. The comparison of the workspace volume between original and optimized designs is shown in Figure 10b. It is demonstrated that the workspace volume of the optimized mechanism has increased by 14.17% compared with the original design. It is indicated that the optimal design of mechanism scale parameters based on kinematic performance is effective.

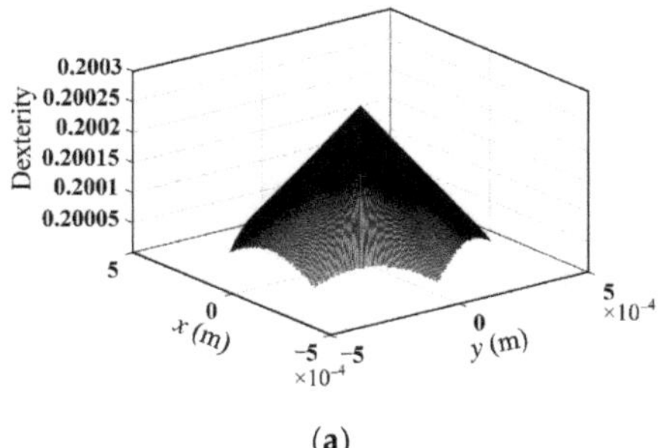

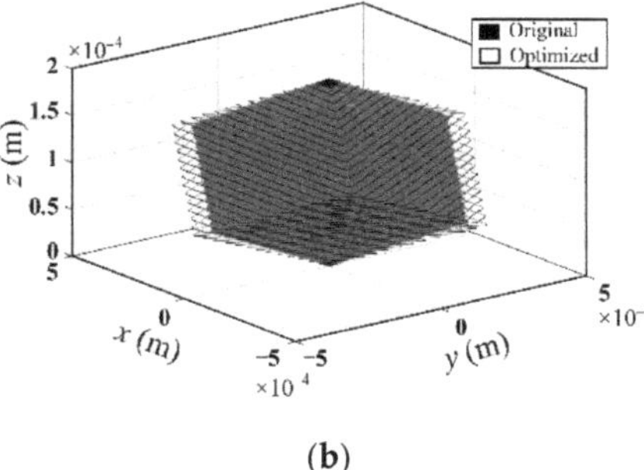

(a) (b)

Figure 10. (**a**) The dexterity within the maximum cross-section of the workspace of the optimized mechanism; (**b**) workspace comparison between original and optimized designs.

3.2. Optimization of Flexure Spherical Hinge Structure Parameters Based on Dynamic Performance

The scale parameters that meet the requirements of the kinematic performance of the micromanipulator were obtained through the above optimization design. However, the dynamic performance of the mechanism is not considered, which has an important impact on the high-frequency control scheme of the mechanism and the selection of the driver. Results from the related research [31] show that the dynamic performance of the flexible parallel micromanipulation mechanism is related to the micromanipulator's scale parameters as well as the structural parameters of the flexure spherical hinge. Among them, the structural parameters of the flexure spherical hinge are the main factors affecting the dynamic performance of the micromanipulator. Therefore, the structural parameters of the flexure spherical hinge are chosen to further optimize its dynamic performance. The flexible hinge of the 3-PSS flexible parallel micromanipulator is a right-circular flexure spherical hinge, and its structural diagram is shown in Figure 11. The main parameters of the flexible hinge are the minimum thickness t_s and the cutting radius R_s. Therefore, the minimum thickness t_s and the cutting radius R_s are selected as the optimization parameters.

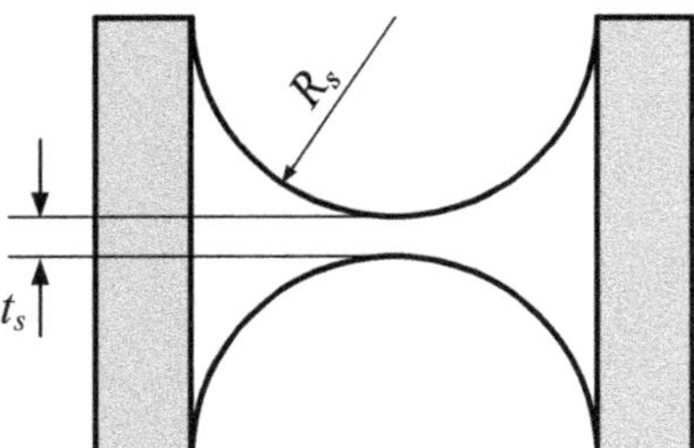

Figure 11. The schematic diagram of flexure spherical hinge structure.

The natural frequency is required to be greater than $\sqrt{2}$ times the fundamental frequency f_b of the driver (piezo stage) for preventing resonance, and f_b of the selected piezo stage is 37 Hz in this study. Meanwhile, the maximum stress σ_{max} of the flexure spherical hinge during the movement of the mechanism should be less than the permissible stress $[\sigma]$ so as to prevent fatigue fracture. Considering the requirements of machining and deformation of the flexible spherical hinge, certain size ranges (unit: mm) are given for the structural parameters t_s and R_s. In summary, the constraint expression for the optimal design can be written as:

$$\begin{cases} f \geq \sqrt{2} f_b \\ \sigma_{max} \leq [\sigma] \\ 0.8 \leq t_s \leq 1.2 \\ 1 \leq R_s \leq 10 \end{cases} \tag{15}$$

The driving force is an important dynamic performance of the micromanipulator and is therefore chosen as the optimization objective. It can be seen from the previous section that the driving force is mainly related to the total mass and stiffness of the mechanism. Generally, the driving force of the micromanipulator is smaller as the mass of the mechanism decreases. For brevity, the mass of the moving platform is set as a constant value, and the mass of the rod and the slider are set to be variable. Since the scale parameters have been determined according to the previous kinematic optimization results, the mass of the rod is only related to the diameter D of the rod. According to the 3D structure shown in Figure 1a, the diameter D of the rod is required to be not less than the sum of the minimum thickness t_s and twice the cutting radius R_s of the flexure spherical hinge. Here we assume that the diameter of rod $D = t_s + 2R_s + 2$ mm. In addition, the size of the slider is also directly related to the diameter of the rod D. In summary, the total mass of the micromanipulator is mainly related to the structural parameters of the flexure spherical hinge. It can be seen from Equations (9) and (11) that under the same motion requirement, the smaller the total mass of the mechanism, the smaller the driving force required. Hence, the minimum total mass M of the mechanism is taken as one of the optimization objectives.

It is known from previous analysis that the bending stiffness k_{bm} of the flexure spherical hinge directly affects the overall stiffness of the mechanism, and further affects the driving force and natural frequencies of the micromanipulator. From the analysis of driving force characteristics in Section 2.2, it can be seen that the less the stiffness of the flexure spherical hinge, the less the required driving force. Therefore, the minimum stiffness is chosen as another optimization objective.

Taking the above two optimization objectives into consideration, in order to make the required driving force small enough, a minimum weighted combination of the total mass of the mechanism and the bending stiffness of the flexible spherical hinge is selected as the overall optimization objective. The overall optimization objective function can be constructed as:

$$\min f(t, R_s) = \alpha \frac{k(t, R_s)}{k_{\min}} + (1 - \alpha) \frac{M(t, R_s)}{M_{\min}} \quad (16)$$

where the weight factor α ($\alpha \in [0, 1]$) represents the proportion of bending stiffness in the optimization. In order to make the two optimization objectives (k and M) in the same order of magnitude, they are divided by k_{min} and M_{min}, respectively, which represent the minimum values of bending stiffness and overall mass under the constraint conditions, respectively.

Governing the constraints in Equation (15) and the optimization objective in Equation (16), optimization is carried out by employing the genetic algorithm toolbox in MATLAB R2018b software. It should be noted that the optimization results of the structural parameters of the flexure spherical hinge are different under different weight factors. Taking the motion trajectory of Equation (17) as an example, we perform the optimization. In order to obtain the structural parameters of the flexure spherical hinge resulting in the smallest maximum absolute value of the driving force, the optimization results of structural parameters of the flexible spherical hinge under different weight factors are first obtained successively. Then we calculate the maximum absolute values F_{bM} of the driving force corresponding to each set of structural parameters. Relevant results are collected in Table 2.

$$\begin{cases} x = 2 \times 10^{-5} \sin(\omega t) \\ y = 2 \times 10^{-5} \cos(\omega t) \\ z = 1 \times 10^{-4} \end{cases} \quad (17)$$

where $\omega = \pi/4$.

Table 2. Optimization results with the weight factors varying from 0.1 to 0.9.

The Weight Factors α	Structural Parameters of Flexure Spherical Hinge (mm)		Mass (kg)	The Bending Stiffness of the Hinge (N/m)	Maximum Absolute Value of Driving Force F_{bM} (N)
	t_s	R_s			
0.1	0.8	1.054	0.1310	4.744	1.8397
0.2	0.8	1.584	0.1379	3.788	1.7304
0.3	0.8	2.073	0.1449	3.279	1.7303
0.4	0.8	2.292	0.1482	3.109	1.7481
0.5	0.8	2.294	0.1483	3.107	1.7483
0.6	0.8	2.294	0.1483	3.107	1.7483
0.7	0.8	2.294	0.1483	3.107	1.7483
0.8	0.8	2.294	0.1483	3.107	1.7483
0.9	0.8	2.294	0.1483	3.107	1.7483

It can be seen from Table 2 that the maximum absolute value F_{bM} of the driving force first decreases, then increases, and then remains unchanged as the stiffness weight factor α increases with an interval of 0.1. After the weight factor is increased to 0.5, as it continues to increase, the optimization results remain unchanged. This is because when the weight α is increased to 0.5, some parameters or properties of the optimized mechanism reach critical values of constraints, so continuing to increase α will not change the optimization results. It is easy to find that the maximum absolute value F_{bM} of the driving force reached the smallest value (1.7303 N) when the weight factor α is 0.3. However, the smallest value is very close to the maximum absolute value (1.7304 N) when the weight factor α is 0.2. Therefore, in order to obtain a more ideal optimization model, another group of weight factors is selected between 0.2–0.3 with the interval of 0.01, and further optimization is carried out according to the above method. Results in Table 3 show that the smallest maximum absolute value (1.7223 N) of the driving force appears as the weight factor α is 0.25. Correspondingly, the structure parameters of the flexure spherical hinge obtained by optimization are t_s = 0.8 mm and R_s = 1.829 mm. In addition, it is interesting to note that no matter the weight factor, α takes any value from 0.1 to 0.9, the optimization results of t_s

always coincide with the minimum value in the constraints, that is 0.8. It demonstrates that the minimum thickness t_s of the flexible hinge has a larger impact on the driving forces compared to the cutting radius R_s.

Table 3. Optimization results with the weight factors varying from 0.21 to 0.29.

The Weight Factors α	Structural Parameters of Flexure Spherical Hinge (mm)		Mass (kg)	The Bending Stiffness of the Hinge (N/m)	Maximum Absolute Value of Driving Force F_{bM} (N)
	t_s	R_s			
0.21	0.8	1.633	0.1385	3.726	1.7272
0.22	0.8	1.682	0.1379	3.788	1.7249
0.23	0.8	1.731	0.1399	3.611	1.7233
0.24	0.8	1.781	0.1406	3.556	1.7225
0.25	0.8	1.829	0.1406	3.557	1.7223
0.26	0.8	1.878	0.1420	3.456	1.7228
0.27	0.8	1.926	0.1427	3.410	1.7238
0.28	0.8	1.976	0.1434	3.363	1.7255
0.29	0.8	2.024	0.1441	3.321	1.7276

The comparison of driving forces before and after optimization is shown in Figure 12. It can be seen that the variation trend of the driving forces before and after optimization is consistent. After optimization, the driving forces are decreased to different degrees on the entire time axis. Moreover, the maximum absolute value F_{bM} of driving forces is reduced by 34.54% compared with that before optimization.

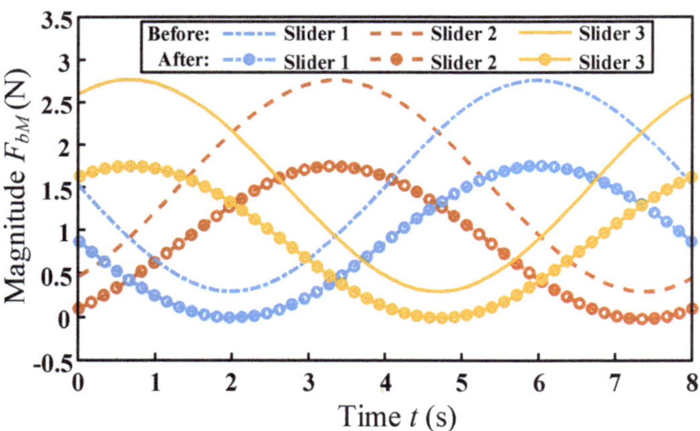

Figure 12. The comparison of driving forces before and after optimization.

The structural parameters of the flexure spherical hinge and the dynamic performance of the mechanism before and after optimization are shown in Table 4. It can be seen from the comparison that the overall mass of the mechanism is reduced by 8.34% after optimization, which greatly reduces the inertial force of the mechanism. At the same time, it is also found that the stiffness and first-order natural frequency of the mechanism are reduced to a certain extent, but they are all within the allowable range. In addition, the ultimate angular displacement of the optimized flexure spherical hinge is slightly reduced, but it is still much larger than the initial set value of 1°. In conclusion, optimization based on dynamic performance is feasible and effective. It can be seen that when the acceleration in Equation (9) is 0, the dynamic equation is transformed into a static equation. Thus, the optimization method is also suitable for optimization design based on a quasi-static performance.

Table 4. Comparison of the structural parameters of flexible spherical hinge and dynamic performance before and after optimization.

Parameters	Before Optimization	After Optimization
Structural parameter (mm)	$t_s = 1$, $R_s = 2.5$	$t_s = 0.8$, $R_s = 1.829$
Total mass (kg)	0.1534	0.1406
Bending stiffness of hinge k_{bm} (N/m)	6.5276	3.557
Ultimate angular displacement (°)	2.625	2.4739
Driving force F_{bM} (N)	2.6329	1.7223
Natural frequency (Hz)	75.67	57.54

4. Synchronous Optimal Design Based on Kinematics and Dynamics

In practical applications, the manufacturing difficulty of flexible spherical hinges is usually much greater than that of other structural parts, and especially the minimum thickness of the spherical hinge is often less than 1 mm, which is often easy to damage by processing. Therefore, when improving the existing 3-PSS flexible parallel micromanipulator to optimize its kinematics and dynamics, sometimes it is not desirable to modify the flexible spherical hinge, and only the scale parameters of the mechanism can be optimized. The following proposes a method to optimize the scale parameters of the mechanism based on both kinematic performance and dynamic performance. In this optimization method, the first-order frequency of the mechanism is used as the optimization objective and the inscribed circle radius at the maximum cross-section of the workspace is selected as constraints. The optimization procedure is shown in Figure 13.

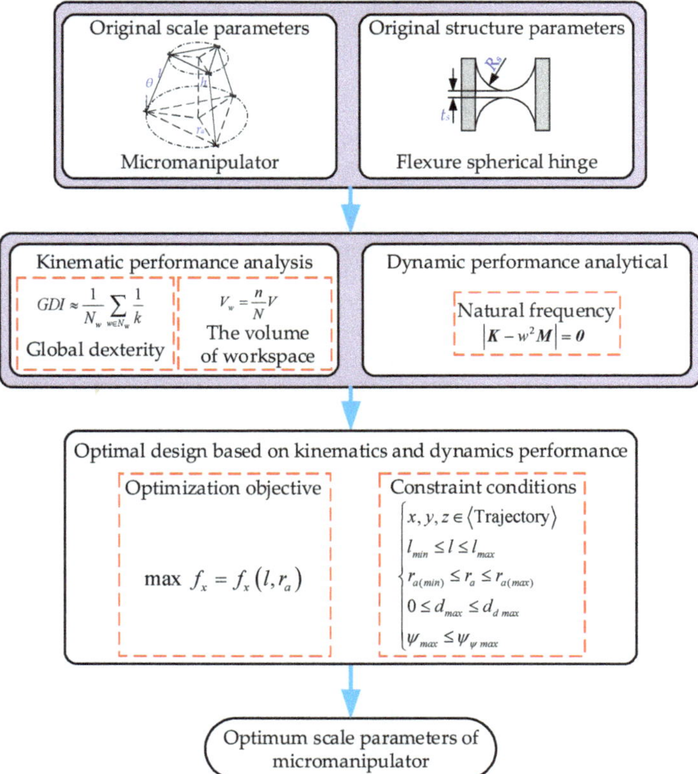

Figure 13. The synchronous optimization procedure based on kinematics and dynamics performance.

Similar to Section 3.1, the moving platform radius r_p of the mechanism is set to be a constant, and the fixed platform radius r_a and the rod length l are used as the optimization parameters. The variation range of the rod length l is selected as approximately ±20% of the original size. The corresponding variation range of the radius r_a of the fixed platform can be obtained through the variation range of the rod length l and the geometry relationship of the mechanism. Considering the maximum cross-section of the workspace is a polygon (see Figure 2d), the workspace of the mechanism is measured by the radius of the inscribed circle in the maximum cross-section. In this study, the radius of the inscribed circle of the maximum cross-section of the workspace is selected to be 400 µm. It is assumed that the maximum stroke d_{max} of the selected piezoelectric stages does not exceed 200 µm, and the ultimate angular displacement $\psi_{\max}$ of the flexure spherical hinge does not exceed 1°. Then the constraint condition can be combined as:

$$\begin{cases} x = 4 \times 10^{-4}\sin(\omega t),\ y = 4 \times 10^{-4}\cos(\omega t),\ z = 1 \times 10^{-4} \\ 50\ \text{mm} \leq l \leq 80\ \text{mm} \\ 26\ \text{mm} \leq r_a \leq 105\ \text{mm} \\ 0 \leq d_{\max} \leq 200\ \mu\text{m} \\ \psi_{\max} \leq 1^{\circ} \end{cases} \tag{18}$$

where $\omega = \pi/4$.

According to the symmetry of the micromanipulator's structure, the first and second-order natural frequencies (corresponding to the motion in the x and y directions, respectively) of the mechanism are equal. Hence, the natural frequency f_x corresponding to the motion in the x direction (or the y direction) is selected as the optimization objective. The optimization objective function is established as:

$$\max f_x = f_x(l, r_a) \tag{19}$$

According to the above optimization parameters (l and r_a), constraints (Equation (18)), and optimization objective (Equation (19)), optimization can thus be carried out by employing the genetic algorithm toolbox in MATLAB R2018b software. The optimized scale parameters are l = 50 mm and r_a = 36.92 mm, respectively. Correspondingly, the natural frequency (x or y direction) of the optimized mechanism is 73.61 Hz.

In order to verify whether the optimization results meet the constraints, the workspace analysis of the optimized mechanism is carried out. As shown in Figure 14, the maximum inscribed circle radius on the maximum cross-section of the workspace of the optimized mechanism is 400 µm, which satisfies the optimization constraints. The comparison of the workspace of the original and optimized mechanisms is given in Figure 15. As can be seen from Figure 15a, the maximum cross-sectional shape of the optimized mechanism's workspace remains a regular hexagon compared with the original mechanism, but the area is increased. The workspace volume of the optimized mechanism increases by 31.93% compared with that of the original mechanism, as shown in Figure 15b.

Since the dexterity of the mechanism directly reflects the motion accuracy of the mechanism, it is necessary to verify the dexterity of the optimized mechanism. According to Equation (8), the global dexterity of the optimized mechanism can be calculated as 0.173 by using MATLAB R2018b. The dexterity distribution within the maximum cross-section of the workspace of the mechanism is shown in Figure 16. It can be seen that compared with the dexterity of the original mechanism (see Figure 4), the dexterity of the optimized mechanism is reduced to a certain extent. It indicates that the proposed method of optimizing the scale parameters of the 3-PSS flexible parallel micromanipulator to maximize the first-order frequency of this mechanism with the specific inscribed circle radius as the constraint is effective.

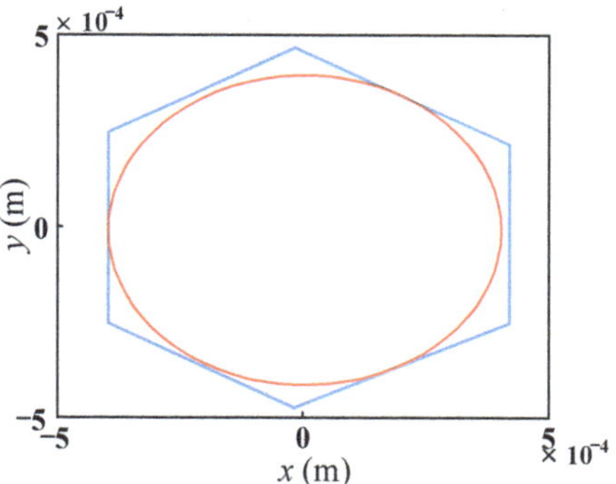

Figure 14. Maximum cross-section of the workspace of optimized mechanism.

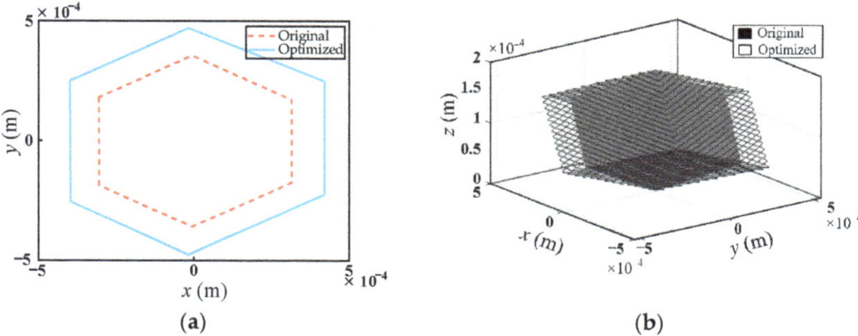

Figure 15. Comparison of the workspace between original and optimized mechanism, (**a**) the maximum cross-section of the workspace of the mechanism; (**b**) micromanipulator workspace.

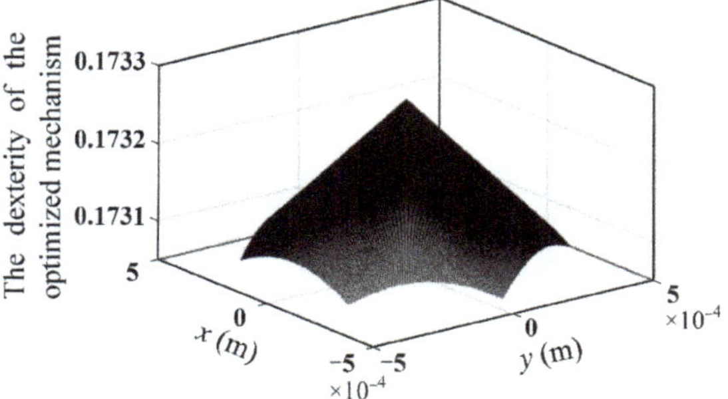

Figure 16. The distribution of dexterity within the maximum cross-section of the workspace of the optimized mechanism.

The comparison of the scale parameters of the micromanipulator, kinematic and dynamic performance before and after optimization, is given in Table 5. It is shown that the first-order natural frequency of the optimized mechanism increased by 31.07% compared with the original mechanism, which shows the effectiveness of the optimization method. The workspace of the optimized mechanism increased by 31.93% (The symbol "↑" in Table 5 indicates increase), however, the global dexterity of the optimized mechanism is reduced. It indicates that the workspace of the optimized mechanism is enlarged, but at the cost of a certain reduction in global dexterity. Therefore, in the optimization process, trade-offs must be made when faced with multiple kinematic and dynamic performance requirements. For example, if the global dexterity is more concerned than the workspace, then in the constraints (Equation (18)) one might consider choosing a smaller inscribed circle radius of the maximum cross-section of the workspace in exchange for higher global flexibility. Appendix A gives the kinematic and dynamic performances of the 3-PSS flexible parallel micromanipulator after optimization with a different inscribed circle radius as the constraints and the results confirm this rule (Table A1).

Table 5. Comparison of the scale parameters of micromanipulator, kinematic and dynamic performances before and after optimization.

Parameters	Before Optimization	After Optimization
Inscribed circle radius (μm)	307	400
Scale parameters (mm)	$l = 65$, $r_a = 45$	$l = 50$, $r_a = 36.92$
Volume of workspace	Original	↑ 31.93%
Global dexterity	0.2287	0.173
Natural frequency (Hz)	56.16	73.61

5. Conclusions

Based on the kinematics and dynamics of the mechanism, this paper proposes two optimal design schemes for the 3-PSS flexible parallel micromanipulator according to different application requirements and conditions. The first is called progressive optimization design, in which the scale parameters (l and r_a) are firstly optimized to maximize the workspace, combining the constraints of the minimum global dexterity of the mechanism. Then, the minimum thickness t_s and the cutting radius R_s of the flexure spherical hinge are further optimized for minimizing the required driving forces, combined with constraints of the minimum first-order natural frequency of the mechanism and maximum stress of the flexure spherical hinge during the movement of the mechanism. The second is called synchronous optimization design, in which the scale parameters (l and r_a) are optimized to maximize the first-order natural frequency of the mechanism, combined with the constraints of a certain inscribed circle radius of the maximum cross-section of the workspace, the maximum stroke of the selected piezoelectric stages, and the maximum ultimate angular displacement of the flexure spherical hinge. A comparison of the kinematic and dynamic characteristics of the original and optimized mechanism demonstrated the effectiveness of both optimization methods.

The advantage of the progressive optimization method is that both the workspace and the driving forces are optimized and the minimum requirements for global dexterity and first-order natural frequency are ensured. Thus, multiple kinematic and dynamic characteristics of the mechanism are taken into account during the optimization process. However, this optimization method needs to optimize the mechanism scale parameters followed by the structural parameters of the flexible spherical hinge in two steps, which is relatively complicated. Especially when the structure of the flexible spherical hinge is inconvenient to change, the employment of this method is limited. The advantage of the synchronous optimization method is that only the scale parameters of the mechanism need to be optimized without changing the structural parameters of the flexible spherical hinge. The optimization process takes only one step and the process is relatively simple. However, this optimization method only optimizes the first-order natural frequency of the mechanism under the premise of the requirement of a certain working space, and does not take the requirement of global dexterity and driving forces of the mechanism into consideration. Therefore, it is suggested that the optimal design scheme be reasonably selected according to different design requirements and the application of the mechanism.

Author Contributions: J.R. is the designer and executor of this study. J.R., Q.L., H.W. and Q.C. completed data analysis, paper writing, and revision. All authors have read and agreed to the published version of the manuscript.

Funding: This research was funded by the Key Projects of Hubei Provincial Department of Education Research Program (Grant numbers D20211401).

Institutional Review Board Statement: Not applicable.

Informed Consent Statement: Not applicable.

Data Availability Statement: Not applicable.

Conflicts of Interest: The authors declare no conflict of interest.

Appendix A. The Scale Parameters of Micromanipulator, Kinematic, and Dynamic Performance of 3-PSS Flexible Parallel Micromanipulator after Optimization with Different Inscribed Circle Radius as Constraints

The scale parameters of micromanipulator, kinematic, and dynamic performance of 3-PSS flexible parallel micromanipulator after optimization with inscribed circle radius varying from 100 μm to 500 μm are as follows (The symbol "↑" and "↓" indicate increase and decrease, respectively):

Table A1. Comparison of the scale parameters of micromanipulator, kinematic and dynamic performances before and after optimization with different inscribed circle radius.

Parameters	After Optimization			
Inscribed circle radius (μm)	100	200	300	500
Scale parameters (mm)	$l = 50$, $r_a = 60.3$	$l = 50$, $r_a = 46.82$	$l = 50$, $r_a = 40.53$	$l = 50$, $r_a = 62.22$
Volume of workspace	↓ 67.46%	↓ 33.27%	↓ 1.03%	↑ 62.22%
Global dexterity	0.7045	0.3428	0.2309	0.1408
Natural frequency (Hz)	77.25	74.29	73.80	73.95

References

1. Howell, L.L. *Compliant Mechanisms*; McCarthy, J.M., Ed.; Springer: London, UK, 2012; pp. 189–216.
2. Speich, J.; Goldfarb, M. A Compliant-Mechanism-Based Three Degree-of-Freedom Manipulator for Small-Scale Manipulation. *Robotica* **2000**, *18*, 95–104. [CrossRef]
3. Kota, S.; Hetrick, J.; Li, Z.; Saggere, L. Tailoring Unconventional Actuators Using Compliant Transmissions: Design Methods and Applications. *IEEE/ASME Trans. Mechatron.* **1999**, *4*, 396–408. [CrossRef]
4. Tian, Y.; Shirinzadeh, B.; Zhang, D.; Alici, G. Development and Dynamic Modelling of a Flexure-Based Scott-Russell Mechanism for Nano-Manipulation. *Mech. Syst. Signal Process.* **2009**, *23*, 957–978. [CrossRef]
5. Zubir, M.N.M.; Shirinzadeh, B.; Tian, Y. A New Design of Piezoelectric Driven Compliant-Based Microgripper for Micromanipulation. *Mech. Mach. Theory* **2009**, *44*, 2248–2264. [CrossRef]
6. Fattah, A.; Angeles, J.; Misra, A.K. Dynamics of a 3-DOF Spatial Parallel Manipulator with Flexible Links. In Proceedings of the 1995 IEEE International Conference on Robotics and Automation, Nagoya, Japan, 21–27 May 1995; pp. 627–633.
7. Hao, F.; Merlet, J.P. Multi-Criteria Optimal Design of Parallel Manipulators Based on Interval Analysis. *Mech. Mach. Theory* **2004**, *40*, 157–171. [CrossRef]
8. Ren, J.; Li, Q. Analysis of Compliance and Kinetostatic of a Novel Class of n-4R Compliant Parallel Micro Pointing Mechanism. *Micromachines* **2022**, *13*, 1014. [CrossRef]
9. Li, C.; Wang, J.; Chen, S. Flexure-Based Dynamic-Tunable Five-Axis Nanopositioner for Parallel Nanomanufacturing. *Precis. Eng.* **2016**, *45*, 423–434. [CrossRef]
10. Berna, Ö.; Raquel, P.; Ahmet, B.; Lucio, P.; Ece, Ö.; Claire, A.; Murat, K.; Francesco, S.; Mooney, D.J.; Selman, S.M. Modular Soft Robotic Microdevices for Dexterous Biomanipulation. *Lab Chip* **2019**, *19*, 778–788.
11. Dsouza, R.D.; Navin, K.P.; Theodoridis, T.; Sharma, P. Design, Fabrication and Testing of a 2 DOF Compliant Flexural Microgripper. *Microsyst. Technol.* **2018**, *24*, 3867–3883. [CrossRef]
12. Giamberardino, P.D.; Bagolini, A.; Bellutti, P.; Rudas, I.J.; Verotti, M.; Botta, F.; Belfiore, N.P. New MEMS Tweezers for the Viscoelastic Characterization of Soft Materials at the Microscale. *Micromachines* **2017**, *9*, 15. [CrossRef]
13. Shao, Z.; Tang, X.; Wang, L.; Sun, D. Atlas based kinematic optimum design of the Stewart parallel manipulator. *Chin. J. Mech. Eng.* **2015**, *28*, 20–28. [CrossRef]
14. Staicu, S.; Zhang, D. Dynamic modelling of a 4-DOF parallel kinematic machine with revolute actuators. *Int. J. Manuf. Res.* **2008**, *3*, 172–187.
15. Ding, B.; Li, Y.; Xiao, X.; Tang, Y.; Li, B. Design and analysis of a 3-DOF planar micromanipulation stage with large rotational displacement for micromanipulation system. *Mech. Sci.* **2017**, *8*, 117–126. [CrossRef]
16. Lu, S.; Li, Y.; Ding, B. Kinematics and dynamics analysis of the 3PUS-PRU parallel mechanism module designed for a novel 6-DOF gantry hybrid machine tool. *J. Mech. Sci. Technol.* **2020**, *34*, 345–357. [CrossRef]

17. Yu, J.; Zhou, Q.; Bi, S.; Zong, G. Optimal Design of a Fully Compliant Mechanism Based on Its Dynamic Characteristics. *Chin. J. Mech. Eng.* **2003**, *39*, 32–36. [CrossRef]
18. He, J.; Li, J.; Sun, Z.; Gao, F.; Wu, Y.; Wang, Z. Kinematic design of a serial-parallel hybrid finger mechanism actuated by twisted-and-coiled polymer. *Mech. Mach. Theory* **2020**, *152*, 103951. [CrossRef]
19. Ding, B.; Li, X.; Li, Y. FEA-based optimization and experimental verification of a typical flexure-based constant force module. *Sens. Actuators A Phys.* **2021**, *332*, 113083. [CrossRef]
20. Yang, M.; Du, Z.; Sun, L.; Dong, W. Optimal design, modeling and control of a long stroke 3-PRR compliant parallel manipulator with variable thickness flexure pivots. *Robot. Comput.—Integr. Manuf.* **2019**, *60*, 23–33.
21. Xu, K.; Liu, H.; Yue, W.; Xiao, J.; Ding, Y.; Wang, G. Kinematic modeling and optimal design of a partially compliant four-bar linkage using elliptic integral solution. *Mech. Mach. Theory* **2021**, *157*, 104214. [CrossRef]
22. Li, Y.; Xu, Q. A novel design and analysis of a 2-DOF compliant parallel micromanipulator for nanomanipulation. *IEEE Trans. Autom. Sci. Eng.* **2006**, *3*, 247–254.
23. Li, Y.; Xu, Q. Design and Optimization of an XYZ Parallel Micromanipulator with Flexure Hinges. *J. Intell. Robot. Syst.* **2009**, *55*, 377–402. [CrossRef]
24. Xu, Q.; Li, Y. Kinematic Analysis and Optimization of a New Compliant Parallel Micromanipulator. *Int. J. Adv. Robot. Syst.* **2006**, *3*, 47. [CrossRef]
25. Jia, X.; Zhang, D.; Tian, Y. Workspace Analysis and Optimization of a 3-DOF Precision Positioning Table. *Acta Armamentarii* **2010**, *31*, 624–630.
26. Wang, W.; Zhang, L.; Bi, S. Dynamic Analysis and Design of Compliant Mechanisms Using the Finite Element Method. *Appl. Mech. Mater.* **2012**, *163*, 111–115. [CrossRef]
27. Li, H.; Yang, Z.; Huang, T. Dynamics and elasto-dynamics optimization of a 2-DOF planar parallel pick-and-place robot with flexible links. *Struct. Multidiscip. Optim.* **2009**, *38*, 195–204. [CrossRef]
28. Du, Z.; Yu, Y.; Su, L. Effects of Inertia Parameters of Moving Platform on the Dynamic Characteristic of Flexible Parallel Mechanism and Optimal Design. *Opt. Precis. Eng.* **2006**, *14*, 1009–1016. [CrossRef]
29. Li, Y.; Xu, Q. Design and Analysis of a Totally Decoupled Flexure-Based XY Parallel Micromanipulator. *IEEE Trans. Robot.* **2009**, *25*, 645–657.
30. Li, Y.; Huang, J.; Tang, H. A Compliant Parallel XY Micromotion Stage with Complete Kinematic Decoupling. *IEEE Trans. Autom. Sci. Eng.* **2012**, *9*, 538–553. [CrossRef]
31. Ren, J.; Cao, Q. Dynamic Modeling and Frequency Characteristic Analysis of a Novel 3-PSS Flexible Parallel Micro-Manipulator. *Micromachines* **2021**, *12*, 678. [CrossRef]

MDPI

Article

Vision Feedback Control for the Automation of the Pick-and-Place of a Capillary Force Gripper

Takatoshi Ito †, Eri Fukuchi †, Kenta Tanaka, Yuki Nishiyama, Naoto Watanabe and Ohmi Fuchiwaki * 

Department of Mechanical Engineering, Yokohama National University, Yokohama 240-8501, Japan
* Correspondence: ohmif@ynu.ac.jp
† These authors contributed equally to this work.

Abstract: In this paper, we describe a newly developed vision feedback method for improving the placement accuracy and success rate of a single nozzle capillary force gripper. The capillary force gripper was developed for the pick-and-place of mm-sized objects. The gripper picks up an object by contacting the top surface of the object with a droplet formed on its nozzle and places the object by contacting the bottom surface of the object with a droplet previously applied to the place surface. To improve the placement accuracy, we developed a vision feedback system combined with two cameras. First, a side camera was installed to capture images of the object and nozzle from the side. Second, from the captured images, the contour of the pre-applied droplet for placement and the contour of the object picked up by the nozzle were detected. Lastly, from the detected contours, the distance between the top surface of the droplet for object release and the bottom surface of the object was measured to determine the appropriate amount of nozzle descent. Through the experiments, we verified that the size matching effect worked reasonably well; the average placement error minimizes when the size of the cross-section of the objects is closer to that of the nozzle. We attributed this result to the self-alignment effect. We also confirmed that we could control the attitude of the object when we matched the shape of the nozzle to that of the sample. These results support the feasibility of the developed vision feedback system, which uses the capillary force gripper for heterogeneous and complex-shaped micro-objects in flexible electronics, micro-electro-mechanical systems (MEMS), soft robotics, soft matter, and biomedical fields.

Keywords: micromanipulation; capillary force; water; vision feedback; non-contact

Citation: Ito, T.; Fukuchi, E.; Tanaka, K.; Nishiyama, Y.; Watanabe, N.; Fuchiwaki, O. Vision Feedback Control for the Automation of the Pick-and-Place of a Capillary Force Gripper. *Micromachines* **2022**, *13*, 1270. https://doi.org/10.3390/mi13081270

Academic Editor: Alessandro Cammarata

Received: 29 June 2022
Accepted: 2 August 2022
Published: 7 August 2022

1. Introduction

Electronics are becoming increasingly sophisticated while decreasing in size. For high-resolution displays, 0201 sized electronic chip parts and micro-LEDs of less than 100 μm have been developed [1]. Microfiber assembly is also one of the most significant technologies for the advancement of information processing instruments [2].

In conventional surface mounting technology (SMT), the suction force generated by compressed air is used mainly to pick up flat components. However, as the size of the chip decreases, a higher mounting speed and accuracy are required to achieve an adequate mounting density. Transfer printing technologies are feasible methods for assembling micro-LEDs [3–5]. Mechanical grippers have important applications, such as medical operations on eyeballs [6]. Shape memory polymers have also been investigated as a feasible micromanipulation method [7,8].

These techniques are based on the contact between the micro-object and the gripper, which may damage the targets. In contact-type grippers, it is necessary to control the gripping force while handling fragile objects. A noncontact-type gripper decreases the possibility of damaging the parts being handled.

In liquids, various non-contact micromanipulation methods have been reported, such as using steady streaming [9,10], electric fields [11], and laser tweezers [12,13] for fragile positioning, such as biological cells and microorganisms.

In the atmosphere, ultrasonic levitation [14,15] and a liquid bridge force (hereinafter referred to as capillary force-based gripper) are categorized as non-contact grasping methods with self-alignment [16,17]. Several studies have been conducted on capillary force grippers for pick-and-place operations [18–22]. Estimation of capillary force [23,24], and image processing technologies [25,26] are also significant research categories for the automation of micromanipulation and the classification of micro fossils [25], micro particles [26], soft robotics [27], soft matters [28], and biomedical fields.

In a previous article, Tanaka et al. reported a double nozzle gripper using two liquid transporting methods: a diaphragm pump and the capillary phenomenon [22]. This study focused on the picking and placing of 1-mm sized objects with various shapes and did not describe a vision feedback control method for decreasing the positioning errors.

In this paper, we describe the details of the image processing procedure and the investigation of a size matching effect that influences final alignment accuracy [28]. We also describe the attitude control by a shape matching effect between the nozzle shape and the picked object [28,29].

The remainder of the paper is structured as follows. We describe the static analysis in Section 2; the mechanical design in Section 3; the experimental setup, as shown in Figure 1, in Section 4; the image processing in Section 5; the success rate, placing error, and attitude control in Section 6; and concluding remarks and future research scope in Section 7. (See also Video S1 "Digest movie" for the outline).

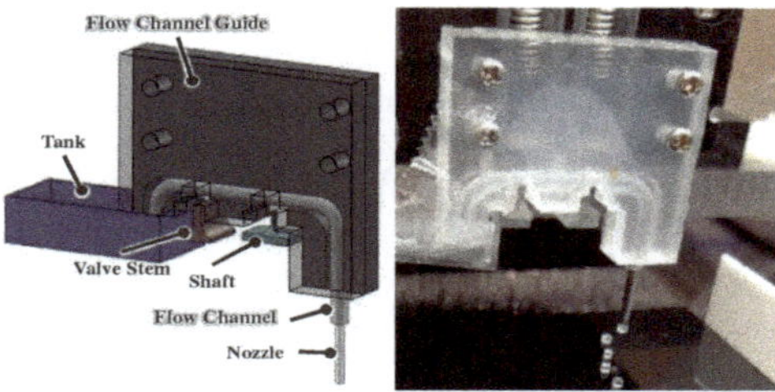

Figure 1. Working area of the pick-and-place of 1-mm cubes.

2. Approximation of the Capillary Force

In this section, we estimate the capillary forces using the geometrical parameters of the objects when the top surface of the picked-up object is a plane.

Figure 2 shows the capillary bridge between the nozzle and the object. Here, the shape of the bottom of the liquid bridge is approximated to be circular. We defined the radius of the interface between the liquid bridge and the nozzle as r_2 and the radius of the interface between the liquid bridge and the object as r_1. In this case, assuming that the water spreads over the nozzle, we could estimate r_2 to be the radius of the nozzle. h is the height of the liquid bridge, as shown in Figure 2. R_1 and R_2 are the arc-approximated meniscus radii of the liquid bridges caused by the interfacial tension, respectively. In this case, it is the length of the wispiest part of the liquid bridge. Moreover, θ_1 and θ_2 are the contact angles between the object and the liquid and the nozzle and the liquid, respectively.

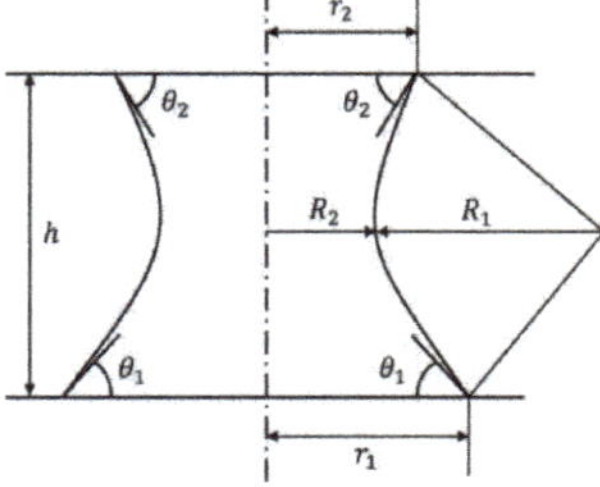

Figure 2. Capillary bridge between two parallel planes.

The capillary force is determined geometrically if the section curves of the meniscus are approximated as parts of the arc [23,24]. In this case, R_1 and R_2 are expressed as follows:

$$R_1 = \frac{h}{\cos\theta_1 + \cos\theta_2} \tag{1}$$

$$R_2 = r_2 - R_1(1 - \sin\theta_2) \tag{2}$$

The mean curvature C of the meniscus is determined as follows:

$$C = \frac{1}{R_2} - \frac{1}{R_1}, \tag{3}$$

where R_2 is positive. When the shape is convex, R_1 is defined as negative. The difference in pressure across the interface, given by Laplace's equation, is as follows:

$$\Delta p = \gamma C \tag{4}$$

Here, γ is the surface tension, and r_1 is given as follows:

$$r_1 = r_2 + R_1(\sin\theta_2 - \sin\theta_1). \tag{5}$$

Thus, the suction force acting on the flat plane (top surface of the cube) by the Laplace pressure F_L is expressed as follows:

$$F_L = \pi r_1^2 \Delta p. \tag{6}$$

The pulling force acting on the flat plane owing to the surface tension is given as follows:

$$F_T = 2\pi r_1 \gamma \sin\theta_1. \tag{7}$$

From (6) and (7), the required capillary force F_C, which is the sum of F_L and F_T, is given as follows:

$$F_C = F_L + F_T = \pi\gamma r_1 \left[r_1 \left(\frac{1}{R_2} - \frac{1}{R_1} \right) + 2\sin\theta_1 \right]. \tag{8}$$

3. Design

The gripper consists of a flow device, flow channel, tank, shaft, and valve stem, all fabricated from acrylic. The flow channel was made of a silicon tube. A stainless-steel tube was used at the end of the flow channel, which is called the nozzle. We used the valve stem and the shaft to refill water and form droplets, and we adopted a spring interlocking mechanism, developed by Hagiwara et al. [21], to move them with the Z-stage.

Figure 3 illustrates the filling of water in the channel through capillary action. As shown in the left panel, even if the water at the tip of the channel disappears owing to evaporation or experiment conduction, water can again be filled to the tip of the channel by pulling down the valve and opening the channel, as shown in the right panel.

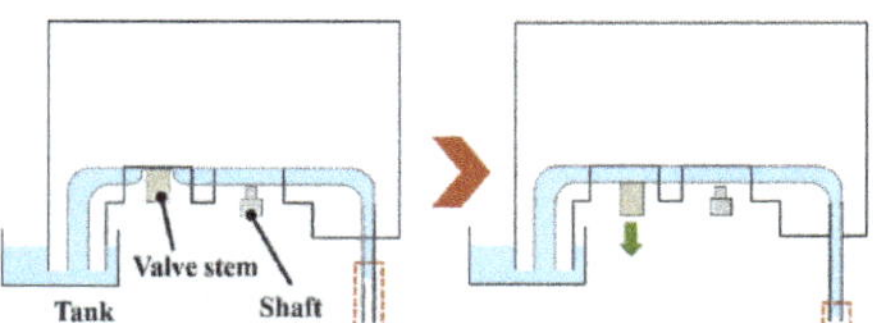

Figure 3. Mechanism of water filling till the tip of the channel by capillary action.

Figure 4 shows the formation of a droplet. When the water channel moves down, the valve closes the channel to prevent backflow (left panel), and the shaft then pushes the channel and forms the droplet at the channel tip using a diaphragm pump mechanism. Since the deformed volume of the diaphragm and the volume displaced from the shaft are almost equal, the volume of the droplet is determined by measuring the length of the shaft pushed in.

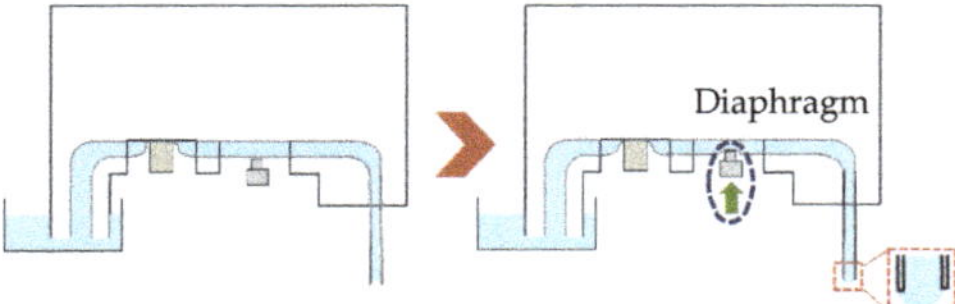

Figure 4. Principle of droplet formation.

The object is picked up by contacting a formed droplet with the object. The object is placed by first contacting and applying a formed droplet to the place surface, and then contacting the bottom surface of the object picked up by the nozzle with the applied droplet. Table 1 shows the parameters of the micro-objects used in the pick-and-place experiments.

Table 1. Specifications of micro-objects.

Name	1 mm Cube	0.5 mm Cube
Image		
Material	Acrylic	
Geometric parameters	Depth: 1 mm Width: 1 mm Height: 1 mm	Depth: 0.5 mm Width: 0.5 mm Height: 0.5 mm
Gravity force	12 μN	1.5 μN

In the experiments, we used various sizes and shapes of nozzles for each object. The experimental conditions and parameters used to calculate the estimated value of the capillary force for each condition are organized and shown in Table 2; the estimated capillary forces are also shown. h, θ_1, and θ_2 were typical values measured from the charge coupled device (CCD) camera image during the experiments. The surface tension γ, measured by the ring method, was 72.9 mN/m; we used purified water as the liquid.

Table 2. Experimental condition and the parameters.

Condition			(C1)	(C2)	(C3)	(C4)	(S)
Picked Object			0.5 mm Cube		1 mm Cube		
Nozzle's scross section							
Parameters	Design parameters	r_2	0.635 mm	0.255 mm	0.635 mm	0.255 mm	0.5 mm
	Experimental parameters	γ	72.9 μN				
		h	0.52 mm	0.25 mm	0.35 mm	0.23 mm	0.50 mm
		θ_1	148.1°	100°	77.3°	73.6°	100.4°
		θ_2	73.7°	109°	82.1°	80°	86.8°
Estimated capillary force from (8)			97.6 μN	163 μN	256.5 μN	197 μN	260 μN
Gravity force			1.5 μN		12 μN		

From Table 2, it can be seen that the estimated capillary force had a much larger value than the corresponding gravitational force, and hence, this gripper could grasp the object under all conditions if there were no additional adhesive or electrostatic forces.

4. Experimental Setup

To increase the success rate of the pick-and-place, we developed an automatic pick-and-place with vision feedback function. The organization of the entire setup is shown in Figure 5.

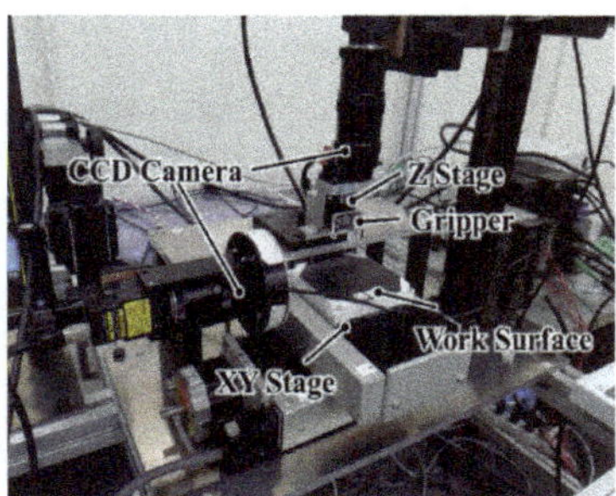

Figure 5. Experimental setup for the automatic pick-and-place.

In this study, the work surface was positioned in the X- and Y-axis with a positioning resolution of 1 μm and a repeatability of ±0.5 μm (YA10A-L1, Kohzu Precision Co., Ltd., Kawasaki City, Japan). The gripper was connected to a linear stage (MMU-40X, Chuo Precision Industrial Co., Ltd., Tokyo, Japan) and moved along the Z-axis. It was also connected to the interlocking mechanism, which controls the shaft and the valve to form a droplet. The traverse limits of the XY- and Z-stages were ±50 mm and ±5 mm, respectively.

To measure the position of the samples, the vision feedback system by OpenCV was used. This system imported the image from the CCD camera set on top of the work bench. The camera had a resolution of 2432 × 2050 pixels (CV-H500M, KEYENCE Corp., Osaka, Japan). The CCD camera also attached to the side of the workbench, and it had a resolution of 1216 × 1025 pixels (CV-200C, KEYENCE Corp.). Each camera had a magnification lens (LA-LM510, Keyence Co.), and the depth of field was 1.28 mm.

The pixel resolutions of the top and side cameras were, approximately, 7 μm/pixel and 9 μm/pixel, respectively. We confirmed that the image processing and gravity center detection process in the vision feedback resulted in a final X- and Y-axis measurement resolution of 1 μm because of the averaging effect.

LabVIEW2018 (National Instruments, Austin, TX, USA) was used as the programming language, and OpenCV was used to conduct image processing. The detailed procedure for the image processing is described in Section 5. The pick-and-place process is as follows:

(1) The object is placed within view of the top camera on the surface to measure the X- and Y- coordinates;
(2) The XY-stage is moved so that the placement position on the work surface is directly below the nozzle;
(3) The gripper is moved up and down by the Z-stage, and a droplet is applied to the work surface;
(4) The XY-stage is moved, and the object is placed directly under the nozzle;
(5) The gripper moves first down and then up to pick up the object;
(6) The XY-stage is moved such that the target placement position is directly below the nozzle grasping the object;
(7) The gripper moves down to place the picked-up object on the pre-applied droplet of (3);

(8) The XY-stage is moved such that the object is within view of the CCD camera to measure the X- and Y- coordinates.

In steps (3), (5), and (7) of the above process, the amount of Z-stage movement is calculated and determined by the image processing program written in Python from images captured by the side camera.

5. Image Processing

In this section, we explain the vision feedback system. The program used Open CV by Python. The system was used when applying place droplets, picking up, and placing objects. Figure 6 shows the image of the process of the vision feedback system in placement, and this process is described as follows:

(1) The side camera captures the image so that both the object and the place droplets are visible (Figure 6a);
(2) The captured image is imported to the PC and the Sobel filter is applied (Figure 6b).
(3) Next, the image is binarized by the threshold (Figure 6c);
(4) The image is then used to detect the contour of the object (Figure 6d);
(5) The distance l_1 between the object and the place surface is calculated from the coordinates of the lowest surface of the object (Figure 6e);
(6) In this experiment, the parameter is the liquid bridge height h_2 between the object and the place surface at placing (Figure 6f). Thus, from h_2 and l_1, the stage descent amount l_2 is determined as follows:

$$l_2 = l_1 - h_2. \quad (9)$$

(7) The amount of Z-stage descent determined in (6) is fed back to the control (Figure 6g).

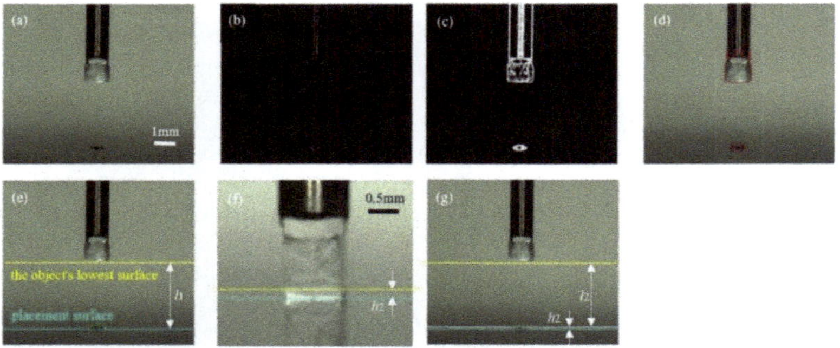

Figure 6. Process of the vision feedback system in placing: (**a**) the capture image, (**b**) the image with the Sobel filter, (**c**) the binarized image, (**d**) detecting contours, (**e**) calculating the distance l_1, (**f**) the definition of h_2, and (**g**) the definition of l_2.

As described in step 6, we used h_2 as a parameter in this experiment. Here, Table 3 shows the set values of h_2 for each experimental condition.

Here, the condition values of (C1)–(C5) are the same as in Table 2. We conducted the pick-and-place experiment 10 times for each condition.

Table 3. Targeted values of h_2.

Condition	**(C1)**	**(C2)**	**(C3)**	**(C4)**	**(S)**
Cube size	0.5 mm			1 mm	
Nozzle shape	Circle				Square
Nozzle size (mm)	1.27	0.51	1.27	0.51	1 × 1
	Diameter				
h_2 (μm)	10 25 50	10 25 50	0 50 70 100 120	25 50	70

6. Experiments

6.1. Comparison of Results with Height h_2

We conducted the pick-and-place operation under the five conditions. Figure 7 shows the sequence of pick-and-place operations under the condition of (C3) for h_2 of 50 μm.

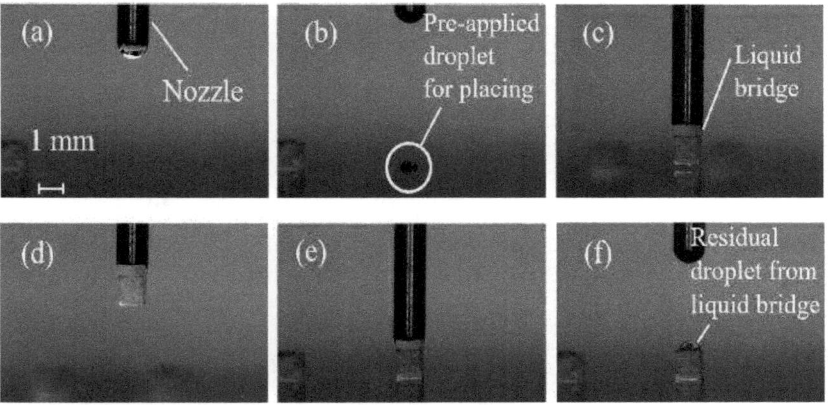

Figure 7. Sequential photograph of the pick-and-place experiment under (C3) for h_2 = 50 μm: (**a**) droplet discharge from the nozzle, (**b**) application of the droplet by stamping the nozzle to the surface, (**c**) forming the liquid bridge between the nozzle and top surface of the cube, (**d**) picking up the cube, (**e**) placing the cube on the droplet on the substrate, and (**f**) releasing the cube by moving up the gripper.

In Table 4, the success rates and positioning errors for h_2 = 0, 50, 70, 100, and 120 μm are compared. From Table 4 and Figure 7, the newly developed vision feedback system confirmed that the gripper could automatically pick-and-place objects without solid contact. Figure 8 shows the relationship between the positioning errors and h_2 under the condition (C3).

Table 4. Comparison of success rate and position errors.

	h_2 (μm)	**0**	**50**	**70**	**100**	**120**
Success rate	Pick-up (%)	100	100	100	100	100
	Place-down (%)	100	90	100	100	100
Positioning error	Average (μm)	140	63	44	77	82
	Standard deviation (μm)	45	28	34	35	50

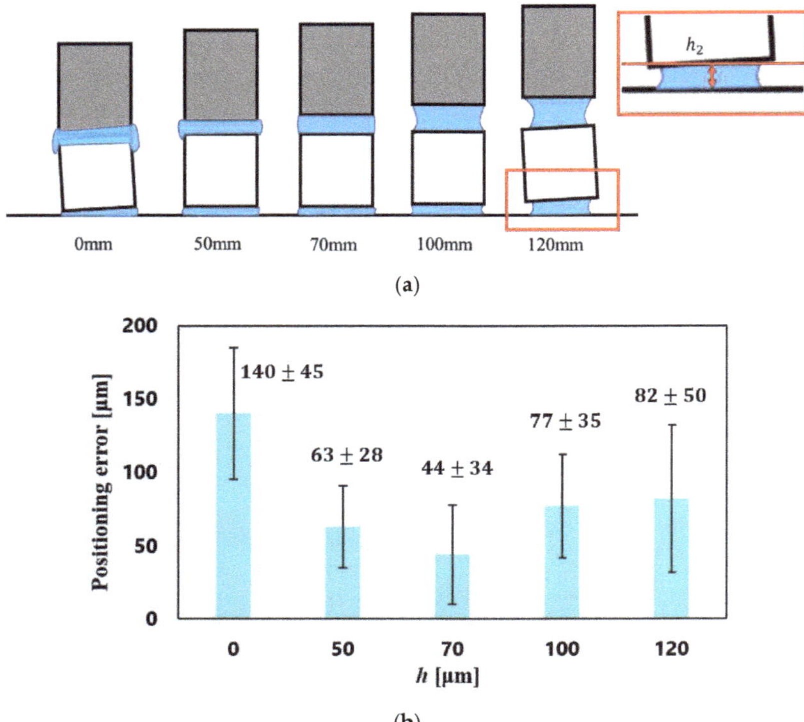

Figure 8. Relationship between the positioning errors and h_2 under (C3). (**a**) Typical schematic diagram for each h_2, (**b**) plots of positioning errors vs. h_2.

The smallest value of the positioning error was 44 ± 34 µm with h_2 = 70 µm (the ratios to the cube side were 4.4 ± 3.4%), although the error with h_2 = 50 µm had almost the same value.

Figure 9 shows the typical examples for a large positioning error. The water spread to the sides of the object when h_2 was too small, as shown in Figure 9a, whereas the cube was easily inclined when h_2 was too large, leading to a large positioning error, as shown in Figure 9b.

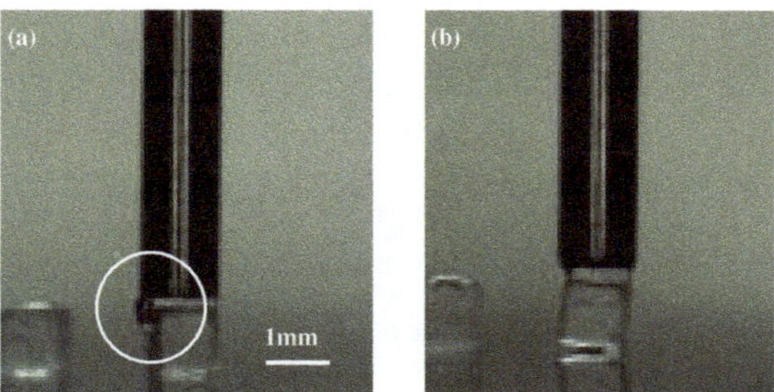

Figure 9. Example of a large positioning error: (**a**) h_2 = 0 µm; (**b**) h_2 = 100 µm.

6.2. Evaluation of Size Matching Effect

The positioning errors for each condition are compared in Table 5. We used the result with the smallest error out of the h_2 values set for each condition.

Table 5. Comparison of the positioning errors for conditions (C1), (C2), (C3), and (C4).

Condition	(C1)	(C2)	(C3)	(C4)
Ratio of nozzle size to object size	254 (%)	102 (%)	127 (%)	51 (%)
h_2	25 (μm)	25 (μm)	70 (μm)	25 (μm)
Average error	80 (μm)	40 (μm)	44 (μm)	154 (μm)
Standard deviation	53 (μm)	17 (μm)	34 (μm)	77 (μm)
Ratio of error to object size	16 ± 10.6 (%)	8.0 ± 3.4 (%)	4.4 ± 3.4 (%)	15.4 ± 7.7 (%)

The cube moves toward the center position of the droplet on the surface due to surface tension as a self-alignment phenomenon. If the average error is only attributed to the center position of the pre-applied droplet, the errors of (C1) and (C3) should be almost equal because the same nozzle is used for applying the droplet to the surface. For the same reason, that of (C2) and (C4) should also be almost equal. However, twice the difference between (C1) and (C3) and quadruple the difference between (C2) and (C4) can be observed from Table 5. We supposed that the size and shape matching effects also influenced the final alignment errors [29–31].

To examine the size matching effect, we defined the diameter of the circular nozzle as D, the positioning error as E, and the side length of the cube as L. We normalized the parameters $(D,\ E)$ by L as follows:

$$D^* \equiv D/L, \tag{10}$$

$$E^* \equiv E/L. \tag{11}$$

We compared the results in Table 5 using these parameters to verify a suitable size of circular shaped nozzle for each object, as shown in Figure 10.

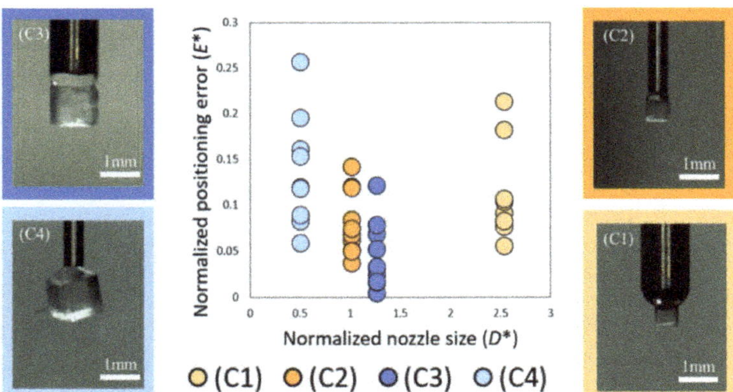

Figure 10. Plots of nozzle size D^* vs. positioning error E^*.

As shown in Figure 10, the closer D^* is to 1, the smaller the positioning error becomes. When the nozzle diameter was approximately twice as large as the object as in condition (C1), the distribution of the final alignment position was larger than those of (C2) and (C3). When the nozzle diameter was approximately half of the object, as in condition (C4), the object was often tilted during pick-up, as shown in the photograph of Figure 10 (C4). That is the main reason why (C4) obtained the largest distribution among the four conditions.

6.3. Attitude Control Using Shape Matching Effect

We considered that the attitude angle of the picked-up object could be controlled by matching the shape of the nozzle to the picked object.

To examine the shape matching effect [30,31], we conducted the pick-and-place experiment as in condition (S) in Table 4 with a square-shaped nozzle and side length of 1 mm for the 1-mm cube. Figure 11 shows a sequential photograph of the pick-and-place operation.

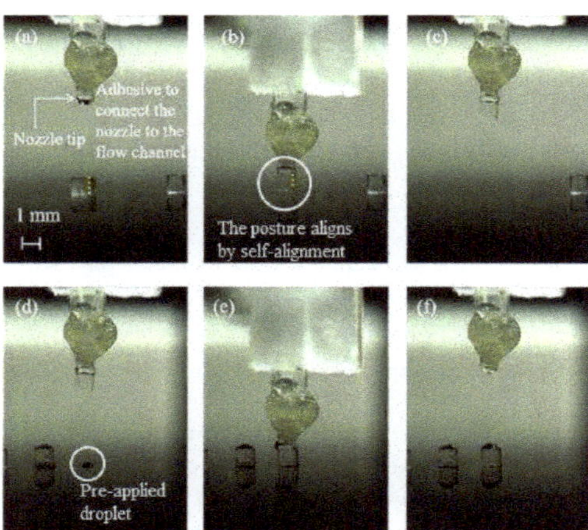

Figure 11. Sequential photograph of the pick-and-place of a 1-mm cube by the square nozzle: (**a**) droplet discharge from the nozzle, (**b**) forming the liquid bridge between the gripper and cube with a self-alignment of the attitude, (**c**) picking up the cube, (**d**) positioning the nozzle just above the pre-applied droplet, (**e**) placing the sample on the droplet on the substrate, and (**f**) releasing the sample by moving the gripper up.

From Figure 11a,b, the object changed its attitude angle for aligning the nozzle shape; we verified that the shape matching effect is useful for controlling the attitude of the sample. The movement of the object is illustrated in Figure 12.

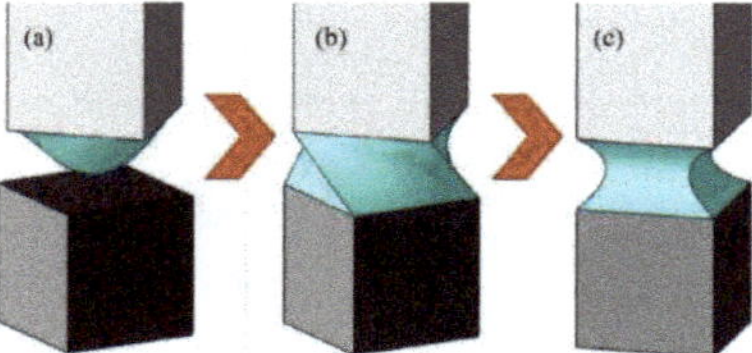

Figure 12. Shape matching sequence between same-sized square nozzle and cube: (**a**) droplet discharge from the square nozzle, (**b**) just before the liquid spreading and making the liquid bridge between two surfaces, and (**c**) liquid bridge deformed into a symmetrical shape such that the surface energy of the liquid bridge is minimized.

We compared the experimental results with condition (C3) using a circular nozzle to conduct a pick-and-place of the same cubed object. The positioning error and the attitude angle of the object after placement are shown in Table 6, and the image of the object after placement is shown in Figure 13.

Table 6. Comparison of the positioning errors and attitude angles for conditions (C3) and (S).

Condition		**(C3)**	**(S)**
Positioning error	Average	44 (μm)	99 (μm)
	Standard deviation	34 (μm)	57 (μm)
Attitude angle	Average	−3.0 (deg.)	−2.8 (deg.)
	Standard deviation	26.9 (deg.)	**7.4** (deg.)

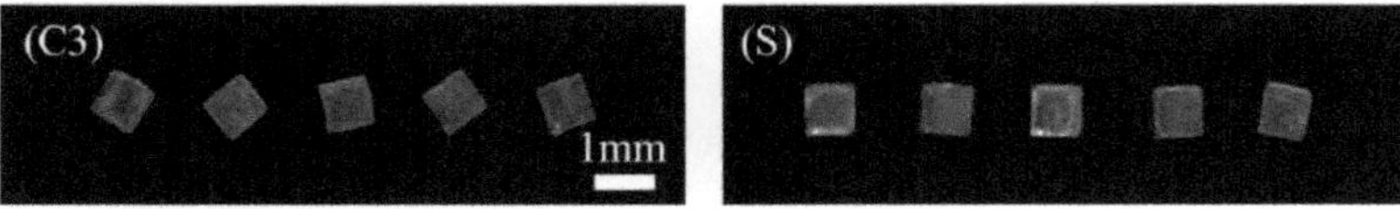

Figure 13. Arrangement of 1-mm cubes under conditions (C3) and (S).

As no target angle was set in these experiments, we focused on the standard deviations (SDs) to discuss the experimental results of the accuracy of the attitude angle. Table 6 shows that the SD of the attitude angle under condition (S) was smaller than condition (C3). Using this nozzle and a rotational stage, we could set the objects at various angles, as shown in Figure 14.

Figure 14. Circular arrangements of 1-mm cubes at various angles.

7. Conclusions and Future Prospects

This paper described a newly developed vision feedback method for improving the placement accuracy for a single nozzle capillary gripper. The developed system was coded in Python using OpenCV.

In the experiment, we automatically picked-and-placed the cubes without any solid contact, and the success rate was 99% (129 out of 130 samples). For the cubes with 1- and 0.5-mm sides, the minimum positioning error $\pm$ SD of each size was 44 $\pm$ 34 μm and 39 $\pm$ 18 μm, respectively; the ratios of the positioning errors to the lengths of the objects were 4.4 $\pm$ 3.4% and 7.8 $\pm$ 3.6%, respectively. We also verified that the size matching effect worked reasonably; the average positioning error minimized when the size of the nozzle was nearly equal to the side of the cube.

In addition, using a square-shaped nozzle with the same shape of the cube, we succeeded in controlling the attitude; the SD of the attitude angle was reduced from a random value to 7.4°.

To reduce the positioning errors, our future plans are as follows. We plan to evaluate the relationship between the surface shape and properties of the objects as well as the radius and arrangement pattern of the droplets, for effectively applying the self-alignment phenomena. Further work will study the accuracy of contour detection in the vision feedback system. The impact of lighting on positioning will also be studied in the future, for example, the effects of droplet size and color and diffraction and refraction of lighting. In addition, the investigation of the release condition is a significant subject from both theoretical and experimental approaches.

Moreover, we wish to take on the challenge of developing a pick-and-place mechanism for smaller complex-shaped objects of less than 200 μm, such as micro-LEDs, helical-shaped coils, thin wires, gears, soft contact lenses, fragile gels, and biomedical cells with the capillary grippers to cultivate the manipulation of heterogeneous and complex-shaped micro-objects in flexible electronics, micro-electro-mechanical systems (MEMS), soft robotics, soft matter, and biomedical fields. In the microminiaturization of an object, we consider the limitations to be determined as follows: the smallest possible droplet size, the sharpest nozzle, the effect of droplet evaporation, and the piezoelectric control of droplet volume.

Supplementary Materials: The following supporting information can be downloaded at: https://www.mdpi.com/article/10.3390/mi13081270/s1, Video S1: Digest movie.

Author Contributions: Conceptualization, T.I.; methodology, T.I., E.F. and K.T.; software, T.I.; validation, T.I., E.F., K.T. and O.F.; formal analysis, T.I., K.T. and E.F.; investigation, T.I., E.F., K.T., Y.N., N.W. and O.F.; resources, O.F.; data curation, T.I., E.F. and O.F.; writing—original draft preparation, T.I., E.F. and O.F.; writing—review and editing, E.F., N.W. and O.F.; visualization, T.I. and E.F.; supervision, Y.N. and O.F.; project administration, O.F.; funding acquisition, O.F. All authors have read and agreed to the published version of the manuscript.

Funding: This research was funded by the Grant-in-Aid for Scientific Research (C) [16K06178], Yokohama Academic Foundation, Yokohama National University Publication Subsidy.

Conflicts of Interest: The authors declare no conflict of interest.

References

1. Templier, F. GaN-based emissive microdisplays: A very promising technology for compact, ultra-high brightness display systems. *J. Soc. Inf. Display* **2016**, *24*, 669–675. [CrossRef]
2. Barwicz, T.; Taira, Y.; Lichoulas, T.W.; Boyer, N.; Martin, Y.; Numata, H.; Nah, J.-W.; Takenobu, S.; Janta-Polczynski, A.; Kimbrell, E.L.; et al. A Novel Approach to Photonic Packaging Leveraging Existing High-Throughput Microelectronic Facilities. *IEEE J. Sel. Top. Quantum Electron.* **2016**, *22*, 455–466. [CrossRef]
3. Henly, F.J. Combining engineered epi growth substrate materials with novel test and mass-transfer equipment to enable MicroLED mass-production. *SID* **2018**, *49*, 688–691. [CrossRef]
4. Feng, X.; Meitl, M.; Bowen, A.; Huang, Y.; Nuzzo, R.; Rogers, J. Competing fracture in kinetically controlled transfer printing. *Langmuir* **2007**, *23*, 12555–12560. [CrossRef]
5. Carlson, A.; Bowen, A.M.; Huang, Y.; Nuzzo, R.G.; Rogers, J.A. Transfer Printing Techniques for Materials Assembly and Micro/Nanodevice Fabrication. *Adv. Mater.* **2012**, *24*, 5284–5318. [CrossRef]
6. Jinno, M.; Iordachita, I. Microgripper using flexible wire hinge for robotic intraocular snake. In Proceedings of the IEEE ICRA2022, Philadelphia, PA, USA, 23–27 May 2022; pp. 6218–6224.
7. Eisenhaure, J.D.; Rhee, S.I.; Al-Okaily, A.M.; Carlson, A.; Ferreira, P.M.; Kim, S. The Use of Shape Memory Polymers for MEMS Assembly. *J. Microelectromechanical Syst.* **2016**, *25*, 69–77. [CrossRef]
8. Xia, Y.; He, Y.; Zhang, F.; Liu, Y.; Leng, J. A review of shape memory polymers and composites: Mechanisms, materials, and applications. *Adv. Mater.* **2020**, *33*, 202000713. [CrossRef]
9. Amit, R.; Abadi, A.; Kosa, G. Characterization of steady streaming for a particle manipulation system. *Biomed. Microdevices* **2016**, *18*, 39. [CrossRef]
10. Fuchiwaki, O.; Tanaka, Y.; Notsu, H.; Hyakutake, T. Multi-axial non-contact in situ micromanipulation by steady streaming around two oscillating cylinders on holonomic miniature robots. *Microfluid. Nanofluidics* **2018**, *22*, 80. [CrossRef]
11. Benhal, P.; Chase, J.G.; Gaynor, P.; Oback, B.; Wang, W. AC electric field induced dipole-based on-chip 3D cell rotation. *Lab Chip* **2014**, *14*, 2717–2727. [CrossRef]
12. Berns, M.W. Laser scissors and tweezers to study chromosomes: A review. *Front Bioeng. Biotechnol.* **2020**, *8*, 721. [CrossRef]
13. Choudhary, D.; Mossa, A.; Jadhav, M.; Cecconi, C. Bio-molecular applications of recent developments in optical tweezers. *Biomolecules* **2019**, *9*, 23. [CrossRef]
14. Lee, V.; James, N.M.; Waitukaitis, S.R.; Jaeger, H.M. Collisional charging of individual submillimeter particles: Using ultrasonic levitation to initiate and track charge transfer. *Phys. Rev. Mater.* **2018**, *2*, 035602. [CrossRef]
15. Shi, Q.; Di, W.; Dong, D.; Yap, L.W.; Li, L.; Zang, D.; Cheng, W. A general approach to free-standing nanoassemblies via acoustic levitation self-assembly. *ACS Nano* **2019**, *13*, 5243–5250. [CrossRef]
16. Mastrangeli, M.; Abbasi, S.; Varel, Ç.; Hoof, C.V.; Celis, C.P.; Böhringer, K.F. Self-assembly from milli to nanoscales: Methods and applications. *J. Micromech. Microeng.* **2009**, *19*, 083001. [CrossRef]
17. Najib, A.M.; Abdullah, M.Z.; Saad, A.A.; Samsudin, Z.; Che Ani, F. Numerical simulation of self-alignment of chip resistor components for different silver content during reflow soldering. *Microelectron. Reliab.* **2017**, *79*, 69–78. [CrossRef]
18. Uran, S.; Šafaric, R.; Bratina, B. Reliable and accurate release of micro-sized objects with a gripper that uses the capillary-force method. *Micromachines* **2017**, *8*, 182. [CrossRef]
19. Fuchiwaki, O.; Kumagai, K. Development of wet tweezers based on capillary force for complex-shaped and heterogeneous micro-assembly. In Proceedings of the 2013 IEEE/RSJ International Conference on Intelligent Robots and Systems, Tokyo, Japan, 3–7 November 2013; pp. 1003–1009.
20. Kato, Y.; Mizuno, T.; Takagi, H.; Ishino, Y.; Takasaki, M. Experimental study on microassembly by using liquid surface tension. *SICE J. Control. Meas. Syst. Integr.* **2010**, *3*, 309–314. [CrossRef]
21. Hagiwara, W.; Ito, T.; Tanaka, K.; Tokui, R.; Fuchiwaki, O. Capillary force gripper for complex-shaped micro-objects with fast droplet forming by on-off control of a piston slider. *IEEE Robot Autom. Lett.* **2019**, *4*, 3695–3702. [CrossRef]

22. Tanaka, K.; Ito, T.; Nishiyama, Y.; Fukuchi, E.; Fuchiwaki, O. Double-nozzle capillary force gripper for cubical, triangular prismatic, and helical 1-mm-Sized-Objects. *IEEE Robot Autom. Lett.* **2022**, *7*, 1324–1331. [CrossRef]
23. Lambert, P. *Capillary Forces in Microassembly*; Springer: New York, NY, USA, 2007.
24. Tselishchev, Y.G.; Val'tsifer, V.A. Influence of the type of contact between particles joined by a liquid bridge on the capillary cohesive forces. *Colloid J.* **2003**, *65*, 385–389. [CrossRef]
25. Itaki, T.; Taira, Y.; Kuwamori, N.; Maebayashi, T.; Takeshima, S.; Toya, K. Automated collection of single species of microfossils using a deep learning–micromanipulator system. *Prog. Earth Planet. Sci.* **2020**, *7*, 19. [CrossRef]
26. Shoji, D.; Noguchi, R.; Otsuki, S.; Hino, H. Classification of volcanic ash particles using a convolutional neural network and probability. *Sci. Rep.* **2018**, *8*, 8111. [CrossRef]
27. Panarello, A.P.; Seavey, C.E.; Doshi, M.; Dickerson, A.K.; Kean, T.J.; Willenberg, B.J. Transforming capillary alginate gel into new 3D-printing biomaterial inks. *Gels* **2022**, *8*, 376. [CrossRef]
28. Ji, X.; Liu, X.; Cacucciolo, V.; Imboden, M.; Civet, Y.; Haitami, A.E.; Cantin, S.; Perriard, Y.; Shea, H. An autonomous untethered fast soft robotic insect driven by low-voltage dielectric elastomer actuators. *Sci. Robot* **2019**, *4*, eaaz6451.
29. Mastrangeli, M.; Ruythooren, W.; Celis, J.; Van Hoof, C. Challenges for capillary self-assembly of microsystems. *IEEE Trans. Compon. Packaging Manuf. Technol.* **2011**, *1*, 133–149. [CrossRef]
30. Stauth, S.A.; Parviz, B. Self-assembled single-crystal silicon circuits on plastic. *Proc. Natl. Acad. Sci. USA* **2006**, *103*, 13922–13927. [CrossRef]
31. Takagi, S.; Shimada, T.; Ishida, Y.; Fujimura, T.; Masui, D.; Tachibana, H.; Eguchi, M.; Inoue, H. Size-matching effect on inorganic nanosheets: Control of distance, alignment, and orientation of molecular adsorption as a bottom-up methodology for nanomaterials. *Langmuir* **2013**, *29*, 2108–2119. [CrossRef]

Article

Dynamic Model of a Conjugate-Surface Flexure Hinge Considering Impacts between Cylinders

Alessandro Cammarata [1,*], Pietro Davide Maddìo [1], Rosario Sinatra [1], Andrea Rossi [2] and Nicola Pio Belfiore [2]

[1] Department of Civil Engineering and Architecture, University of Catania, Viale Andrea Doria 6, 95123 Catania, Italy; pietro.maddio@unict.it (P.D.M.); rosario.sinatra@unict.it (R.S.)
[2] Department of Industrial, Electronic and Mechanical Engineering, University of Roma Tre, Via Vito Volterra 62, 00154 Rome, Italy; andrea.rossi@uniroma3.it (A.R.); nicolapio.belfiore@uniroma3.it (N.P.B.)
* Correspondence: alessandro.cammarata@unict.it; Tel.: +39-095-738-2403

Abstract: A dynamic model of a Conjugate-Surface Flexure Hinge (CSFH) has been proposed as a component for MEMS/NEMS Technology-based devices with lumped compliance. However, impacts between the conjugate surfaces have not been studied yet and, therefore, this paper attempts to fill this gap by proposing a detailed multibody system (MBS) model that includes not only rigid-body dynamics but also elastic forces, friction, and impacts. Two models based on the Lankarani-Nikravesh constitutive law are first recalled and a new model based on the contact of cylinders is proposed. All three models are complemented by the friction model proposed by Ambrosìo. Then, the non-smooth Moreau time-stepping scheme with Coulomb friction is described. The four models are compared in different scenarios and the results confirm that the proposed model outcomes comply with the most reliable models.

Keywords: multibody systems; CSFH; event-driven scheme; non-smooth contact; LN-model; Moreau time-stepping scheme

Citation: Cammarata, A.; Maddìo, P.D.; Sinatra, R.; Rossi, A.; Belfiore, N.P. Dynamic Model of a Conjugate-Surface Flexure Hinge Considering Impacts between Cylinders. *Micromachines* **2022**, *13*, 957. https://doi.org/10.3390/mi13060957

Academic Editor: Alberto Corigliano

Received: 16 May 2022
Accepted: 14 June 2022
Published: 16 June 2022

1. Introduction

During the last decades, the development of both MEMS (micro electro-mechanical systems) and NEMS (nano electro-mechanical systems) technology-based devices encountered several technological issues [1]. As far as the mechanical structure is concerned, the micromachining methods and the available materials inevitably restricted the mobility of most micro/nanosystems mobility to a plane motion with a few degrees of freedom (DoF), when not even down to a single DoF only.

The appearance of flexure hinges, together with lumped compliant structure, disclosed new ways to design. As a consequence, several new devices, obtained by means of micromachining, were proposed in the literature. For example, biosensing acoustic wave based devices [2], CMOS-MEMS resonators [3], microgrippers [4], drug delivery micropumps [5], surgery [6], micromirror platforms [7], and more generally actuators [8,9] and sensors [10].

A peculiar hinge, called Conjugate-Surface Flexure Hinge (CSFH), has been successfully proposed as a component for MEMS/NEMS Technology-based devices with lumped compliance [11,12] and the next section will be dedicated to some important details on CSFHs.

Although several aspects of CSFH equipped microsystems have been already studied, such as, adaptability to precision mechanism [13], kinetostatics [14], operational in aqueous environment [15], vibrations [16] and technological issues [17], the dynamical behavior of a CSFH still remains unexplored. For example, the impacts and their consequences on dynamics have not been studied yet. Contact is an inherent feature of the CSFH but can yield wear [18].

The contact in mechanical joints with clearance has been largely discussed in the literature [19]. Since the CSFH conjugate surfaces can be described as a journal-bearing, here, we refer only to this class of joint. The two most commonly used approaches to describe the phenomenon of impact see a continuous regularized approach [20–22] versus a non-smooth approach [23,24]. The regularized approach derives the contact laws using geometric and material parameters of the contacting surfaces, the non-smooth approach does not require impact laws and the impact is instantaneous. Although both approaches are valid, the dynamics of impact can be quite different, especially in the presence of external forces capable of amplifying the small differences coming from different contact dynamic responses.

For this reason, we will attempt to fill this gap by proposing a detailed multibody system (MBS) model that includes rigid-body dynamics, elastic forces, friction, and impacts. This complex mathematical tool must be flexible and provide reliable results. Motivated by this reason, starting from models widely employed in multibody systems with impacts and experimentally validated such as those reported in [20–22], we propose a novel model in which the generalized stiffness obtained considering the impact of two cylinders and not two spheres. Results will demonstrate that our model is consistent with other experimentally verified models.

In Section 3, the circular beam flexure hinge is described and the generalized elastic forces to include in the dynamic model are obtained. Section 4 introduces continuous impact models based on regularized approaches. Hertz contact theory is first recalled, then the Lankarani-Nikravesh model and its modified version with a non-constant generalized contact stiffness are described. Starting from the classic Lankarani-Nikravesh constitutive law, a novel method based on the contact of two cylindrical surfaces is presented. All three continuous impact models are complemented with the friction model proposed by Ambrosìo. In Section 5, the non-smooth Moreau time-stepping scheme with Coulomb friction is described. Section 6 compares the methods of the previous sections considering different impact scenarios. First, a central and an asymmetrical impact are described. Then, a complete CSFH dynamics is simulated. Section 7 deals with the influence of model parameters on system dynamics. Finally, Sections 8 and 9 summarize the final comments, future developments and conclusions.

2. Motivations and Contributions

The main motivations that guided the study are listed below:

1. absence of a complete dynamic model of a CSFH,
2. create a specific impact model for CSFH that is a valid alternative to those most commonly used in the literature,
3. provide some guidelines for proper CSFH modeling and design.

These motivations have led to the following main contributions:

1. developing a detailed multibody model of a CSFH including rigid-body dynamics, elastic forces, friction, and impacts,
2. developing a novel event-driven model considering impacts between cylinders,
3. conducting a parametric analysis to understand the influence of each parameter on the CSFH dynamics.

3. Flexure Description

The Conjugate-Surface Flexure Hinge (CSFH) is composed of two parts: a flexure beam connecting two bodies and a conjugate-surface area where the two bodies can either slide or collide. The curved beam leads to nonlinear motion characteristics [25]. While the kineto-static analysis of CSFH has been already described in detail [26], this paragraph recalls some relevant outlines that will be useful for the sake of the present investigation.

3.1. Kinematics

The basic layout illustrated in Figure 1 will be the reference model for the dynamic analysis. Accordingly, the flexure hinge is connected to a rigid body, where both undeformed and deformed configurations have been displayed.

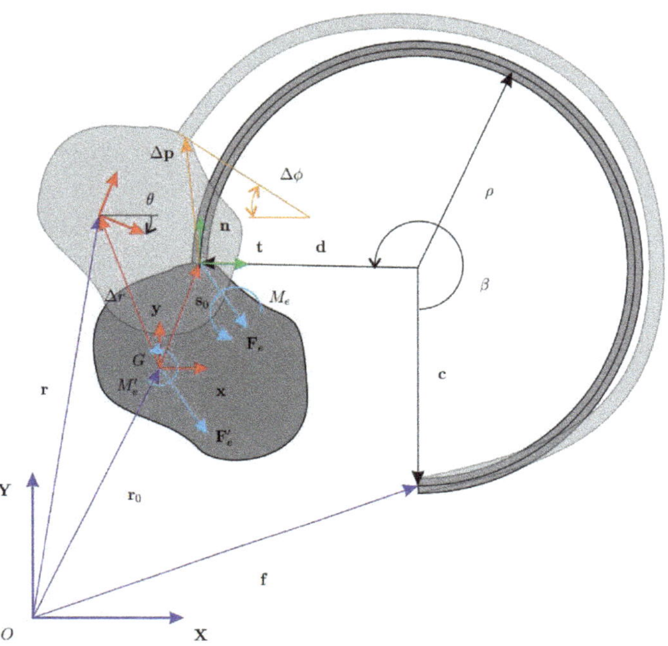

Figure 1. Layout of a CSFH.

Kinetostatic analysis will be herein introduced through the vector

$$\mathbf{r}_0 = \mathbf{f} - \mathbf{c} + \mathbf{d} - \mathbf{s}_0 \tag{1}$$

that defines the mass-center position of the rigid body in the undeformed configuration, where:

- $\mathbf{f}$ stands for the position of the beam root,
- $\mathbf{c}$ and $\mathbf{d}$, respectively, denote the vectors from the center of the circular hinge to the endpoints of the curved beam,
- $\mathbf{s}_0$ is the vector connecting the mass-center to the body-hinge attachment point in the undeformed configuration.

Being ρ and β the flexure radius and opening angle, respectively, the vector $\mathbf{h} = \mathbf{d} - \mathbf{c}$, can be expressed as

$$\mathbf{h} = \rho \begin{bmatrix} c(\beta - \frac{\pi}{2}) \\ 1 + s(\beta - \frac{\pi}{2}) \end{bmatrix} \tag{2}$$

where the compact notation $s \equiv \sin$ and $c \equiv \cos$ has been employed. Vector $\mathbf{s}_0$ connects two rigid-body points and can be expressed considering the rotation matrix $\mathbf{A}$ of the body in its undeformed configuration and the vector $\bar{\mathbf{s}}$ relative to the body-frame $(\mathbf{x}, \mathbf{y})$, i.e.

$$\mathbf{s}_0 = \mathbf{A}(\theta_0)\bar{\mathbf{s}} \Rightarrow \begin{bmatrix} s_{0x} \\ s_{0y} \end{bmatrix} = \begin{bmatrix} c(\theta_0) & -s(\theta_0) \\ s(\theta_0) & c(\theta_0) \end{bmatrix} \begin{bmatrix} \bar{s}_x \\ \bar{s}_y \end{bmatrix} \tag{3}$$

where θ is the angle between the axes $\mathbf{x}$ and $\mathbf{X}$. If $\theta_0 = 0$, it follows that $\mathbf{s}_0 \equiv \bar{\mathbf{s}}$. The expression of $\mathbf{r}_0$ becomes

$$\begin{bmatrix} r_{0x} \\ r_{0y} \end{bmatrix} = \begin{bmatrix} f_x + \rho c(\beta - \frac{\pi}{2}) - \bar{s}_x \\ f_y + \rho + \rho s(\beta - \frac{\pi}{2}) - \bar{s}_y \end{bmatrix} \tag{4}$$

being f_x and f_y the components of $\mathbf{f}$ in the reference frame.

Consider the deformed configuration and taking $\mathbf{p} = \mathbf{r} + \mathbf{A}\bar{\mathbf{s}}$ as the positioning vector of the body-hinge attachment point, the displacement $\Delta\mathbf{p}$ is simply expressed as

$$\Delta\mathbf{p} = \mathbf{p} - \mathbf{p}_0 \equiv \mathbf{r} + \mathbf{A}\bar{\mathbf{s}} - \mathbf{r}_0 - \mathbf{s}_0 \equiv \Delta\mathbf{r} + (\mathbf{A} - \mathbf{1})\bar{\mathbf{s}} \tag{5}$$

in which $\Delta\mathbf{r}$ is the mass-center displacement. As it can be observed in Figure 1, the rotation $\Delta\phi$ of the attachment section is equal to θ. The standard stiffness model for this case [26] has been obtained by considering the local frame $(\mathbf{n}, \mathbf{t})$, respectively, composed of the vectors normal and tangent to the attachment cross-section. This frame moves with the body and has a constant orientation with respect to the body-frame $(\mathbf{x}, \mathbf{y})$. The rotation matrix $\mathbf{R}$ expressing this constant orientation has the following expression

$$\mathbf{R} = \begin{bmatrix} c(\psi) & -s(\psi) \\ s(\psi) & c(\psi) \end{bmatrix}, \quad \psi = 2\pi - \beta \tag{6}$$

Composing $\mathbf{R}$ and $\mathbf{A}$ is possible to pass from the local frame $(\mathbf{n}, \mathbf{t})$ to the global frame $(\mathbf{X}, \mathbf{Y})$. In the following, the steps necessary to find the generalized elastic force vector are detailed.

3.2. Elastic Force

The above-mentioned stiffness model can be summarized through the following expression:

$$\begin{bmatrix} \Delta\bar{\mathbf{p}} \\ \Delta\phi \end{bmatrix} = \hat{\mathbf{C}} \begin{bmatrix} \hat{\mathbf{F}}_e \\ M_e \end{bmatrix} \tag{7}$$

where $\hat{\mathbf{C}}$ is the compliance matrix [26]. The force vector $\hat{\mathbf{F}}_e$ contains the normal and tangential forces expressed in the frame $(\mathbf{n}, \mathbf{t})$ and applied at the attachment section while M_e is the moment. These components can be gathered into the generalized vector $\hat{\mathbf{w}} = [\hat{\mathbf{F}}_e^T, M_e]^T$.

Given the generic configuration of the rigid body expressed through the three-elements array $\mathbf{q} = [\mathbf{r}^T, \theta]^T$ the following procedure will be used.

1. find $\Delta\mathbf{p}$ using Equation (5),
2. find the vector $\mathbf{w}_e = [\mathbf{F}_e^T, M_e]^T$ defined in $(\mathbf{X}, \mathbf{Y})$:

$$\mathbf{w}_e = \mathbf{K}\Delta\mathbf{p}, \quad \mathbf{K} = \mathbf{A}\mathbf{R}\hat{\mathbf{K}}\mathbf{R}^T\mathbf{A}^T \tag{8}$$

   where $\mathbf{K}$ is the stiffness matrix expressed in the reference frame $(\mathbf{X}, \mathbf{Y})$ and $\hat{\mathbf{K}} \equiv \hat{\mathbf{C}}^{-1}$ is the local stiffness matrix in $(\mathbf{n}, \mathbf{t})$,
3. transport the vector $\mathbf{w}_e$ to the mass-centre point G to find the generalized force vector $\mathbf{Q}_e = [\hat{\mathbf{F}}'^T_e, M'_e]^T$, defined as

$$\mathbf{Q}_e = \mathbf{T}_e\mathbf{w}_e \equiv \begin{bmatrix} \mathbf{1} & 0 \\ \mathbf{s}^T\tilde{\mathbf{1}} & 1 \end{bmatrix} \begin{bmatrix} \mathbf{F}_e \\ M_e \end{bmatrix}, \quad \tilde{\mathbf{1}} = \begin{bmatrix} 0 & 1 \\ -1 & 0 \end{bmatrix} \tag{9}$$

   where $\mathbf{T}_e$ is the 3×3 rigid-body transformation matrix needed to transport $\mathbf{w}_e$ and $\tilde{\mathbf{1}}$ is necessary to consider the cross-product in the planar case [27].

4. Event-Driven Scheme with Regularized Approach

We first describe an event-driven scheme with a contact detection algorithm. These schemes integrate the equations of motion until a slip-stick transition or an impact is detected.

4.1. Contact Kinematics

The CSFH limits the hinge deformation by introducing two conjugate surfaces where the two bodies can collide. This solution has positive effects on motion accuracy and improves resistance to yielding as well. Figure 2 shows the undeformed and deformed

CSFH. The conjugate surfaces, represented through two circles of radii R_1 and R_2, are separated by a radial clearance $\delta = R_2 - R_1$ in the undeformed configuration. The local vector $\bar{\mathbf{s}}_1$ denotes the position of the center O_1 with respect to the body frame. During motion, the hinge deforms and the bodies collide at one point C. Observing the Figure 2, the following closure equation can be written

$$\mathbf{c}_1 = \mathbf{c}_2 \Rightarrow \mathbf{r}_1 + \mathbf{A}\bar{\mathbf{s}}_1 - R_1\mathbf{n}_c = \mathbf{r}_2 - R_2\mathbf{n}_c \tag{10}$$

where point C is thought to belong to the two bodies. Here, the second body is fixed for convenience. Equation (10) provides the unit vector $\mathbf{n}_c$ normal to the conjugate surfaces at point C, i.e.,

$$\mathbf{n}_C = -\frac{\mathbf{e}}{\|\mathbf{e}\|} \qquad \mathbf{e} = \mathbf{r}_1 + \mathbf{A}\bar{\mathbf{s}}_1 - \mathbf{r}_2 \tag{11}$$

where $\mathbf{e}$ is the eccentricity vector expressing the position of O_1 with respect to O_2. The tangent unit vector $\mathbf{t}_c$ is calculated rotating $\mathbf{n}_c$ 90° counter-clockwise. Time-differentiating the expression of $\mathbf{c}_1$, the velocity $\dot{\mathbf{c}}_1$ is obtained as

$$\dot{\mathbf{c}}_1 = \dot{\mathbf{r}}_1 + \mathbf{\Omega}(\mathbf{s}_1 - R_1\mathbf{n}_c) \tag{12}$$

being $\mathbf{\Omega}$ the angular-velocity matrix of the first body. The expression of $\dot{\mathbf{c}}_1$ is required to calculate the tangent velocity, that is the component of $\dot{\mathbf{c}}_1$ along $\mathbf{t}_c$.

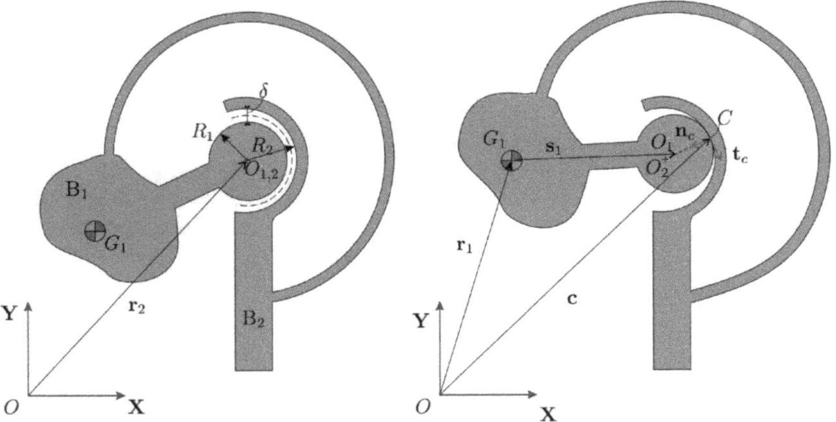

Figure 2. Undeformed CSFH (**left**). Deformed CSFH with impact (**right**).

4.2. Contact Model with Friction

In this subsection, different Hertzian contact models are first described. Then, the static friction force model of Ambrósio is recalled.

4.2.1. Impact Models

The regularized approach to contact starts from the work by Hertz on the theory of elasticity [28,29]. Hertz introduced a non-linear law between the normal contact force F_N and the indentation ϱ, i.e.,

$$F_N = K\varrho^n \tag{13}$$

where K is the generalized stiffness parameter and n is the nonlinear exponent factor. The stiffness K is evaluated following Hertz contact theory. Considering two spheres, the following expression is obtained

$$K = \frac{4}{3(\sigma_1 + \sigma_2)}\sqrt{\frac{R_1R_2}{R_2 + R_1}} \tag{14}$$

where σ_i, $i = 1,2$, are material parameters expressed in terms of Young's modulus E and Poisson's ratio η, defined by

$$\sigma_i = \frac{1-\nu_i^2}{E_i}, \quad i = 1,2 \tag{15}$$

Considering two cylinders of length L with parallel axes, we have a contact on a rectangular area. In this case, Hertz proposed to use

$$K = \frac{\pi}{4}E^* \tag{16}$$

where E^* is an equivalent Elastic modulus, defined as

$$\frac{1}{E^*} = \frac{1-\nu_1^2}{E_1} + \frac{1-\nu_2^2}{E_2} \tag{17}$$

The ESDU-78035 Tribology Series [30] proposed to use the following implicit law instead

$$\varrho = F_N\left(\frac{\sigma_1+\sigma_2}{L}\right)\left[\ln\left(\frac{4L(R_i-R_j)}{F_N(\sigma_1+\sigma_2)}\right)+1\right] \tag{18}$$

In this case, the stiffness K is not constant and can be obtained numerically from the derivative of the force-indentation curve.

Hertz law does not include energy dissipation due to internal damping. Therefore, following the work by Kelvin and Voigt, [31], different viscoelastic models have been proposed in the literature, [32]. The generic viscoelastic model has the following form

$$F_N = K\varrho^n + D\varrho^m\dot{\varrho} \tag{19}$$

where D is the damping coefficient representing the dissipative term proportional to the relative normal contact velocity $\dot{\varrho}$ and m is a non-linear coefficient making the dissipation dependent on the indentation. The coefficient m can be empirical or based on dissipation models.

Hunt and Crossley [33] proposed a dissipative contact force model adding a non-linear viscoelastic term to Hertz law, thus coming to the following force-penetration law

$$F_N = K\varrho^n + D\dot{\varrho} \tag{20}$$

Hunt and Crossley law can be written in terms of the coefficient of restitution c_r, i.e.

$$F_N = K\varrho^n\left[1+\frac{3(1-c_r)}{2}\frac{\dot{\varrho}}{\dot{\varrho}^{(-)}}\right] \tag{21}$$

being $\dot{\varrho}^{(-)}$ the initial contact velocity. Following Hunt and Crossley law, Lankarani-Nikravesh proposed an impact model to be applied in multibody systems [20]. The normal force is written as

$$F_N = K\varrho^n\left[1+\frac{3(1-c_r^2)}{4}\frac{\dot{\varrho}}{\dot{\varrho}^{(-)}}\right] \tag{22}$$

Since the Lankarani-Nikravesh model applies to stiff materials with the coefficient of restitution greater than 0.9, Flores et al. proposed an equivalent model for soft materials [34], i.e.,

$$F_N = K\varrho^n\left[1+\frac{8(1-c_r)}{5c_r}\frac{\dot{\varrho}}{\dot{\varrho}^{(-)}}\right] \tag{23}$$

In this case, the impact law considers different energy dissipation between compression and restitution of the contact phases.

In [22] the authors proposed to use a constitutive law of type

$$F_N = K\varrho^n + D_{LN}\frac{K_{ESDU}}{K}\dot{\varrho} \tag{24}$$

to describe the journal-bearing contact. While K follows from Equation (14), K_{ESDU} is derived from the ESDU law in Equation (18). Finally, D_{LN} is the damping obtained using the Lankarani-Nikravesh (LN) model (22), i.e.,

$$D_{LN} = \frac{3K(1 - c_r^2)}{4\dot{\varrho}^{(-)}} \tag{25}$$

This model hereafter referred to as the LNA-ESDU, has been experimentally verified for a slider-crank mechanism involving contact events at low or moderate impact velocities. Compared to the classic LN model, LNA-ESDU provides lower impact forces and is capable to accurately reproduce the experimental results probably because the actual impact forces are distributed on an wider area due to the plasticity deformation of the bristles.

Considering this literature, in this work we broaden the classic LN model of Equation (22) to also include the generalized stiffness obtained by the contact of two cylinders as in Equation (16).

In the numerical part, this model will be compared to the other LNA-based models for validation.

4.2.2. Friction Model

Static friction force models start with the work of Coulomb [35]. Modified Coulomb's laws such as those proposed by Threlfall [36] or Bo and Pavelescu [37] included the Stribeck effect, i.e., the transition from static to dynamic friction.

In most of these static models, the friction force can have a discontinuity at zero velocity. To solve this issue, Karnopp [38], Leine et al. [39], Bengisu and Akay [40] proposed static models with finite slope at zero velocity.

Nevertheless, in stick conditions, the relative tangent velocity should be zero but it does not occur due to numerical issues. Rather, it maintains close to zero and switches its sign with high frequency introducing numerical instability in the system's response. To prevent this unwanted behavior, the Ambrósio static friction model put the friction force to zero for low velocities [41]. The friction force becomes

$$\mathbf{F}_T = \begin{cases} \mathbf{0} & |v_T| \leq v_0 \\ -\frac{|v_T| - v_0}{v_1 - v_0} f_d F_N \mathrm{sgn}(\mathbf{v}_T) & v_0 < |v_T| < v_1 \\ -f_d F_N \mathrm{sgn}(\mathbf{v}_T) & |v_T| \geq v_1 \end{cases} \tag{26}$$

in which f_d is the kinetic coefficient of friction, v_0 is the stiction velocity, v_1 is the slip velocity [38], and $\mathbf{v}_T = (\mathbf{t}_c^T \mathbf{v})\mathbf{t}_c$ and v_T are the relative tangential velocity and its module, respectively. In this case, $\mathbf{v} \equiv \dot{\mathbf{c}}_1$ reported in Equation (12).

For its stability, the Ambrósio static friction model is often used in multibody systems with frictional joints. Many other friction models could be coupled to the impact law and several empirical models have been presented in the literature. We refer to [42,43] for further details.

4.3. *Equations of Motion*

During the motion, the first body B_1 is subject to different generalized forces: the generalized elastic force $\mathbf{Q}_e$, the gravitational force $\mathbf{Q}_g$, and the generalized contact force $\mathbf{Q}_c$. The latter is present only if the contact is triggered. Therefore, the undamped equations of motion are

$$\mathbf{M}\ddot{\mathbf{q}} = \mathbf{Q} \tag{27}$$

where the generalized force $\mathbf{Q}$ changes its expression according to the contact condition $||\mathbf{e}|| \geq \delta$, i.e.,

$$\mathbf{Q} = \mathbf{Q}_e + \mathbf{Q}_g, \quad \text{free motion} \tag{28a}$$

$$\mathbf{Q} = \mathbf{Q}_e + \mathbf{Q}_g + \mathbf{Q}_c, \quad \text{contact} \tag{28b}$$

The generalized gravitational force is given by $\mathbf{Q}_g = [m_1\mathbf{g}^T, 0]^T$, being m_1 the mass of B_1 and $\mathbf{g}$ the gravity acceleration vector.

The contact is triggered when $\|\mathbf{e}\| \geq \delta$. If this condition is met, a contact force $\mathbf{F}_c$ is generated in the contact area. This force has two components along $\mathbf{n}_c$ and $\mathbf{t}_c$, respectively, i.e.,

$$\mathbf{F}_c = F_N\mathbf{n}_c + F_T\mathbf{t}_c \tag{29}$$

where F_N is the normal force and F_T is the friction force calculated following contact and friction models presented in the previous subsection. When carried to the mass center position G_1, the contact force $\mathbf{F}_c$ yields a moment M_c, thus the 3-dimensional generalized contact force $\mathbf{Q}_c$ can be expressed as

$$\mathbf{Q}_c \equiv \begin{bmatrix} \mathbf{F}_c \\ M_c \end{bmatrix} = \begin{bmatrix} 1 \\ \mathbf{s}_1^T\bar{\mathbf{1}} \end{bmatrix}\mathbf{F}_c \tag{30}$$

In Figure 3 the flowchart of procedure for dynamic analysis including impact is shown, [21,32]. When an impact condition is fulfilled, trying to get closer to the instant of the impact, a maximum tolerance δ_{tol} activates the bisection of the time interval. The generalized force applied to the system changes according to Equation (28).

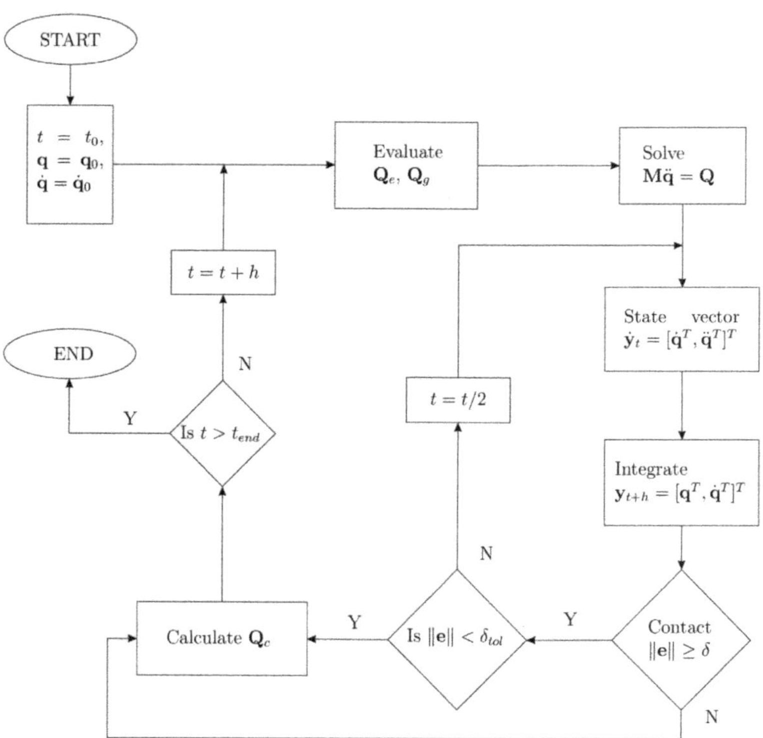

Figure 3. Flowchart of the event-driven scheme.

5. Time-Stepping Method: Moreau's Scheme

In contrast to the previous method, time-stepping methods do not require the detection of impact points and the consequent change of state. Rather, a discrete state is determined and held over the entire time step. These methods are less accurate than the previous event-driven schemes but are more robust and easy-to-implement. Here, Moreau's scheme with the midpoint rule has been implemented [23,24]. Readers interested to time-stepping methods are referred to specialized bibliography [44–46]. In Moreau's scheme, the equations of dynamics (27) are written as an index-2 DAE (differential algebraic equation) system, i.e.,

$$\begin{cases} \mathbf{M}\dot{\mathbf{u}} - \mathbf{Q}_e - \mathbf{Q}_g - \mathbf{D}^T\boldsymbol{\lambda} = \mathbf{0} \\ \mathbf{u} = \dot{\mathbf{q}} \\ -\boldsymbol{\gamma} \equiv -\mathbf{D}\mathbf{u} \in N_A(\boldsymbol{\lambda}) \end{cases}. \tag{31}$$

where $\mathbf{u}$ is the generalized velocity vector, $\mathbf{D}$ is the Jacobian matrix, $\boldsymbol{\lambda}$ is the vector of Lagrangian multipliers, γ is the time-derivative of the unilateral constraints, and N_A is the cone of inclusion defined on the set A. The third equation is a set-valued law on velocity level that can be activated or not depending on a geometric gap function able to trigger the unilateral contact. Here, the gap function g can be defined as

$$g(\mathbf{q}) = \delta - \sqrt{\mathbf{e}^T\mathbf{e}} \tag{32}$$

where $\mathbf{e}$ has been defined in Equation (11). If $g(\mathbf{q}) \leq 0$ the unilateral contact is active and the normal force modulus is

$$F_N = \lambda_1, \quad \lambda_1 \in \mathbb{R}_0^+ \equiv A_1 \tag{33}$$

Now, the normal contact is defined at position level and not at velocity level, as required by Moreau's time-stepping scheme. The time-derivative of the unilateral constraint $g(\mathbf{q}) = 0$ yields

$$\gamma_1 = \begin{bmatrix} \mathbf{n}_c^T & \mathbf{n}_c^T\tilde{\mathbf{I}}\mathbf{s}_1 \end{bmatrix}\dot{\mathbf{q}} \tag{34}$$

where $\gamma_1 = dg(\mathbf{q})/dt$.

The same gap function g activates also the tangential or friction force F_T. Once the impact has been triggered, the friction force can be either in impressed or constrained mode. In impressed mode there is slip between the surfaces in contact and F_T is defined as

$$F_T = -f_d\lambda_1\frac{\mathbf{v}_T}{v_T} \tag{35}$$

Notice that here F_T is not following the smooth transition provided by Equation (26). In constrained mode there is stiction and F_T belongs to the interval

$$F_T \in [-f_a\lambda_1, +f_a\lambda_1] \tag{36}$$

The same expression can be written in terms of the law of inclusion using the relative tangential velocity module v_T, i.e.,

$$-v_T \in N_{A_2}(\lambda_2), \quad A_2 = \{\lambda_2 \in \mathbb{R} \mid \ |\lambda_2| \leq f_a\lambda_1\} \tag{37}$$

The latter expression allows defining a kinematic set-valued law in which $\gamma_2 \equiv v_T$, i.e.,

$$\gamma_2 = \begin{bmatrix} \mathbf{t}_c^T & \mathbf{t}_c^T\tilde{\mathbf{I}}(\mathbf{s}_1 - R_1\mathbf{n}_c) \end{bmatrix}\dot{\mathbf{q}} \Rightarrow \gamma_2 = \begin{bmatrix} \mathbf{t}_c^T & \mathbf{t}_c^T\tilde{\mathbf{I}}\mathbf{s}_1 \end{bmatrix}\dot{\mathbf{q}} \tag{38}$$

Combining the Equations (34) and (38), we write

$$\boldsymbol{\gamma} \equiv \mathbf{D}\mathbf{u}, \quad \mathbf{D} = \begin{bmatrix} \mathbf{n}_c^T & \mathbf{n}_c^T\tilde{\mathbf{I}}\mathbf{s}_1 \\ \mathbf{t}_c^T & \mathbf{t}_c^T\tilde{\mathbf{I}}\mathbf{s}_1 \end{bmatrix} \tag{39}$$

in which $\mathbf{D}$ is the 2×3 Jacobian matrix of system (31).

6. Numerical Simulations

The proposed formulation has been tested considering the CSFH displayed in Figure 4. Without any loss of meaning, we suppose that the mass centers G_1 and G_2 are, respectively, located at the geometric centers O_1 and O_2 of the conjugate cylindrical surfaces. Geometrical and structural parameters are reported in Table 1. Even if impact and friction follow different models, some parameters such as the restitution coefficient c_r, and the dynamic friction coefficient f_d are common to both event-driven and time-stepping models.

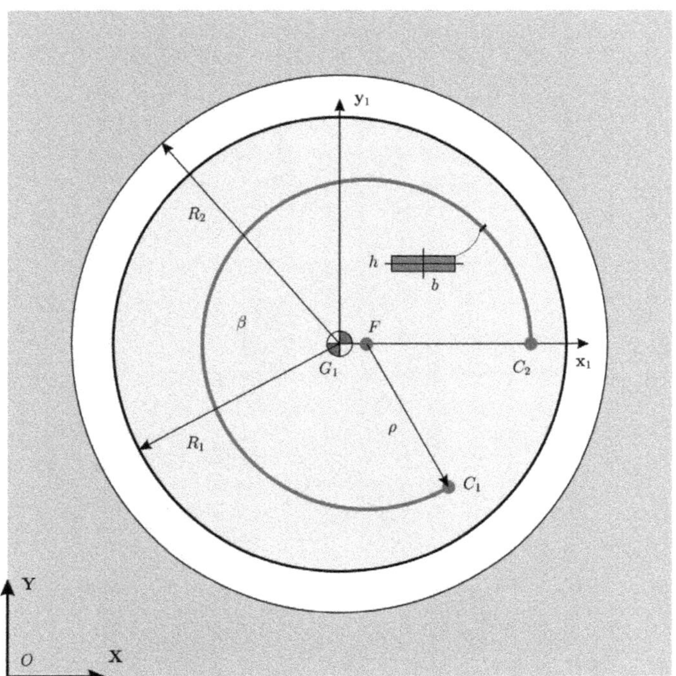

Figure 4. Layout of the CSFH used in the numerical simulations. Initial undeformed configuration.

Table 1. Common geometric, inertial, and structural parameters of the CSFH used in the numerical simulations.

Notation	Description	Value	Unit
β	hinge opening angle	300	(°)
ρ	hinge radius	0.144	(m)
$\overrightarrow{OF}$	hinge center coordinates	$[0.22, 0.1]^T$	(m)
$\overrightarrow{G_1C_1}$	hinge attachment point to body 1	$[0.2920, -0.02471]^{(1)^T}$	(m)
$\overrightarrow{OC_2}$	hinge attachment point to body 2	$[0.364, 0.1]^T$	(m)
h	cross-section height of the hinge	0.005	(m)
b	cross-section width of the hinge	0.025	(m)
R_1	body 1 radius	0.18	(m)
R_2	body 2 radius	0.20	(m)
ν_h	hinge Poisson's ratio	0.3	(-)
E_h	hinge Young's modulus	100	(GPa)
m_1	body 1 mass	10	(kg)
I_1	body 1 moment of inertia	0.1617	(kg m^2)
ν_1, ν_2	body 1, 2 Poisson's ratio	0.3	(-)
E_1, E_2	body 1, 2 Young's modulus	100	(GPa)
f_a	adherence coefficient	0.11	(-)
f_d	dynamic friction coefficient	0.055	(-)
c_r	restitution coefficient	0.9	(-)

In the following, four models will be compared:

- the classic Lankarani-Nikravesh impact model with generalized stiffness K as in Equation (14) + the modified Ambrósio friction model (LNA K spheres),
- the novel Lankarani-Nikravesh impact model with generalized stiffness K as in Equation (16) + the modified Ambrósio friction model (LNA K cylinders) proposed in this paper,
- the Lankarani-Nikravesh/ESDU impact model [22] + the modified Ambrósio friction model (LNA-ESDU),
- the Moreau time-stepping scheme.

As recalled, the first model is the classic LN impact model described in [20]. The second one is a modified version proposed in this paper that takes into account the contact of two cylinders modifying the generalized stiffness K through Equation (16). The LNA-ESDU has been validated experimentally for a journal-bearing contact with clearance of a slider-crank mechanism in [22]. Finally, the fourth model is the Moreau time-stepping scheme. It is noteworthy that the first three models are continuous models with event-driven schemes while the fourth is a time-stepping method. This article offers a first numerical comparison to understand if the proposed method is valid and comparable with methods widely accepted by the scientific community. The choice of using three models for comparison is linked to their importance in the multibody field. In fact, the LN model and the Moreau model are the most popular models, each for its own category. The ESDU model is more recent and does not have the same notoriety as the previous methods; however, it proved to be very reliable from an experimental point of view.

The parameters used in the three continuous models are reported in Tables 2 and 3. It can be observed that the generalized stiffness parameter used for the classic LN impact model, i.e., considering the contact of two spheres, is one order of magnitude stiffer than that employed in our modified version in which the contact of two cylinders is considered. The same feature can be observed in Figure 5 for the LNA-ESDU where the generalized stiffness parameter is not constant but grows with the indentation [22]. The dynamic simulation has been performed using the explicit Runge-Kutta 4th-order method. The initial time step has been set to $h_0 = 2 \times 10^{-4}$ (s) while the tolerance of the event-driven scheme is $\delta_{tol} = 1 \times 10^{-4}$ (m).

The Moreau model employs the parameters reported in Table 1. The time-stepping method is based on the midpoint rule with a fixed time step $h_0 = 1\times 10^{-5}$ (s).

Table 2. Lankarani-Nikravesh impact model [20].

Notation	Description	Value	Unit
K (spheres)	generalized stiffness parameter for Equation (14)	47.35	(GPa)
K (cylinders)	generalized stiffness parameter for Equation (16)	2.27	(GPa)
n	nonlinear exponent factor	1.5	(-)

Table 3. Modified Ambrósio friction model [41].

Notation	Description	Value	Unit
v_0	lower tolerance for the tangential velocity	1×10^{-4}	(m/s)
v_1	upper tolerance for the tangential velocity	1×10^{-2}	(m/s)

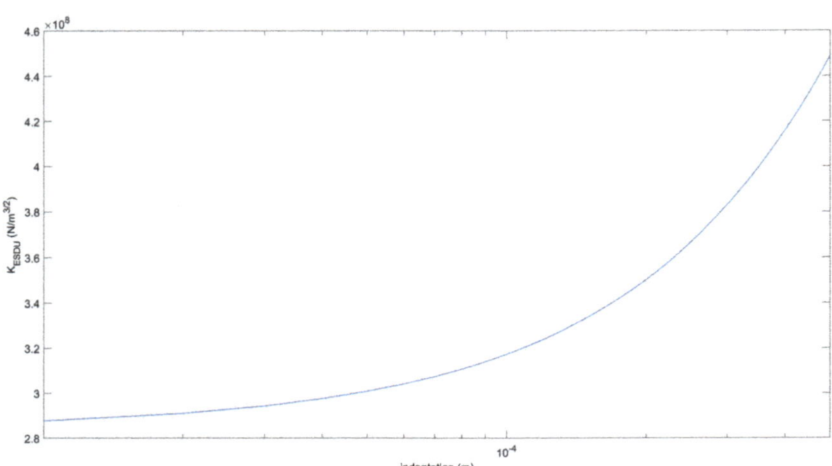

Figure 5. Generalized stiffness evaluated using the Lankarani–Nikravesh/ESDU impact model [22].

6.1. Central Impact

First, we considered the simplest case of a central impact on the two conjugate surfaces. In this particular scenario, the friction force is zero making it possible to evaluate the differences between the four models in the impact process only. The initial state of body 1 is $\mathbf{q}_0 = [0, 0.01, 0]^T$. The results of the simulation are displayed in Figure 6 where the vertical displacement of body 1 and the impact force are plotted. It can be observed that, after free-flying, body 1 impacts at the same instant for the four models, and the bounce height gradually decreases in time due to the dissipative effects. Since the Moreau model is non-smooth, its contact is impulsive and the impact force is the highest among the four models. The three smooth continuous models have the common feature that the impact is spread over a finite, albeit very small, time interval, therefore reducing the contact force peaks. Moreover, it can be observed that the dissipated energy is lower than in the Moreau model and that the sequence of impacts is dilated. Comparing the three continuous LNA-based models, we realize that the LNA (K cylinders) is the most rigid model with lower indentation. The remaining two models have similar characteristics with lower contact forces and greater penetration depths. From this simple experiment, we understand that although some common parameters employed in the models are the same, the comparison of the results shows dynamics that gradually become different. The discrepancies between regularized smooth LNA-based methods and non-smooth methods should not be surprising as they are inherent in different formulations. The contact in the LNA-based models is divided into two phases, the impact and the restitution phase while in the Moreau scheme the contact is non-smooth and impulsive and the rebound height is predominantly influenced by the restitution coefficient, here considered equal for all models. The discrepancies between the models are reduced when the materials are sufficiently rigid since the contact phase is reduced, tending to the limit case of impulsive contact.

To better understand the differences between the three continuous models, let's analyze the first contact. Figure 7 shows the indentation curves and the hysteresis cycles. It can be observed that the three models have maximum indentation decreasing with the generalized stiffness. The LNA-ESDU is the model with a wider curve and with a flatter hysteresis cycle while the classic LNA model is the most rigid one. Finally, the LNA model with K calculated for two cylinders in contact is placed between the two. It should also be observed that the longer duration of the contact for the LNA-ESDU entails a greater computational burden that gradually decreases up to the classic LNA model.

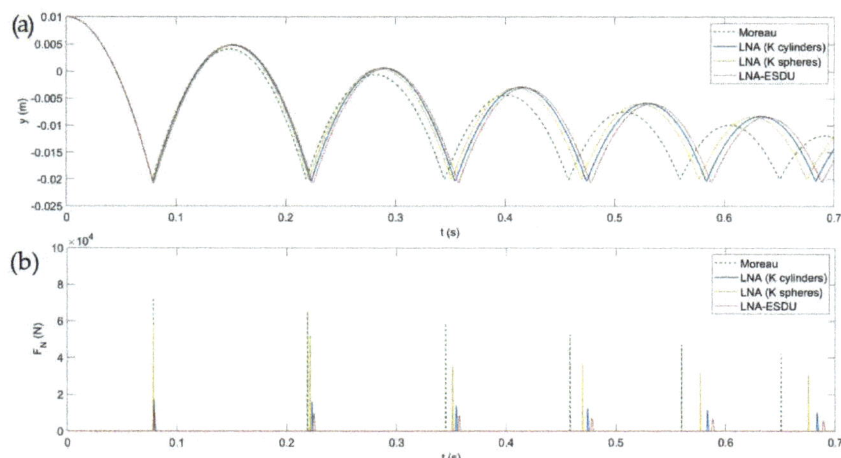

Figure 6. Model comparison considering a central impact: (**a**) y-coordinate of G_1, (**b**) impact force F_N.

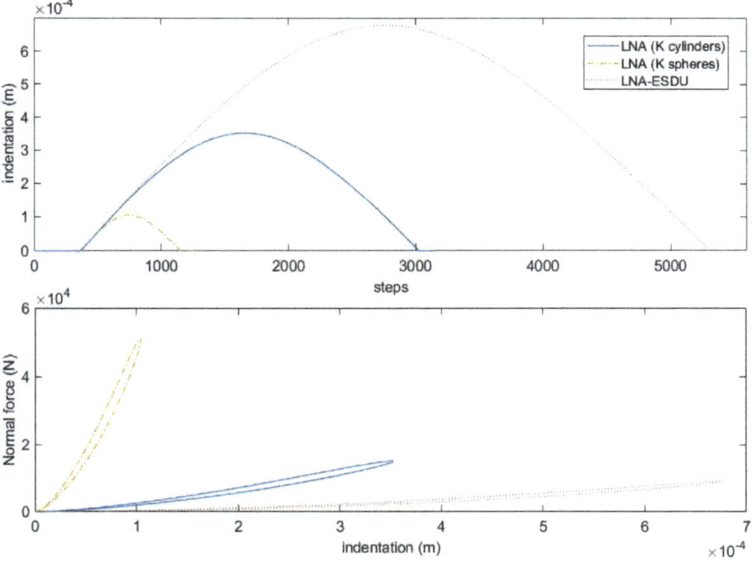

Figure 7. Model comparison considering an asymmetrical impact: from the top to the bottom: indentation *vs.* number of steps; hysteresis cycles.

6.2. Asymmetrical Impact

Let us now consider an asymmetrical collision where the friction force also comes into play. In this second scenario the initial state vector is $\mathbf{q}_0 = [0.01, 0.01, 0]^T$.

Observing Figure 8, many of the conclusions of the previous case are also valid in this scenario. The Moreau model is confirmed as the most rigid while the models LNA (K cylinders) and LNA-ESDU are those with less stiffness. While the three continuous models show similar dynamics, the non-smooth Moreau model presents evident differences not only in terms of contact forces but also in terms of gross motion quantities such as position and rotation. This is due to the friction force that amplifies the contact law differences leading to a chaotic behavior that is difficult to predict.

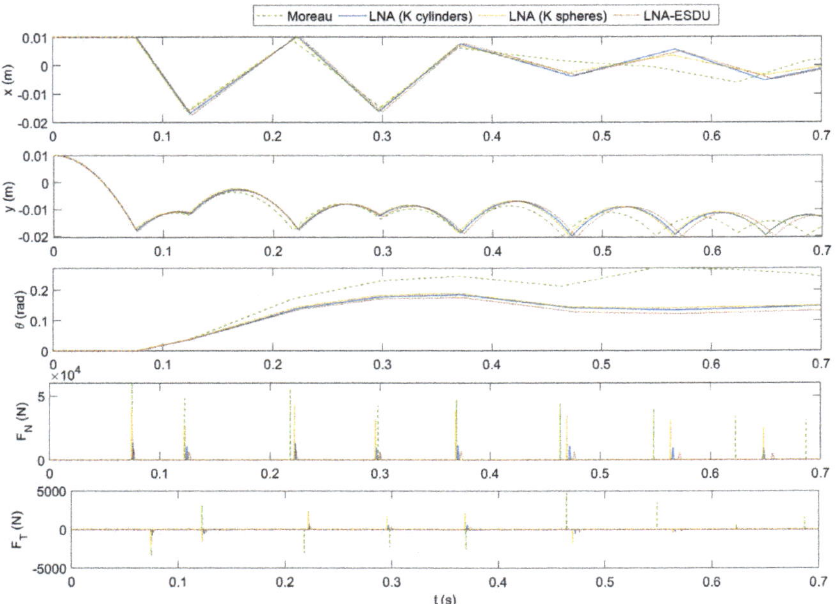

Figure 8. Model comparison considering an asymmetrical impact: from the top to the bottom: x-coordinate of G_1, y-coordinate of G_1, rotation angle θ of body 1, normal contact force F_N, tangential friction force F_T.

To better understand the influence of friction on the dynamics, in Figure 9 the relative tangential velocity v_T and the friction force F_T are displayed in terms of the time steps for the LNA-based models. Since the number of steps depends on contact duration, stiffer models produce a lower number of iterations and are computationally less expensive. It can be observed that most of the computational time is spent during contact phases since the event-driven scheme reduces the time step, therefore leading to a higher number of steps. Vice-versa, the algorithm increases the time step when no contact is detected. Observing the friction force F_T, the plots are similar but the peaks grow proportionally to the stiffness of the model. Furthermore, the proposed LNA (K cylinders) generates friction forces compared to the LNA-ESDU. This feature is promising as the LNA (K cylinders) has lower complexity than LNA-ESDU while providing similar results.

Figure 10 reports the trajectories of the center G_1 of body 1 for the four models. Due to the presence of contacts, all trajectories are inside the circle of clearance, i.e., a circular region with a radius equal to $\delta = R_2 - R_1$. Points inside this region are subject only to gravity and inertia forces. Points on the boundary or outside are subject also to impact forces.

It can be noted that the less rigid models, namely LNA (K cylinders) and LNA-ESDU, have greater impact depths, highlighted by the points outside the circle of clearance. Besides, LNA (K cylinders) confirms to be closer to LNA-ESDU than to LNA (K spheres).

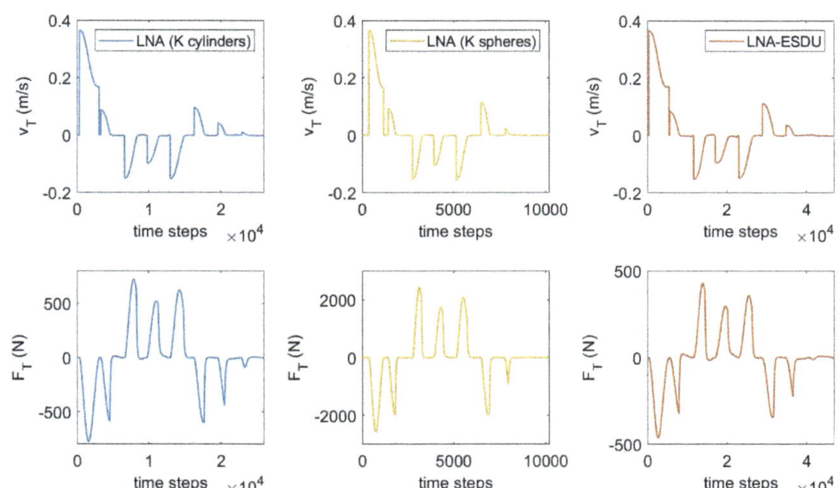

Figure 9. Relative tangential velocity v_T and friction force F_T in terms of the number of time steps.

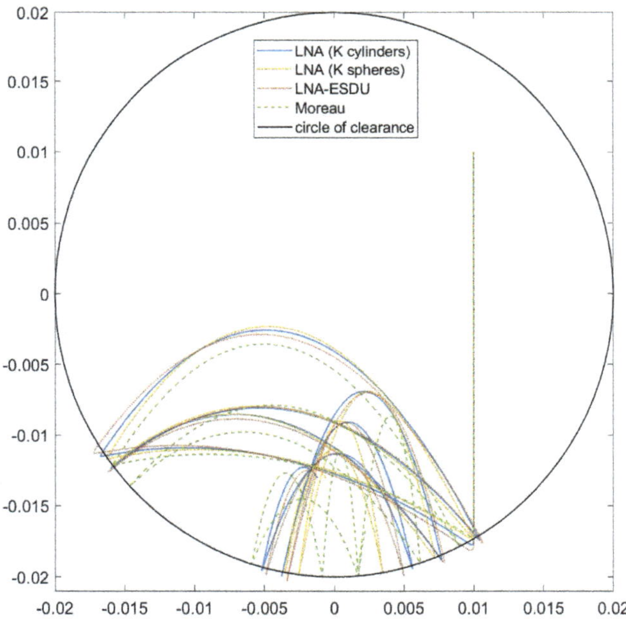

Figure 10. Trajectory of the body–center G_1 for the four models considering the asymmetrical impact. Starting point at $\mathbf{q}_0 = [0.01, 0.01, 0]^T$.

6.3. CSFH Simulation

Finally, the complete simulation of a CSFH is presented. Figure 11 reports some relevant results of the numerical simulations. As for the previous case, the models start with similar dynamics from the state $\mathbf{q}_0 = [0.01, 0.01, 0]^T$. Comparing the x-coordinate of G_1 in Figures 8 and 11, we can distinguish the influence of the flexure hinge on the horizontal dynamics of the circular flexure pushes the body 1 towards the center G_2 of body 2. The first impact generates different contact forces and the trajectories rapidly change amplified by the influence of the circular flexure beam. Furthermore, the friction force affects both the translational and the rotational motion. In turn, this modifies the status of the body and therefore the flexure's response.

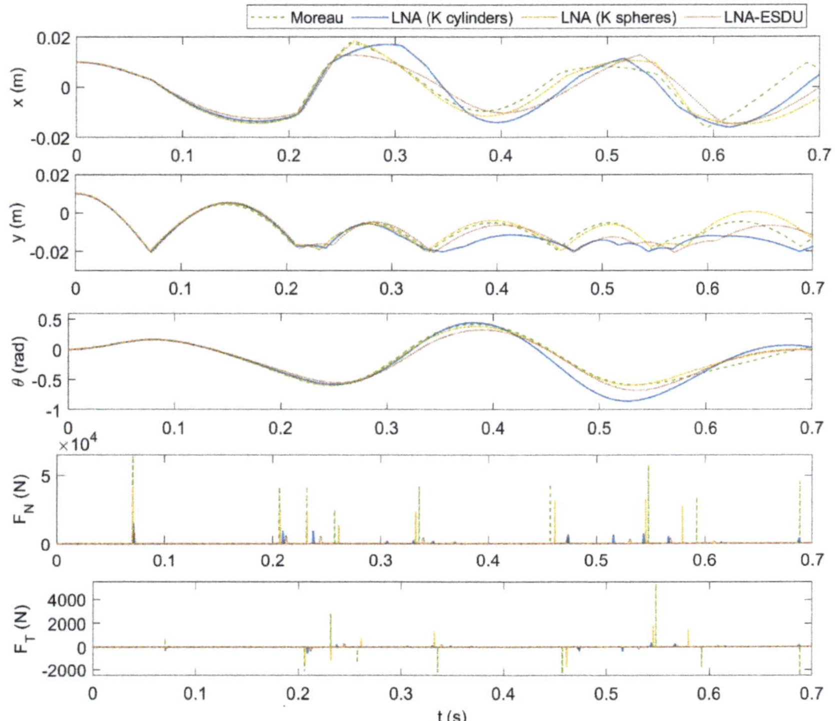

Figure 11. CSFH simulation—comparison of the four models: from the top to the bottom: x-coordinate of G_1, y-coordinate of G_1, rotation angle θ of body 1, normal contact force F_N, tangential friction force F_T.

Finally, Figure 12 reports the trajectories of the center G_1 of body 1 for the four models. As already pointed out, the elastic force pulls body 1 towards the center deviating the vertical fall that is observed in Figure 10. It can be seen that, from the first impact, the trajectories of the models begin to deviate and the dynamics are chaotic. Even for this case, the less rigid models, namely LNA (K cylinders) and LNA-ESDU, have greater impact depths, highlighted by the points outside the circle of clearance.

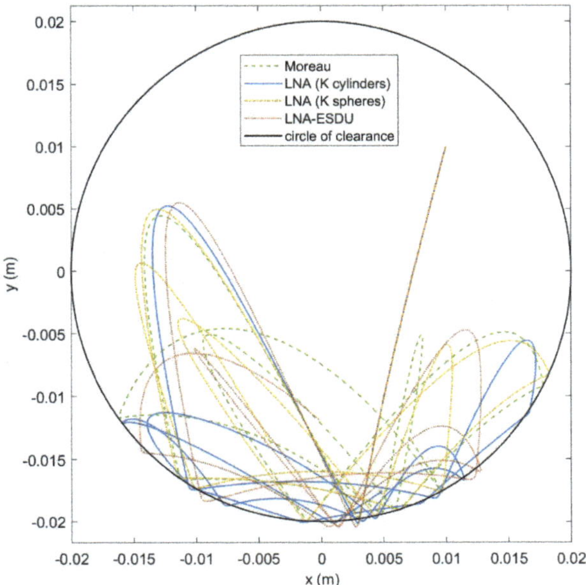

Figure 12. Trajectory of the body–center G_1 for the four models. Starting point at $\mathbf{q}_0 = [0.01, 0.01, 0]^T$.

7. Parametric Analysis

To understand how the parameters of the models influence the CSFH dynamic response we conducted a parametric analysis by varying all the parameters of the model one at a time. The results for the event-driven models are reported in Table 4 where the nominal values are those of Table 1. We have grouped all three event-driven models into a single table because the behavior is similar. The analysis is qualitative and the number of arrows, increasing from one to three, indicates the degree of influence of a parameter on an output variable. Low influence (one arrow) means that the changes in the dynamic response are limited. Medium influence (two arrows) means that the differences gradually amplify as the simulation proceeds. Finally, high influence (three arrows) implies that the differences appear already in the early stages of the simulation generating completely different dynamics.

Table 4. Parametric analysis for the event-driven models. One arrow (low influence), two arrows (medium influence), three arrows (high influence).

	x	y	θ	F_N	F_T
$R_1 \pm 1$ (mm)	↑↑↑	↑↑↑	↑↑↑	↑↑	↑
$m_1 \pm 0.1$ (kg)	↑	↑	↑	↑	↑
$I_1 \pm 0.001$ (kg m^2)	↑	↑	↑	↑	↑
$E_{1,2} \pm 10$ (GPa)	↑	↑	↑	↑	↑
$c_r \pm 0.01$ (-)	↑	↑	↑	↑	↑
$f_d \pm 0.005$ (-)	↑	↑	↑	↑	↑
$\rho \pm 1$ (mm)	↑↑	↑↑	↑	↑↑	↑↑
$b \pm 1$ (-)	↑	↑	↑	↑	↑
$h \pm 1$ (mm)	↑↑	↑↑	↑↑	↑↑	↑↑
$\beta \pm 1$ (°)	↑↑↑	↑↑↑	↑↑	↑↑↑	↑↑
$E_h \pm 10$ (GPa)	↑↑↑	↑↑↑	↑↑↑	↑↑	↑↑

It can be observed that the radius R_1, and therefore the radial clearance δ, is a critical parameter for the model. Increasing or decreasing the clearance modifies the kinematics of the impact, anticipating or delaying it, and implies important changes especially on the gross motion (x, y, θ). Other parameters related to the impact model produce small deviations in the dynamic response. A very different thing happens with regard to the flexure parameters whose modification has strong repercussions on the system dynamics.

Finally, the Table 5 shows the parametric analysis for the time-stepping Moreau's method. It should be noted that, while following the trend of the event-driven methods, the model is more sensitive to changes in the parameters. Probably, this behavior derives from the higher stiffness of the Moreau's method.

Table 5. Parametric analysis for the time-stepping Moreau's method. One arrow (low influence), two arrows (medium influence), three arrows (high influence).

	x	y	θ	F_N	F_T
$R_1 \pm 1$ (mm)	↑↑↑	↑↑↑	↑↑↑	↑↑	↑↑
$m_1 \pm 0.1$ (kg)	↑↑	↑↑	↑↑	↑↑	↑↑
$I_1 \pm 0.001$ (kg m^2)	↑	↑	↑	↑	↑
$c_r \pm 0.01$ (-)	↑↑	↑↑	↑↑	↑↑	↑↑
$f_d \pm 0.005$ (-)	↑↑	↑↑	↑	↑	↑
$f_a \pm 0.01$ (-)	↑	↑	↑	↑	↑
$\rho \pm 1$ (mm)	↑↑	↑↑	↑	↑↑	↑↑
$b \pm 1$ (-)	↑	↑	↑	↑	↑
$h \pm 1$ (mm)	↑↑↑	↑↑↑	↑↑↑	↑↑	↑↑
$\beta \pm 1$ (°)	↑↑↑	↑↑↑	↑↑	↑↑	↑↑
$E_h \pm 10$ (GPa)	↑↑↑	↑↑↑	↑↑↑	↑↑↑	↑↑↑

8. Discussion

Comparing the four different models reported in the previous sections revealed important insights. The contact problem with friction was confirmed to be tough. The strong coupling between stiffness and hysteresis loss creates highly non-linear dynamics making the system's evolution chaotic. This tendency is further amplified by the flexure, being dependent on the position and orientation of the attached bodies. Considering these premises, comparing different impact models would seem useless. Experimentally, it is preferred to quantify the extent of the collision by monitoring, for example, the accelerations produced on bumping bodies. The acceleration is correlated to the impact forces and by observing the first one can quantify the second, which is much more difficult to observe directly. Monitoring the levels of impact forces has important repercussions on various issues of industrial interest such as wear and durability.

To study microcontacts in MEMS application a nanoindenter based experimental setup similar to that proposed in [47] could be designed. A piezoelectric transducer could push the body 1 to touch the body 2. Then, a microprobe, linked to the body 2, could measure forces and displacements.

From the previous numerical results, we can state that the proposed LNA (K cylinders) is very close to the experimentally verified LNA-ESDU. Considering the computational efficiency, the Moreau time-stepping method is the fastest. Nevertheless, the impulsive nature of impact forces makes the Moreau model too stiff. The classic LNA method is less stiff than Moreau but it needs more computational resources to resolve the continuous contact. The LNA-ESDU is the most reliable but the slowest method at the same time. Compared to the latter, LNA (K cylinders) has the advantage of being simpler by using a constant stiffness instead of a variable one.

The results look promising and worthy of further future developments as the LNA (K cylinders) seems to be a good compromise in terms of efficiency and reliability.

9. Conclusions

The dynamic model of a CSFH including impacts and friction has been described. First, the flexure hinge elasto-kinematic model has been recalled. Then, the impact kinematics of two conjugate cylindrical surfaces has been formulated and the event-driven contact models have been introduced. These models are based on the detection of the impact instants in correspondence of which a switch among different dynamic models or states is imposed. Based on this class of impact models, we proposed to modify the classic Lankarani-Nikravesh impact model using a generalized stiffness derived from the contact of two cylinders. This solution seemed well-suited to describe the CSFH where two cylindrical conjugate surfaces collide. The proposed model has been equipped with the modified Ambrósio friction model. Then, the non-smooth Moreau's scheme has been recalled. The latter is a time-stepping method that does not require the detection of impact points and the consequent change of state.

In the numerical part, the proposed method, the Moreau scheme, and two continuous event-driven schemes including the classic Lankarani-Nikravesh-Ambrósio model and its evolution obtained by the ESDU constitutive law have been compared in different impact scenarios. All numerical simulations revealed that the dynamics are strongly influenced by the impact formulation and its parameters. Furthermore, due to the high non-linearity of the problem, the differences are further amplified by the presence of the flexure. The proposed method not only provided indentation, impact, and friction force values comparable to those provided by the experimentally verified LNA-ESDU model but has the advantage of requiring lower computational resources equal to those needed by the classic LNA method.

The parametric analysis revealed that some parameters such as the radial clearance or flexure parameters have a strong influence on system dynamics. This can be very useful in helping designers to establish the right manufacturing processes and dimensional tolerances.

The present investigation is expected to be a first step towards the understanding of the dynamic behavior of CSFH. Furthermore, the benefits introduced by the method could be important in developing control strategies. For example, the impact model could characterize the displacements of a micro-gripper equipped with CSFHs in function of the comb-drives actuation voltage. In this way, it would be possible to fully exploit the potential of the CSFH by taking into account the contact dynamics among the conjugated surfaces and ensuring, at the same time, control over the maximum stresses that the materials can withstand during the impact phase.

Author Contributions: Conceptualization, A.C. and N.P.B.; methodology, A.C. and N.P.B.; software, A.C. and A.R.; validation, A.C. and A.R.; formal analysis, A.C. and N.P.B.; investigation, P.D.M.; resources, R.S.; data curation, P.D.M.; writing—original draft preparation, A.C. and A.R.; writing—review and editing, R.S.; visualization, A.C. and A.R.; supervision, R.S. All authors have read and agreed to the published version of the manuscript.

Funding: This research received no external funding.

Conflicts of Interest: The authors declare no conflict of interest.

References

1. Gad-el Hak, M. (Ed.) *The MEMS Handbook*, 2nd ed.; CRC Press: Boca Raton, FL, USA, 2006; Volume 2. [CrossRef]
2. Voiculescu, I.; Nordin, A.N. Acoustic wave based MEMS devices for biosensing applications. *Biosens. Bioelectron.* **2012**, *33*, 1–9. [CrossRef] [PubMed]
3. Uranga, A.; Verd, J.; Barniol, N. CMOS-MEMS resonators: From devices to applications. *Microelectron. Eng.* **2015**, *132*, 58–73. [CrossRef]
4. Dochshanov, A.; Verotti, M.; Belfiore, N.P. A comprehensive survey on microgrippers design: Operational strategy. *J. Mech. Des.* **2017**, *139*, 070801. [CrossRef]
5. Nisar, A.; Afzulpurkar, N.; Mahaisavariya, B.; Tuantranont, A. MEMS-based micropumps in drug delivery and biomedical applications. *Sens. Actuators B Chem.* **2008**, *130*, 917–942. [CrossRef]
6. Belfiore, N.P.; Scorza, A.; Ursi, P. Engineering-aided inventive surgery. *Appl. Sci.* **2020**, *10*, 3957. [CrossRef]
7. Pengwang, E.; Rabenorosoa, K.; Rakotondrabe, M.; Andreff, N. Scanning micromirror platform based on MEMS technology for medical application. *Micromachines* **2016**, *7*, 24. [CrossRef]
8. Belfiore, N.P. Micromanipulation: A challenge for actuation. *Actuators* **2018**, *7*, 85. [CrossRef]
9. Belfiore, N.P.; Bagolini, A.; Rossi, A.; Bocchetta, G.; Vurchio, F.; Crescenzi, R.; Scorza, A.; Bellutti, P.; Sciuto, S.A. Design, fabrication, testing and simulation of a rotary double comb drives actuated microgripper. *Micromachines* **2021**, *12*, 1263. [CrossRef]
10. Bogue, R. Recent developments in MEMS sensors: A review of applications, markets and technologies. *Sensor Rev.* **2013**, *33*, 300–304. [CrossRef]

11. Crescenzi, R.; Balucani, M.; Belfiore, N.P. Operational characterization of CSFH MEMS technology based hinges. *J. Micromech. Microeng.* **2018**, *28*, 055012. [CrossRef]
12. Verotti, M.; Dochshanov, A.; Belfiore, N.P. Compliance Synthesis of CSFH MEMS-Based Microgrippers. *J. Mech. Des.* **2017**, *139*, 022301. [CrossRef]
13. Sanò, P.; Verotti, M.; Bosetti, P.; Belfiore, N.P. Kinematic Synthesis of a D-Drive MEMS Device with Rigid-Body Replacement Method. *J. Mech. Des.* **2018**, *140*, 075001. [CrossRef]
14. Belfiore, N.P.; Emamimeibodi, M.; Verotti, M.; Crescenzi, R.; Balucani, M.; Nenzi, P. Kinetostatic optimization of a MEMS-based compliant 3 DOF plane parallel platform. In Proceedings of the ICCC 2013—IEEE 9th International Conference on Computational Cybernetics, Tihany, Hungary, 8–10 July 2013; pp. 261–266. [CrossRef]
15. Vurchio, F.; Ursi, P.; Buzzin, A.; Veroli, A.; Scorza, A.; Verotti, M.; Sciuto, S.; Belfiore, N.P. Grasping and releasing agarose micro beads in water drops. *Micromachines* **2019**, *10*, 436. [CrossRef] [PubMed]
16. Botta, F.; Verotti, M.; Bagolini, A.; Bellutti, P.; Belfiore, N.P. Mechanical response of four-bar linkage microgrippers with bidirectional electrostatic actuation. *Actuators* **2018**, *7*, 78. [CrossRef]
17. Bertini, S.; Verotti, M.; Bagolini, A.; Bellutti, P.; Ruta, G.; Belfiore, N.P. Scalloping and stress concentration in DRIE-manufactured comb-drives. *Actuators* **2018**, *7*, 57. [CrossRef]
18. Kim, H.J.; Yoo, S.S.; Kim, D.E. Nano-scale wear: A review. *Int. J. Precis. Eng. Manuf.* **2012**, *13*, 1709–1718. [CrossRef]
19. Tian, Q.; Flores, P.; Lankarani, H.M. A comprehensive survey of the analytical, numerical and experimental methodologies for dynamics of multibody mechanical systems with clearance or imperfect joints. *Mech. Mach. Theory* **2018**, *122*, 1–57. [CrossRef]
20. Lankarani, H.M.; Nikravesh, P.E. A contact force model with hysteresis damping for impact analysis of multibody systems. *J. Mech. Des.* **1990**, *112*, 369–376. [CrossRef]
21. Flores, P.; Ambrósio, J. On the contact detection for contact-impact analysis in multibody systems. *Multibody Syst. Dyn.* **2010**, *24*, 103–122. [CrossRef]
22. Koshy, C.S.; Flores, P.; Lankarani, H.M. Study of the effect of contact force model on the dynamic response of mechanical systems with dry clearance joints: Computational and experimental approaches. *Nonlinear Dyn.* **2013**, *73*, 325–338. [CrossRef]
23. Moreau, J.J. Unilateral contact and dry friction in finite freedom dynamics. In *Nonsmooth Mechanics and Applications*; Moreau, J., Panagiotopoulos, P., Eds.; Springer: Vienna, Austria, 1988; Volume 302, pp. 1–82. [CrossRef]
24. Moreau, J.J. Some numerical methods in multibody dynamics: Application to granular materials. *Eur. J. Mech.-A Solids* **1994**, *13*, 93–114.
25. Ghayesh, M.H.; Farokhi, H.; Alici, G. Size-dependent electro-elasto-mechanics of MEMS with initially curved deformable electrodes. *Int. J. Mech. Sci.* **2015**, *103*, 247–264. [CrossRef]
26. Verotti, M.; Crescenzi, R.; Balucani, M.; Belfiore, N.P. MEMS-based conjugate surfaces flexure hinge. *J. Mech. Des.* **2015**, *137*, 012301. [CrossRef]
27. Angeles, J. *Fundamentals of Robotic Mechanical Systems*, 4th ed.; Springer: Cham, Switzerland, 2002; Volume 124. [CrossRef]
28. Hertz, H. Ueber die Berührung fester elastischer Körper. *J. Reine Angew. Math.* **1882**, *1882*, 156–171. [CrossRef]
29. Johnson, K.L. One hundred years of Hertz contact. *Proc. Inst. Mech. Eng.* **1982**, *196*, 363–378. [CrossRef]
30. ESDU. *Contact Phenomena. I: Stresses, Deflections and Contact Dimensions for Normally-loaded Unlubricated Elastic Components*; ESDU Tribology Series; ESDU International PLC: London, UK, 1978.
31. Goldsmith, W. *Impact: The Theory and Physical Behavior of Colliding Solids*; Dover Civil and Mechanical Eng, Dover Publications Inc.: Mineola, NY, USA, 2001.
32. Flores, P.; Lankarani, H.M. *Contact Force Models for Multibody Dynamics*; Springer: Berlin/Heidelberg, Germany, 2016; Volume 226.
33. Hunt, K.H.; Crossley, F.R.E. Coefficient of restitution interpreted as damping in vibroimpact. *J. Appl. Mech.* **1975**, *42*, 440–445. [CrossRef]
34. Flores, P.; Machado, M.; Silva, M.T.; Martins, J.M. On the continuous contact force models for soft materials in multibody dynamics. *Multibody Syst. Dyn.* **2011**, *25*, 357–375. [CrossRef]
35. Coulomb, C.A. Théorie des Machines Simples en Ayant Égard au Frottement de Leurs Parties et à la Roideur des Cordages. Bachelor's Thesis, Charles-Louis-Étienne, Paris, France, 1821.
36. Threlfall, D. The inclusion of Coulomb friction in mechanisms programs with particular reference to DRAM au programme DRAM. *Mech. Mach. Theory* **1978**, *13*, 475–483. [CrossRef]
37. Bo, L.C.; Pavelescu, D. The friction-speed relation and its influence on the critical velocity of stick-slip motion. *Wear* **1982**, *82*, 277–289. [CrossRef]
38. Karnopp, D. Computer simulation of stick-slip friction in mechanical dynamic systems. *J. Dyn. Syst. Meas. Control* **1985**, *107*, 100–103. [CrossRef]
39. Leine, R.; Van Campen, D.; De Kraker, A.; Van Den Steen, L. Stick-slip vibrations induced by alternate friction models. *Nonlinear Dyn.* **1998**, *16*, 41–54. [CrossRef]
40. Bengisu, M.; Akay, A. Stability of friction-induced vibrations in multi-degree-of-freedom systems. *J. Sound Vib.* **1994**, *171*, 557–570. [CrossRef]
41. Ambrósio, J.A.C. Impact of rigid and flexible multibody systems: deformation description and contact models. In *Virtual Nonlinear Multibody Systems*; NATO ASI Series; Schiehlen, W., Valášek, M., Ed.; Springer: Dordrecht, The Netherlands, 2003; Volume 103, pp. 57–81. [CrossRef]

42. Marton, L.; Lantos, B. Modeling, identification, and compensation of stick-slip friction. *IEEE Trans. Ind. Electron.* **2007**, *54*, 511–521. [CrossRef]
43. Marques, F.; Flores, P.; Pimenta Claro, J.; Lankarani, H.M. A survey and comparison of several friction force models for dynamic analysis of multibody mechanical systems. *Nonlinear Dyn.* **2016**, *86*, 1407–1443. [CrossRef]
44. Pfeiffer, F.; Glocker, C. *Multibody Dynamics with Unilateral Contacts*; John Wiley & Sons Inc.: Hoboken, NJ, USA, 1996. [CrossRef]
45. Acary, V.; Brogliato, B. *Numerical Methods for Nonsmooth Dynamical Systems: Applications in Mechanics and Electronics*, 1st ed.; Springer: Berlin/Heidelberg, Germany, 2008. [CrossRef]
46. Studer, C. *Numerics of Unilateral Contacts and Friction: Modeling and Numerical Time Integration in Non-Smooth Dynamics;* Part of the Book Series: Lecture Notes in Applied and Computational Mechanics; Springer: Berlin/Heidelberg, Germany, 2009; Volume 47. [CrossRef]
47. Gilbert, K.W.; Mall, S.; Leedy, K.D.; Crawford, B. A nanoindenter based method for studying MEMS contact switch microcontacts. In Proceedings of the 2008 Proceedings of the 54th IEEE Holm Conference on Electrical Contacts, Orlando, FL, USA, 27–29 October 2008; pp. 137–144.

Article

Micromanipulation and Automatic Data Analysis to Determine the Mechanical Strength of Microparticles

Zhihua Zhang [1,2], Yanping He [3] and Zhibing Zhang [1,*]

1 School of Chemical Engineering, University of Birmingham, Birmingham B15 2TT, UK; zxz836@student.bham.ac.uk
2 Changzhou Institute of Advanced Manufacturing Technology, Changzhou 213164, China
3 School of Chemical Engineering, Kunming University of Science and Technology, Chenggong Campus, Kunming 650504, China; grace.he1985@hotmail.com
* Correspondence: z.zhang@bham.ac.uk

Abstract: Microparticles are widely used in many industrial sectors. A micromanipulation technique has been widely used to quantify the mechanical properties of individual microparticles, which is crucial to the optimization of their functionality and performance in end-use applications. The principle of this technique is to compress single particles between two parallel surfaces, and the force versus displacement data are obtained simultaneously. Previously, analysis of the experimental data had to be done manually to calculate the rupture strength parameters of each individual particle, which is time-consuming. The aim of this study is to develop a software package that enables automatic analysis of the rupture strength parameters from the experimental data to enhance the capability of the micromanipulation technique. Three algorithms based on the combination of the "three-sigma rule", a moving window, and the Hertz model were developed to locate the starting point where onset of compression occurs, and one algorithm based on the maximum deceleration was developed to identify the rupture point where a single particle is ruptured. Fifty microcapsules each with a liquid core and fifty porous polystyrene (PS) microspheres were tested in order to produce statistically representative results of each sample, and the experimental data were analysed using the developed software package. It is found that the results obtained from the combination of the "3σ + window" algorithm or the "3σ + window + Hertz" algorithm with the "maximum-deceleration" algorithm do not show any significant difference from the manual results. The data analysis time for each sample has been shortened from 2 to 3 h manually to within 20 min automatically.

Keywords: micromanipulation; automatic data analysis; mechanical strength; microparticles; algorithms

Citation: Zhang, Z.; He, Y.; Zhang, Z. Micromanipulation and Automatic Data Analysis to Determine the Mechanical Strength of Microparticles. *Micromachines* **2022**, *13*, 751. https://doi.org/10.3390/mi13050751

Academic Editor: Alessandro Cammarata

Received: 22 March 2022
Accepted: 6 May 2022
Published: 10 May 2022

1. Introduction

Microparticles are widely used in many functional products in the industry [1]. Measuring their mechanical strength is essential to optimizing their performance during manufacturing, processing, and end-use applications [2]. For example, microcapsules with self-sensing agents used to produce smart structural composites [3–5] should be mechanically strong enough to survive different engineering processing steps leading to their incorporation into the composites but weak enough to break after mechanical damage is occurring to the composites so that the need for repair can be indicated quickly. Understanding the mechanical strength of the self-sensing microcapsules plays a crucial role in ensuring the functionalities of the composites. Furthermore, characterizing the mechanical strength of other microparticles, e.g., perfume microcapsules for fabric softeners and detergents [6], and microspheres for chromatography media for bio-separation [7], can also provide essential technical data for new product development and production as well as help to optimize their functionality and performance in end-use applications.

Experimental techniques to determine the mechanical strength of microparticles can be classified as ensemble test methods and single-particle test methods [1,2,8]. The former methods are relatively quick as a group of particles are tested simultaneously, but only the average mechanical strength values can be obtained. The latter methods test particles one by one; thus, their mechanical strength distribution can be obtained, which is crucial in many applications to optimize their functionality and performance. Several techniques have been developed to determine the mechanical strength of single particles, including optical/magnetic tweezers [9], pressure probe [10,11], micropipette aspiration [12,13], atomic force microscopy (AFM) [14,15], nanoindentation [16,17], and micromanipulation based on diametrical compression [18]. The main difference among them lies in the different deformations, which can be generated, and magnitudes of forces, which can be measured. For example, the typical force measured by micromanipulation is from μN to N, while the force by AFM is from pN to μN [2]. Consequently, micromanipulation can provide the rupture strength parameters by compressing particles to break, while it is difficult for other techniques to do so [1].

The micromanipulation technique was firstly developed to test the rupture strength of single mammalian cells [18] and since then has been modified to test the mechanical and surface properties, including the elasticity, plasticity, viscoelasticity, adhesion, and cohesion of a variety of biological and non-biological micro-materials [1], e.g., microcapsules [19–21], microspheres [7,22,23], microbeads [24], pollen grains [25], yeast cells [26,27], chondrocytes and chondrons [28,29], biofilms [30,31], fouling deposits [32–35], and microneedles [36]. It has provided essential technical data to a number of global companies to assist their micro-product development and also played a very important role in academic research to develop new applications of various micro-materials [1,37].

The micromanipulation technique involves sample preparation, compression of single particles, and data analysis to obtain the mechanical properties of microparticles. The raw data from a micromanipulation test are a series of voltage data as shown in Figure 1. The main task of the data analysis is to identify the starting point M, where the onset of loading occurs, and rupture point R, where the tested particle ruptured, from which the rupture strength parameters and force-displacement data can be obtained. Unlike some commercial or open-source software packages to analyse the force-displacement data from AFM experiments [38,39], it was carried out manually by interacting with the raw data and template spreadsheets to obtain the results from the micromanipulation experiments, which is quite laborious and time-consuming.

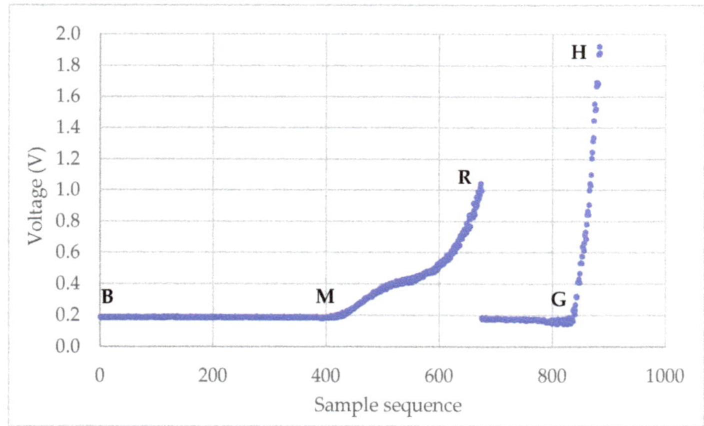

Figure 1. Typical curve of voltage versus sampling sequence from compression of single particles.

The software packages for AFM are not easy to be adapted to process the data from the micromanipulation technique because of the differences in data formats, mechanical property parameters to be obtained, and specific mathematical model formulas required to be used. However, similar to the starting point M in the micromanipulation tests, the contact point (CP) is also crucial to analysing the force-displacement data from AFM experiments. Several algorithms have been developed to locate the CP of the force-displacement data obtained from AFM experiments. A simple algorithm with a threshold (typically 0.1%) was used to estimate the CP from the approach curve above the baseline [40]. However, this threshold needs to be modified according to the baseline value and noise level manually, which is not suitable for automatic data analysis. A local regression-based algorithm was then introduced to determine the CP by slope changes [41]. Three parameters, including the number of data points for regression, and two thresholds need to be properly set to locate the CP. An algorithm was developed to estimate the CP by fitting the data in a liner elastic region to a Hertz-like model for the nano indentation data [42]. The algorithm worked well but requires new sets of parameters for other materials of different mechanical behaviours. Moreover, in AFM force data analysis, usually a force map, e.g., 64 × 64 force curves, are obtained for a single particle to yield a spatial distribution of the mechanical strength parameters. These algorithms above are aimed to locate the CPs for the force curves for a single particle so that the parameters set for the algorithms may not need to be adjusted frequently for every force curve. In contrast, from micromanipulation measurements, a single voltage (force) curve is obtained for a single particle, and usually, the particles in a sample have different sizes and mechanical strength values; therefore, the parameters set may need to be modified frequently for each dataset to ensure the above algorithms can work properly for every tested single particle in a sample. Consequently, the algorithms used in AFM data analysis cannot be applied directly to automatic analysis of the micromanipulation data.

The aim of this study is to develop a software package to analyse the experimental data obtained from using the micromanipulation technique to automatically obtain the mechanical strength parameters of microparticles to simplify the procedure, save time and labour, and enhance the capability of the micromanipulation technique.

In this paper, three algorithms are presented to identify the starting point M, and an algorithm is introduced to locate the rupture point R from the raw voltage data of micromanipulation. Two samples of microparticles, i.e., the microcapsules for self-sensing and the porous PS microspheres with various potential applications, have been tested using the micromanipulation technique, and the experimental data analysed using the developed software package are compared to the manual results to validate the algorithms developed.

2. Materials and Methods

2.1. Microparticles for Micromanipulation

2.1.1. Microcapsules for Self-Sensing

The microcapsules for self-sensing were a very robust type of double-walled microcapsules made by interfacial polymerization. The detailed fabrication methods are described in [5]. The outer and inner shells were made from urea formaldehyde (PUF) and polyurethane (PU), respectively. The core is oil with a fluorophore substance.

2.1.2. Porous Polystyrene Microspheres

The porous polystyrene (PS) microspheres with various potential applications were fabricated via a novel solvent evaporation methodology based on foaming transfer. The detailed fabrication process is reported in [43]. Specifically, the porous PS microspheres obtained by introducing 20 wt% ethanol concentration to the continuous phrase were used in this paper.

2.2. Micromanipulation of the Microparticles

2.2.1. Micromanipulation Rig

The principle of the micromanipulation technique is to compress single particles to different deformations or rupture between two parallel surfaces, and the force versus displacement data are obtained simultaneously. The schematic diagram of the micromanipulation rig used in this work is illustrated in Figure 2, which is also reported elsewhere [19–22]. Single microparticles are placed on the glass slide, which is fixed on the sample stage of a three-dimensional micromanipulator, and then compressed by the output probe (with flat end) of the force transducer that is mounted to the one-dimensional fine micromanipulator. The corresponding compression force is acquired by a data acquisition device (USB-201-OEM, Measurement Computing Corporation, Norton, MA, USA) in the control and acquisition box and the data is saved in the computer for post processing. The fine micromanipulator is driven by a servo motor. The power of the servo motor is 24 V DC. Before compression, single microparticles are moved to just below the force probe by operating the sample-stage micromanipulator. Using the sideview camera, the video images of the compression procedure can be displayed by the industrial computer monitor and saved in the computer. The force transducer can be changed according to the mechanical strength scale of the microparticles to be measured.

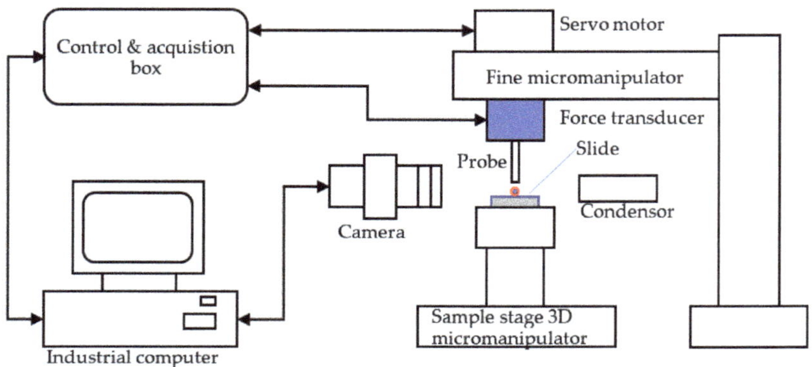

Figure 2. Schematic diagram of the micromanipulation rig.

2.2.2. Micromanipulation of the Microcapsules for Self-Sensing

Dry microcapsules were placed onto a glass slide, and single microcapsules were compressed to rupture using the micromanipulation rig at a compression speed of 2.0 μm/s. The sampling time was 0.01887 s, and the force transducer model was GS0-10 (Transducer Techniques, LLC, Temecula, CA, USA) with a pre-calibrated sensitivity of 8.674 mN/V. In total, 50 microcapsules were tested at ambient temperature of 26 ± 2 °C.

2.2.3. Micromanipulation of the Porous PS Microspheres

The micromanipulation procedure of the porous PS microspheres was the same as the measurement of the self-sensing microcapsules. The transducer used was GS0-10 with a pre-calibrated sensitivity of 7.423 mN/V. In total, 50 PS microspheres were tested under ambient temperature of 16 ± 2 °C.

Figure 3 illustrates the procedure to compress a porous PS microsphere between the two parallel surfaces, i.e., the probe end and the glass surface. The diameter of the transducer probe was around 50 μm, and the diameter of the particle was 16.3 μm.

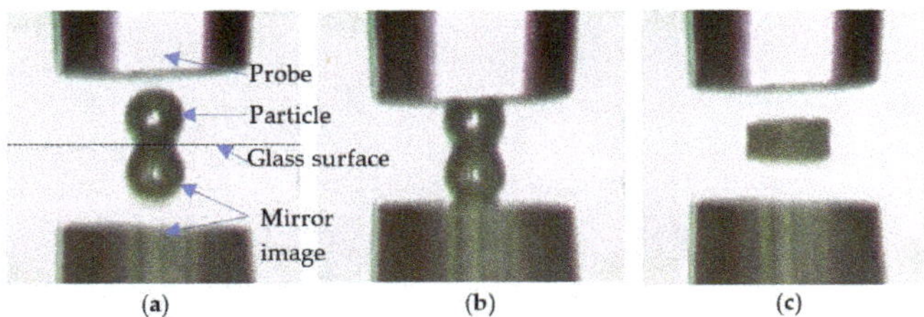

Figure 3. A porous PS microsphere before (**a**), during (**b**), and after (**c**) compression.

2.3. Rutpure Strength of Microparticles

The raw data from a micromanipulation test is a series of voltage versus sample sequence data $(V_1, V_2, \ldots, V_n)$, where n is the number of the voltage data points. A typical curve is shown in Figure 1. At the beginning, the voltage remains stable along the baseline as the probe moves in the air due to the initial gap between the probe and the microparticle. Then, it starts to increase at M when the probe begins to touch the particle. The voltage keeps rising until R and drops suddenly when the particle is ruptured. After that, the voltage rises again from G as the probe compresses the debris of the particle on the hard bottom surface and stops at H when the voltage limit is reached, or the movement is stopped manually. Point M is named as the starting point and R as the rupture point. The line segment BM is termed as "baseline". The main task of the data analysis is to identify the starting point M and rupture point R from which the rupture strength parameters, including displacement at rupture δ_r, rupture force F_r, fractional deformation at rupture ε_r, nominal rupture stress σ_r, nominal rupture tension T_r, and toughness T_C, can be calculated using the following equations

$$F_r = s(V_r - V_B) \tag{1}$$

$$\delta_r = vT_s(r - m) - cF_r \tag{2}$$

$$\varepsilon_r = \frac{\delta_r}{D}, \tag{3}$$

$$\sigma_r = \frac{4F_r}{\pi D^2}, \tag{4}$$

$$T_r = \frac{F_r}{D}, \tag{5}$$

and

$$T_C = \int_0^{\varepsilon_r} \sigma d\varepsilon, \tag{6}$$

where m is the starting point index, r is the rupture point index, V_r is the voltage corresponding to rupture, V_B is the average voltage of the baseline, v is the compression speed, T_s is the sampling time, s is the sensitivity of the force transducer, c is the compliance of the force transducer, D is the initial diameter of the single microparticle, σ is the nominal stress, and ε is the fractional deformation.

The force-displacement data can be obtained using the following two equations:

$$F_i = s(V_{i+m} - V_B), \tag{7}$$

and

$$\delta_i = ivT_s - cF, \tag{8}$$

where i $(1 \le i \le n - m)$ is the index, F is the compression force, and δ is the displacement. Then, the nominal stress and fractional deformation can be calculated using

$$\sigma_i = \frac{4F_i}{\pi D^2}, \tag{9}$$

and

$$\varepsilon_i = \frac{\delta_i}{D}, \tag{10}$$

In practice, the microparticle toughness in Equation (6) can be determined using the trapezoidal numerical integration as Equation (11).

$$T_C = \frac{1}{2}\sum_{i=1}^{r}(\sigma_i + \sigma_{i+1})(\varepsilon_{i+1} - \varepsilon_i), \tag{11}$$

2.4. Algorithms to Locate the Starting Point

2.4.1. "3σ" Algorithm

During the micromanipulation test, the voltage $V(t)$ can be expressed as follows:

$$V(t) = \frac{1}{s}F(t) + e(t), \tag{12}$$

where $F(t)$ is the true value of the compression force, s is the sensitivity of the force transducer, and $e(t)$ is a random noise. Before the onset of compression, $F(t)$ is constant (zero), thus $V(t)$ and $e(t)$ have the same distribution during this period. Assuming the distribution is a normal distribution (the most common distribution [44] for noise), according to the three-sigma rule [45], the possibility (Pr) of $V(t)$ falling away from the mean value (μ) of the baseline by more than three standard deviations (3σ) is at most 0.27%,

$$\Pr(|V(t) - \mu| \geq 3\sigma) \leq 0.27\%, \tag{13}$$

Thus, if the voltage value at a point starts to deviate from the baseline mean value by three standard deviations, it has a high possibility (99.73%) that the onset of compression begins; i.e., the first point when the voltage deviates from the baseline by three standard deviations can be located as the starting point. In practice, the μ and σ can be estimated by the average (V_B) and standard deviation (S_B) of the voltage data of the baseline. Then, a criterion is obtained to determine the starting point.

$$|V_m - V_B| > 3S_B, \tag{14}$$

where m is the index of the starting point. The flowchart of the "3σ" algorithm is illustrated in Figure 4.

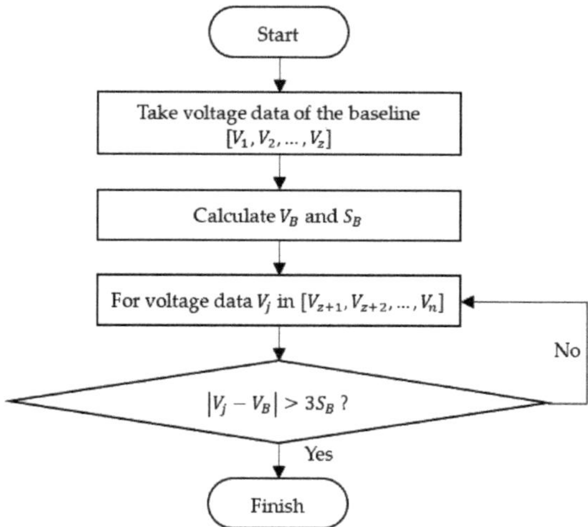

Figure 4. Flowchart of the "3σ" algorithm.

After initialization, the first z points of voltage $(V_1, V_2, \ldots, V_z)$ are taken from the raw voltage data series $(V_1, V_2, \ldots, V_n)$ as the baseline, from which the average V_B and standard deviation S_B are calculated using the following equation.

$$V_B = \frac{1}{z}\sum_{i=1}^{z} V_i, \quad S_B = \sqrt{\frac{\sum_{i=1}^{z}(V_i - V_B)}{z-1}} \tag{15}$$

Then, the voltage data after z, $(V_{z+1}, V_{z+2}, \ldots, V_n)$, are looked through for the first point when Inequation (14) is satisfied, whereafter the algorithm is stopped. The value of z can be estimated by the compression speed, sampling time, and the initial gap between the probe and the particle. Usually, $z = 20$ is used, which is sufficiently accurate to determine V_B and S_B.

2.4.2. "3σ + Window" Algorithm

Normally, the "3σ" algorithm can locate the starting point successfully. However, if a pulse noise exists, the starting point may be determined incorrectly as illustrated in Figure 5. The point m_1 rather than m_2 will be misidentified as the starting point because of the impulse noise around m_1. Although smoothing the raw data by filtering can deal with the impulse noise, other key points such as the rupture value will be evened.

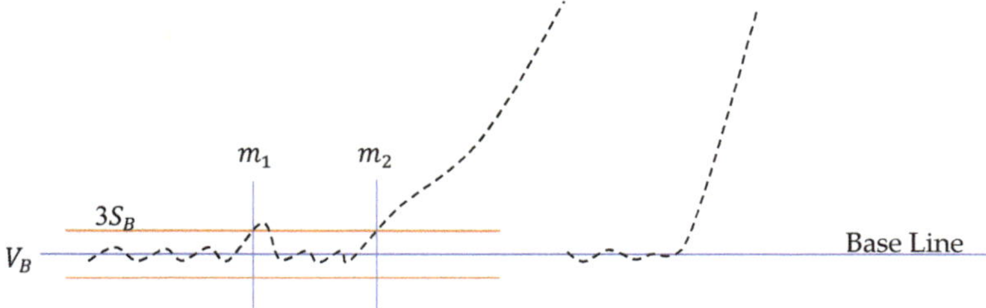

Figure 5. Problem of the "3σ" algorithm when impulse noise exists.

To tackle this problem, the "3σ" algorithm was modified by introducing a moving window with width w. A point m can be identified as the starting point only if all the points from it in the moving window fall away from V_B by $3S_B$, which leads to the following criterion:

$$\Lambda_{i=0}^{w-1}(|V_{m+i} - V_B| > 3S_B) == true, \tag{16}$$

where Λ is the logical "and" Boolean operator. The width of the moving window w can be estimated as an integer corresponding to a percent of the diameter of the microparticle. As some brittle capsules and biological cells may rupture at a fractional deformation as small as 0.06 [8], a percent of 5% can ensure w less than the rupture deformation of most microparticles. Thus, w can be estimated using the following equation:

$$w = \frac{0.05D}{vT_s}, \tag{17}$$

2.4.3. "3σ + Window + Hertz" Algorithm

The "3σ + window" algorithm can deal with most cases including those with random noise and impulse noise. However, it may underestimate the displacement when the voltage corresponding to three standard deviations of the baseline is just chosen as the starting point. This can result in a bigger value of the starting point index (m) so that the displacement will be underestimated, as it is related to the starting point by Equations (2), (7), and (8). The underestimation will be even worse when the signal to noise ratio is low. Following the same strategy described in [42], a mathematical model such as the Hertz model can be used to estimate the starting point (m) from the force-displacement data calculated using the "3σ + window" algorithm.

For diametrical compression of purely linear elastic microspheres, the Hertz model [1] relates the force to the displacement by the following equation:

$$F = \frac{E\sqrt{D}}{3(1-v^2)}\delta^{\frac{3}{2}}, \tag{18}$$

where E is the Young's modulus, and v is the Poisson's ratio. Assume the force and displacement obtained from the "3σ + window" is F' and δ', respectively, and the difference between the true displacement and the one obtained from the "3σ + window" is $\Delta\delta$; then, Equation (18) can be written as

$$F' = k'\left(\delta' + \Delta\delta\right)^{\frac{3}{2}}, \tag{19}$$

Equation (19) can be transformed to

$$\delta' = k\left(F'\right)^{\frac{2}{3}} - \Delta\delta, \tag{20}$$

where $k = 1/(k')^{2/3}$. Although the Hertz model is for purely linear elastic microspheres, it can be used to evaluate the true starting point by fitting into the initial compression data, such as within 5% deformation of the force-displacement data [23] obtained using the "3σ + window" algorithms. The flowchart of the algorithm can be illustrated by Figure 6. Firstly, a starting point index m' is estimated using the "3σ + window" algorithm, and the force-displacement data series $\left((F_1', \delta_1'), (F_2', \delta_2'), \ldots,)\right]$ are calculated using Equations (7) and (8). Then, the force-displacement data within 5% deformation $\left((F_1', \delta_1'), (F_2', \delta_2'), \ldots, \left(F_q', \delta_q'\right)\right)$ are fit using Equation (20), and thus, $\Delta\delta$ is obtained, from which the different number Δm is estimated by Equation (21) to compensate the starting point.

$$\Delta m = \frac{(CoD) \cdot \Delta\delta}{vT_s}, \tag{21}$$

CoD is often explained as the proportion of the variance in the dependent variable that is predictable from the independent variable [46]. It also indicates the extent to which the dependent variable is predictable by the fitting model. In our case, the CoD represents how well the Hertz model can be used to present the relationship between the force and displacement data up to 5% fractional deformation. A value of 1.0 indicates a perfect fit, whilst a value of 0.0 would indicate that the Hertz model fails to model the data. Multiplying $\Delta\delta$ by CoD is expected only to use the predictable percent of $\Delta\delta$ to compensate the starting point. In other words, the compensated number Δm is not only calculated from the $\Delta\delta$ value estimated by the Hertz model but also from the "goodness" of the fit, i.e., how well the Hertz model can fit the data. In this way, Equation (21) adjusts the compensation extent automatically according to the goodness of fit (CoD), which makes the compensation algorithm intelligent.

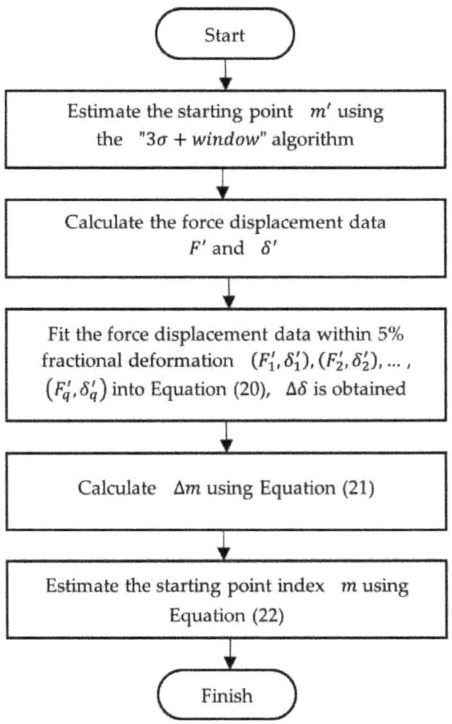

Figure 6. Flowchart of the "3σ + window + Hertz" algorithm.

Finally, the index of the starting point can be obtained by Equation (22).

$$m = m' - \Delta m, \tag{22}$$

2.5. *Algorithms to Locate the Rupture Point*

Maximum-Deceleration Algorithm

Normally, the voltage drops most dramatically just after the rupture point so that it can be identified by looking for the maximum deceleration through the voltage series. Practically, the deceleration is calculated from the following equation:

$$\Delta V_i = \frac{V_{i+1} + V_{i+2}}{2} - V_i,\ i = 1, 2, \cdots, n-2, \tag{23}$$

where $(V_{i+1} + V_{i+2})/2$ rather than V_{i+1} is used to filter the data slightly to reduce the possible impact of random noise.

The flowchart of the algorithm is illustrated in Figure 7. Initially, the drop (deceleration) series is calculated from the voltage series. Then, the point with the maximum drop (point p) is found, and the rupture point (r) is located as the peak point before p.

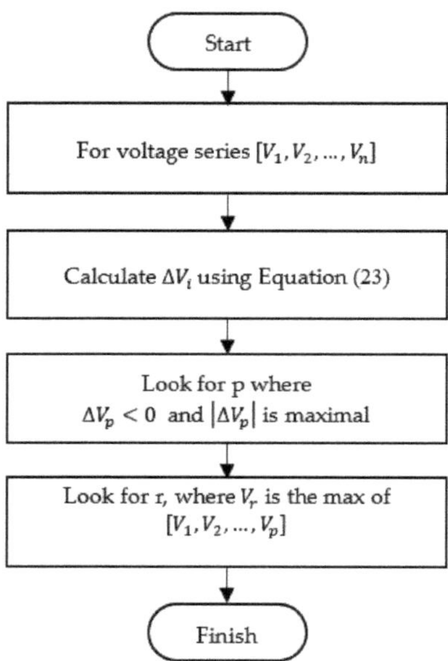

Figure 7. Flowchart of the maximum-deceleration algorithm.

2.6. Development of the Software Package

Visual Studio 2017 Community and .NET from Microsoft were chosen as the main development platform to develop the automatic data analysis software package. The user interface (UI) module, report-generating module, and main program module are mainly developed with the C# language, and the data read and conversion module, data processing and analysis module are mainly developed with the F# language. Besides, some open-source software libraries, such as Math.Net and EEPlus, are used to facilitate the software development. The open-source libraries used are listed in Table 1.

Table 1. Open-source libraries used in the development of the software package.

Library	Version	License
Daria	2.0.2	MIT
EEPlus	4.5.3.2	LGPL-3.0-or-later
ExcelDataReader	3.6.0	MIT
ExcelDataReader.DataSet	3.6.0	MIT
Math.NET Numerics	4.8.0	https://numerics.mathdotnet.com/License.html (accessed on 8 May 2022).
Accord.NET	3.8.0	http://accord-framework.net/license.txt (accessed on 8 May 2022).

3. Results and Discussion

3.1. Performance of the Algorithms

For the experimental raw voltage data of a microcapsule for self-sensing shown in Figure 8a, the starting point m1, m2, and m3 found by the "3σ", "3σ + window", and "3σ + window + Hertz" algorithms, respectively, are shown in Figure 9a. The diameter of the microcapsule was 87.6 μm. It can be seen that for this set of experimental data, the result of the "3σ" algorithm seems to underestimate the starting point because of the impulse noise, whilst the result of the "3σ + window" appears to overestimate the starting point. The starting point found by the "3σ + window + Hertz" algorithm looks more reasonable as the voltage starts to increase around this point. However, when no impulse noise exists, the starting point values obtained from the "3σ" (m1) and "3σ + window" (m2) algorithms are the same. For instance, for the raw voltage data of a porous PS microsphere in Figure 8b, the starting point is m1 = m2 = 213 as shown in Figure 9b. In both cases, the rupture points are successfully identified by the maximum-deceleration algorithm.

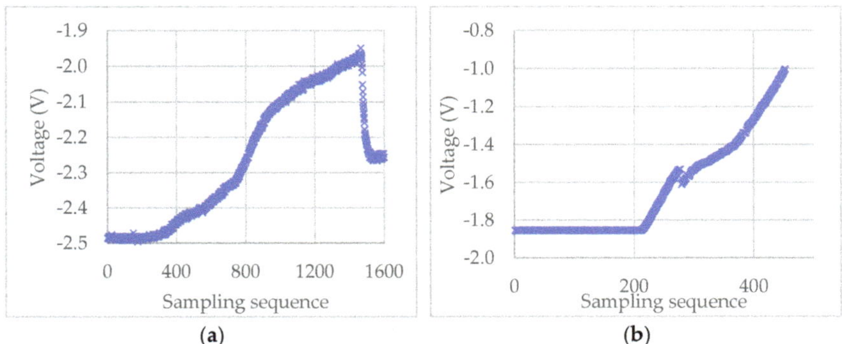

Figure 8. Experimental voltage-sampling sequence curves of a self-sensing microcapsule (**a**) and a porous PS microsphere (**b**).

In the following analysis, the staring point (m3) was found by the "3σ + window + Hertz" algorithm, as it is more reasonable as discussed above. The force-displacement data of the microcapsule in Figure 8a and microsphere in Figure 8b were calculated using Equations (7) and (8), and their curves are shown in Figure 10a,b, respectively. It can be seen from Figure 10a that the force-displacement curve of the self-sensing microcapsule is not very smooth, with some local peaks before rupture that might be due to the roughness of the out-layer PUF [5,47]. The out layer could crack several times before the rupture point shown in Figure 10a, where the inner shell was ruptured, and the force dropped sharply. In contrast, the force-displacement curve of the PS microsphere is quite smooth before the rupture, as its shell was smooth [43]. However, the force at the rupture point did not drop as dramatically as the self-sensing microcapsule since there was no release of any mateiral from the PS microsphere at the rupture point.

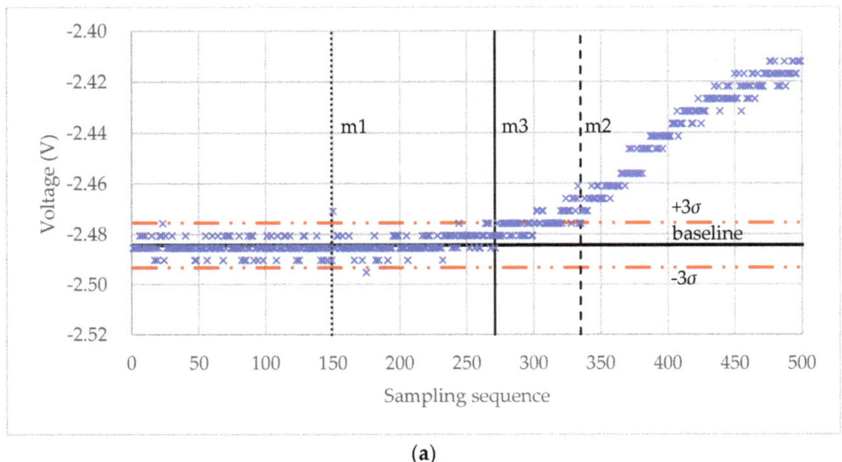

Figure 9. *Cont.*

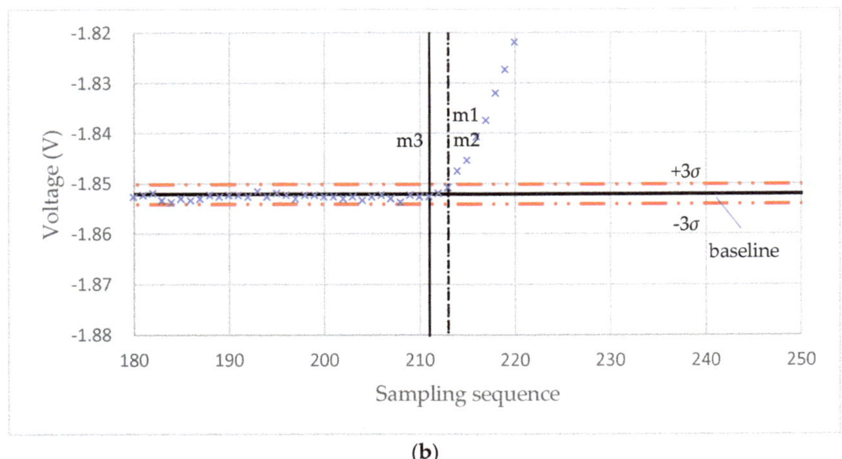

(b)

Figure 9. Starting point values found by the three algorithms. (**a**) Results for a self-sensing microcapsule and (**b**) results for a porous PS microsphere.

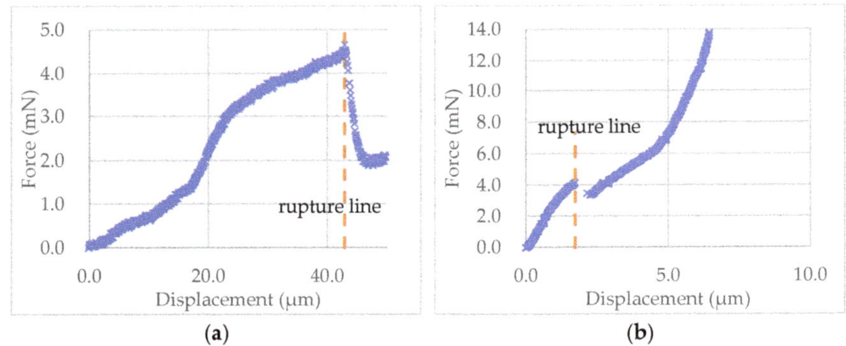

Figure 10. The force-displacement curves obtained using the starting point of m3 and the rupture point determined automatically for the experimental data of the self-sensing microcapsule (**a**) and the PS microsphere (**b**) in Figure 8a,b, respectively.

The nominal stress-fractional deformation data up to rupture of the PS microsphere in Figure 8b was calculated using Equation (9) and (10), and its curve is shown in Figure 11. The starting point was m3 in Figure 9b, found using the "3σ + window + Hertz" algorithm. The toughness of the particle was 1.15 MPa, calculated using the trapezoidal numerical integration in Equation (11), corresponding to the area under the curve up to rupture.

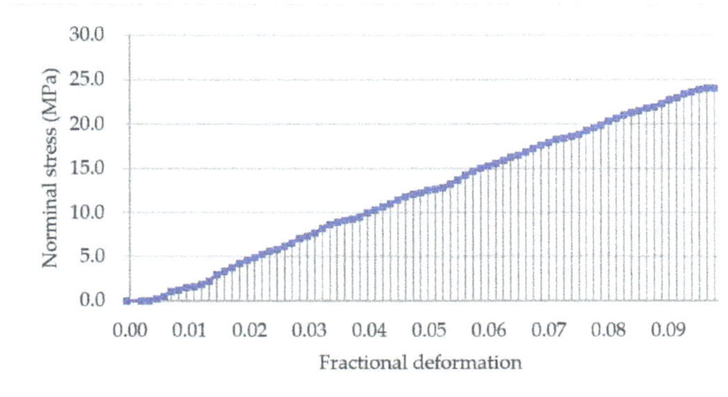

Figure 11. Nominal stress versus fractional deformation up to rupture of the PS microsphere in Figure 8b. The toughness corresponds to the area under the curve, i.e., the integration of the nominal rupture stress over the fractional deformation using Equation (11). The starting point M was found using the "3σ + window + Hertz" algorithm (m3 in Figure 8b).

The experimental data of 50 self-sensing microcapsules and 50 PS microspheres were analysed utilizing the developed software package, and a manual analysis was also carried out for comparison. The average and standard error of the calculated rupture strength parameters for the two samples are shown in Tables 2 and 3. It appears that for the two samples, the average rupture force values from the automatic data analyses are all the same as those from the manual analysis, which shows that the "maximum-deceleration" algorithm is very robust to locate the rupture point. So are the average values of nominal rupture stress, nominal rupture tension and the toughness as the former two parameters are calculated from the rupture force and the diameter of the microparticle. Although the toughness is related to the fractional deformation, which depends on the starting point, the force changes little around the starting point, so the effect of the initial integration of the nominal stress over the fractional deformation on the toughness value is negligible. Thus, the average values of the toughness from the four analyses show the same results. The values of the displacement at rupture from "3σ + window" and "3σ + window + Hertz" overlap with the results from the manual analysis. Because of the appearance of impulse noises, the values of displacement at rupture and deformation at rupture from "3σ" algorithm appear to be different significantly from the manual analysis results. It was found that the starting points for nearly half (24/50) of the tested self-sensing microcapsules and 17/50 of the porous PS microspheres were not correctly identified using the "3σ" algorithm.

Table 2. Rupture strength of the self-sensing microcapsules obtained from different algorithms.

Algorithm	Diameter (μm)	Displacement at Rupture (μm)	Rupture Force (mN)	Deformation at Rupture (%)	Nominal Rupture Stress (MPa)	Nominal Rupture Tension (μN/μm)	Toughness (MPa)
Manual	86.2 ± 3.1	40.5 ± 1.4	4.61 ± 0.22	47.7 ± 1.1	0.85 ± 0.05	53.8 ± 2.2	0.19 ± 0.01
3σ	86.2 ± 3.1	44.7 ± 1.7	4.61 ± 0.22	52.6 ± 1.5	0.85 ± 0.05	53.8 ± 2.2	0.19 ± 0.01
3σ + Window	86.2 ± 3.1	39.0 ± 1.4	4.61 ± 0.22	45.7 ± 1.1	0.85 ± 0.05	53.8 ± 2.2	0.19 ± 0.01
3σ + Window + Hertz	86.2 ± 3.1	41.0 ± 1.4	4.61 ± 0.22	48.2 ± 1.1	0.85 ± 0.05	53.8 ± 2.2	0.19 ± 0.01

Table 3. Rupture strength of the porous PS microspheres obtained from different algorithms.

Algorithm	Diameter (μm)	Displacement at Rupture (μm)	Rupture Force (mN)	Deformation at Rupture (%)	Nominal Rupture Stress (MPa)	Nominal Rupture Tension (μN/μm)	Toughness (MPa)
Manual	11.1 ± 0.4	1.3 ± 0.1	2.53 ± 0.15	12.0 ± 0.4	26.4 ± 1.2	223.0 ± 9.9	1.73 ± 0.11
3σ	11.1 ± 0.4	2.5 ± 0.3	2.53 ± 0.15	21.9 ± 2.3	26.4 ± 1.2	223.0 ± 9.9	1.73 ± 0.11
3σ + Window	11.1 ± 0.4	1.3 ± 0.1	2.53 ± 0.15	11.9 ± 0.4	26.4 ± 1.2	223.0 ± 9.9	1.73 ± 0.11
3σ + Window + Hertz	11.1 ± 0.4	1.4 ± 0.1	2.53 ± 0.15	12.7 ± 0.4	26.4 ± 1.2	223.0 ± 9.9	1.73 ± 0.11

Based on the data of these two samples, the results obtained from using "3σ + window" and "3σ + window + Hertz" algorithms have no significant difference from the manual results so that they both can be used in the automatic analysis of the rupture strength of microparticles.

3.2. Further Discussion

From the mean values in Tables 2 and 3, it appears that the fractional deformation at rupture of the self-sensing microcapsules is quite big (nearly 50%) in comparison with that of the porous PS microspheres (just around 12%). This indicates that the self-sensing microcapsules with double PUF-PU shells showed a ductile failure behaviour, while the porous PS microspheres showed a brittle failure behaviour [8]. However, the nominal rupture stress of the former (0.85 MPa) is much smaller than the latter (26.4 MPa). It is the same with the toughness, as it is related to the nominal stress versus fractional deformation up to rupture. This may result from the large difference in the particle sizes between the two samples since the nominal rupture stress normally decreases with the increasing particle diameter [48]. The values of the diameter for the self-sensing microcapsules and PS microspheres are 86.2 ± 3.1 μm and 11.1 ± 0.4 μm, respectively.

Moreover, the nominal rupture tension of the self-sensing microcapsules (53.9 μN/μm) is also much smaller than that of the porous PS microspheres (227.9 μN/μm). This is reasonable, as the former had a liquid core surrounded by a solid shell with thickness between 200 and 500 nm [5], whilst the latter were solid with a few pores on the surface [43].

The nominal rupture tension and the toughness versus diameter of the two samples of individual microspheres are illustrated in Figure 12. Statistical analysis of the data shows that the nominal rupture tension does not change with diameter for each sample significantly (Figure 12a,b), which can be used to compare the mechanical strength between samples with particles of different sizes. In contrast, the toughness decreases with the diameter, which indicates bigger particles were weaker than smaller ones (Figure 12c,d), similar to the nominal rupture stress [48].

3.3. Comparison with Other Algorithms

The standard deviation was used in several algorithms to evaluate the noise level of the raw data and to help estimate the parameters of the algorithms to identify CP for AFM force data [38,41,42,49]. A moving window was also introduced to help the identification of the CP [41,42]. However, it was only used for local regression rather than dealing with the impulse noise as addressed by the "3σ+ window" algorithm. Besides, the width of the moving window needs to be set manually in the reported algorithms, whilst it is estimated automatically by Equation (17) in the "3σ + window" and "3σ + window + Hertz" algorithms developed in this work. Furthermore, the algorithm in [42] pre-estimates a CP* with a threshold of five standard deviations of the baseline values and then determines the CP by fitting force-displacement data into a Hertz-like model from CP* to an indentation depth empirically determined by the stiffness of the force curve. The "3σ + window + Hertz" algorithm also pre-estimates a prone starting point m' followed by the regression of the force-displacement data within 5% fractional deformation to the Hertz model to determine the real starting point m. However, these two algorithms have two main differences. One is that the "3σ + window" algorithm is used to estimate the prone starting point, which can well deal with the impulse noises in the "3σ + window + Hertz" algorithm, whereas CP* is just estimated with a threshold of five standard deviations of the baseline values [42] that may result in a wrong value when the impulse noise greater than the threshold exists before the real CP. The other difference is that after the Hertz regression, the *CoD* is used in Equation (21) to adjust the degree of the compensation automatically so that when the tested material is not linear ealstic, fewer points will be compensated to m'. In contrast, the algorithm reported in [42] was designed for linear elastic materials and cannot adjust automatically for other mechanical behaviours of the tested materials. Besides, using the "maximum-deceleration" algorithm for the detection of rupture point in this work requires no parameter to be adjusted and is fully automatic, which is advantageous.

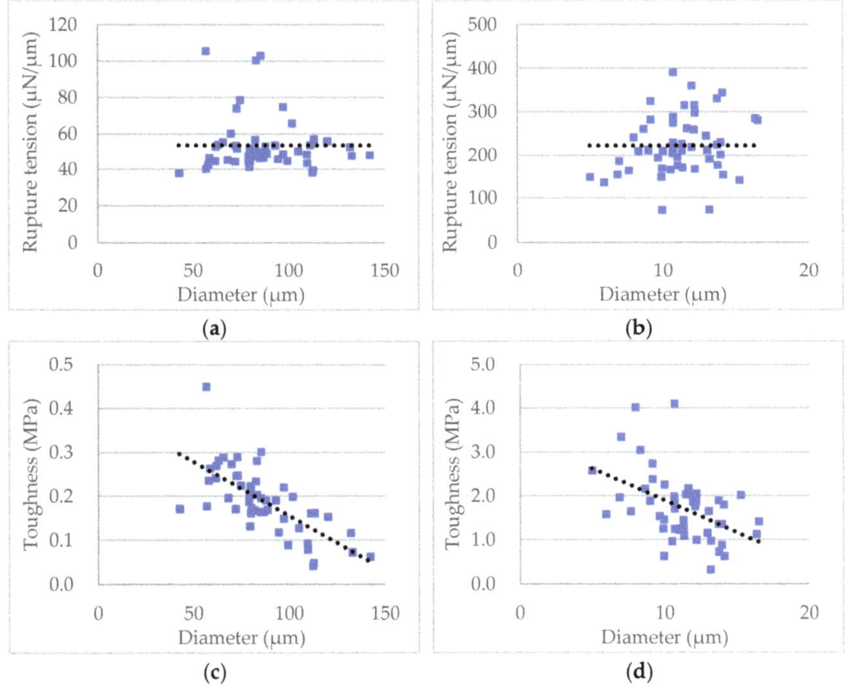

Figure 12. Data of the nominal rupture tension and toughness versus diameter of the two samples. (**a**) Nominal rupture tension of the self-sensing microcapsules. (**b**) Nominal rupture tension of the porous PS microspheres. (**c**) Toughness of the self-sensing microcapsules. (**d**) Toughness of the porous PS microspheres. Each fitted line (dotted) only indicates the trend, with 95% confidence.

4. Conclusions

In this study, a data analysis software package was developed to analyse the rupture strength of microparticles automatically from the experimental data of micromanipulation measurements. Three algorithms were developed to find the starting point of the compression data, i.e., the "3σ", "3σ + window" and "3σ + window + Hertz". The "3σ" algorithm determines the starting point where the voltage of a point deviates from the mean of baseline (V_B) by three standard deviations ($3S_B$), whilst in the "3σ + window" algorithm, a point is determined as the starting point only if the following w points (including this point) all deviate from V_B by $3S_B$. In the "3σ + window + Hertz" algorithm, the starting point is further adjusted by fitting the force-displacement data corresponding to very small deformations (up to 5% fractional deformation) into the Hertz model to compensate the underestimation of the displacement corresponding to the three standard deviations. One algorithm based on the maximum deceleration of the voltage series was developed to determine the rupture point. The results show that the combination of the "3σ + window" or "3σ + window + Hertz" algorithm with the "maximum-deceleration" algorithm can produce results that are in excellent agreement with those obtained manually, and there is no significant difference between them. Moreover, all the developed algorithms work fully automatically without any parameter modification.

For analysing 50 microparticles in a typical sample, the time spent on analysing the rupture strength parameters manually was from 2 to 3 h. In contrast, it took less than 20 min to analyse the same data automatically using the software package developed in this work. It is believed that this software package can also be used to analyse the force-displacement data obtained using conventional mechanical testing machines for macro-scale materials, which can have a wide range of applications.

Author Contributions: Conceptualization, Z.Z. (Zhibing Zhang); methodology, Z.Z. (Zhibing Zhang) and Z.Z. (Zhihua Zhang); software Z.Z. (Zhihua Zhang); validation, Z.Z. (Zhibing Zhang), Z.Z. (Zhihua Zhang) and Y.H.; formal analysis, Z.Z. (Zhihua Zhang); investigation, Z.Z. (Zhihua Zhang); resources, Z.Z. (Zhibing Zhang), Z.Z. (Zhihua Zhang) and Y.H.; data curation, Z.Z. (Zhihua Zhang); writing—original draft preparation, Z.Z. (Zhihua Zhang); writing—review and editing, Z.Z. (Zhibing Zhang); visualization, Z.Z. (Zhihua Zhang); supervision, Z.Z. (Zhibing Zhang); project administration, Z.Z. (Zhibing Zhang); funding acquisition, Z.Z. (Zhibing Zhang). All authors have read and agreed to the published version of the manuscript.

Funding: This research was funded by the University of Birmingham (UK), Jiangsu Industrial Technology Research Institute (JITRI) (China), and Changzhou Institute of Advanced Manufacturing Technology (China).

Data Availability Statement: Not applicable.

Acknowledgments: We acknowledge Zhenxue Zhang in School of Metallurgy and Materials, University of Birmingham, Birmingham, UK, for providing the sample of self-sensing microcapsules.

Conflicts of Interest: The authors declare no conflict of interest.

Nomenclature

c	Compliance of the force transducer (mN^{-1})
CoD	Coefficient of determination
D	Initial diameter of the single microparticle (m)
$e(t)$	Random noise
E	Young's modulus (Pa)
F, F_i, $F(t)$	Compression force (N)
F_r	Rupture force (N)
F'	Estimated force in the "3σ + window + Hertz" algorithm (N)
$((F'_1, \delta'_1), (F'_2, \delta'_2), \ldots, (F'_p, \delta'_p))$	Estimated force-displacement series in the "3σ + window + Hertz" algorithm
k, k'	$k = 1/(k')^{2/3}$, used in the "3σ + window + Hertz" algorithm
m	Starting point index
m_1	Starting point index found by the "3σ" algorithm
m_2	Starting point index found by the "3σ + window" algorithm
m_3	Starting point index found by the "3σ + window + Hertz" algorithm
m'	Estimated starting point index in the "3σ + window + Hertz" algorithm
Δm	Values to compensate starting point index in the "3σ + window + Hertz" algorithm
Pr	Possibility
r	Rupture point index
s	Sensitivity of the force transducer (NV^{-1})
S_B	Standard deviation of baseline voltage (V)

T_C	Particle toughness (Pa)
T_r	Nominal rupture tension (Nm^{-1})
T_s	Sampling time (s)
v	Compression speed (ms^{-1})
$V, V_i, V_m, V(t)$	Voltage (V)
$(V_1, V_2, \ldots, V_z)$	Voltage series of the baseline (V)
$(V_1, V_2, \ldots, V_n)$	Raw voltage data series (V)
V_B	Average voltage of the baseline (V)
V_r	Voltage corresponding to rupture (V)
$\Delta V, \Delta V_i$	Voltage deceleration in the "maximum-deceleration" algorithm (V)
w	Width of moving window in the "3σ + window" algorithm
z	Number of initial voltage points to estimate baseline values
Greek letters	
δ, δ_i	Displacement (m)
δ_r	Displacement at rupture (m)
δ'	Estimated displacement in the "3σ + window + Hertz" algorithm (m)
$\Delta\delta$	Difference between the true displacement and the one obtained from the "3σ + window" algorithm (m)
$\varepsilon, \varepsilon_i$	Fractional deformation
ε_r	Fractional deformation at rupture
μ	Mean value
v	Poisson's ration
3σ	Three sigma, three standard deviations
σ, σ_i	Nominal stress (Pa)
σ_r	Nominal rupture stress (Pa)

References

1. Zhang, Z.; Stenson, J.D.; Thomas, C.R. Micromanipulation in mechanical characterisation of single particles. In *Advances in Chemical Engineering*; Li, J., Ed.; Academic Press: Cambridge, MA, USA, 2009; Chapter 2; pp. 29–85.
2. Gray, A.; Egan, S.; Bakalis, S.; Zhang, Z. Determination of microcapsule physicochemical, structural, and mechanical properties. *Particuology* **2016**, *24*, 32–43. [CrossRef]
3. Maia, F.; Tedim, J.; Bastos, A.C.; Ferreira, M.G.S.; Zheludkevich, M.L. Active sensing coating for early detection of corrosion processes. *RSC Adv.* **2014**, *4*, 17780–17786. [CrossRef]
4. Aniskevich, A.; Kulakov, V.; Bulderberga, O.; Knotek, P.; Tedim, J.; Maia, F.; Leisis, V.; Zeleniakiene, D. Experimental characterisation and modelling of mechanical behaviour of microcapsules. *J. Mater. Sci.* **2020**, *55*, 13457–13471. [CrossRef]
5. Roche, S.; Ibarz, G.; Crespo, C.; Chiminelli, A.; Araujo, A.; Santos, R.; Zhang, Z.; Li, X.; Dong, H. Self-sensing polymeric materials based on fluorescent microcapsules for the detection of microcracks. *J. Mater. Res. Technol.* **2021**, *16*, 505–515. [CrossRef]
6. He, Y.; Bowen, J.; Andrews, J.W.; Liu, M.; Smets, J.; Zhang, Z. Adhesion of perfume-filled microcapsules to model fabric surfaces. *J. Microencapsul.* **2014**, *31*, 430–439. [CrossRef]
7. Müller, E.; Chung, J.-T.; Zhang, Z.; Sprauer, A. Characterization of the mechanical properties of polymeric chromatographic particles by micromanipulation. *J. Chromatogr. A* **2005**, *1097*, 116–123. [CrossRef]
8. Mercadé-Prieto, R.; Zhang, Z. Mechanical characterization of microspheres—Capsules, cells and beads: A review. *J. Microencapsul.* **2012**, *29*, 277–285. [CrossRef]
9. Neuman, K.C.; Nagy, A. Single-molecule force spectroscopy: Optical tweezers, magnetic tweezers and atomic force microscopy. *Nat. Methods* **2008**, *5*, 491–505. [CrossRef]
10. Zimmermann, U.; Steudle, E. Kontinuierliche Druckmessung in Pflanzenzellen. *Naturwissenschaften* **1969**, *56*, 634. [CrossRef]
11. Hüsken, D.; Steudle, E.; Zimmermann, U. Pressure Probe Technique for Measuring Water Relations of Cells in Higher Plants. *Plant Physiol.* **1978**, *61*, 158–163. [CrossRef]
12. Mitchison, J.M.; Swann, M.M. The Mechanical Properties of the Cell Surface. *J. Exp. Biol.* **1954**, *31*, 443–460. [CrossRef]
13. Mitchison, J.M.; Swann, M.M. The mechanical properties of the cell surface: II. The unfertilized sea-urchin egg. *J. Exp. Biol.* **1954**, *31*, 461–472. [CrossRef]
14. Binnig, G.; Quate, C.F.; Gerber, C. Atomic force microscope. *Phys. Rev. Lett.* **1986**, *56*, 930–933. [CrossRef] [PubMed]
15. Bennig, G.K. Atomic force microscope and method for imaging surfaces with atomic resolution. U.S. Patent RE33387, 4 August 1986.
16. McConnaughey, W.B.; Petersen, N.O. Cell poker: An apparatus for stress-strain measurements on living cells. *Rev. Sci. Instrum.* **1980**, *51*, 575–580. [CrossRef] [PubMed]
17. Broitman, E. Indentation Hardness Measurements at Macro-, Micro-, and Nanoscale: A Critical Overview. *Tribol. Lett.* **2016**, *65*, 23. [CrossRef]

18. Zhang, Z.; Ferenczi, M.; Lush, A.C.; Thomas, C.R. A novel micromanipulation technique for measuring the bursting strength of single mammalian cells. *Appl. Microbiol. Biotechnol.* **1991**, *36*, 208–210. [CrossRef]
19. Zhang, Z.; Saunders, R.; Thomas, C.R. Mechanical strength of single microcapsules determined by a novel micromanipulation technique. *J. Microencapsul.* **1999**, *16*, 117–124. [CrossRef]
20. Stenekes, R.J.H.; De Smedt, S.C.; Demeester, J.; Sun, G.; Zhang, Z.; Hennink, W.E. Pore Sizes in Hydrated Dextran Microspheres. *Biomacromolecules* **2000**, *1*, 696–703. [CrossRef]
21. Sun, G.; Zhang, Z. Mechanical properties of melamine-formaldehyde microcapsules. *J. Microencapsul.* **2001**, *18*, 593–602.
22. Xue, J.; Zhang, Z. Preparation and characterization of calcium-shellac spheres as a carrier of carbamide peroxide. *J. Microencapsul.* **2008**, *25*, 523–530. [CrossRef]
23. Yan, Y.; Zhang, Z.; Stokes, J.; Zhou, Q.-Z.; Ma, G.-H.; Adams, M.J. Mechanical characterization of agarose micro-particles with a narrow size distribution. *Powder Technol.* **2009**, *192*, 122–130. [CrossRef]
24. Olderøy, M.; Xie, M.; Andreassen, J.-P.; Strand, B.L.; Zhang, Z.; Sikorski, P. Viscoelastic properties of mineralized alginate hydrogel beads. *J. Mater. Sci. Mater. Electron.* **2012**, *23*, 1619–1627. [CrossRef] [PubMed]
25. Liu, T.; Zhang, Z. Mechanical properties of desiccated ragweed pollen grains determined by micromanipulation and theoretical modelling. *Biotechnol. Bioeng.* **2004**, *85*, 770–775. [CrossRef] [PubMed]
26. Mashmoushy, H.; Zhang, Z.; Thomas, C.R. Micromanipulation measurement of the mechanical properties of baker's yeast cells. *Biotechnol. Tech.* **1998**, *12*, 925–929. [CrossRef]
27. Smith, A.; Zhang, Z.; Thomas, C. Wall material properties of yeast cells: Part Cell measurements and compression experiments. *Chem. Eng. Sci.* **2000**, *55*, 2031–2041. [CrossRef]
28. Wang, Q.G.; Magnay, J.L.; Nguyen, B.; Thomas, C.R.; Zhang, Z.; El Haj, A.J.; Kuiper, N.J. Gene expression profiles of dynamically compressed single chondrocytes and chondrons. *Biochem. Biophys. Res. Commun.* **2009**, *379*, 738–742. [CrossRef]
29. Nguyen, B.V.; Wang, Q.G.; Kuiper, N.J.; El Haj, A.J.; Thomas, C.R.; Zhang, Z. Biomechanical properties of single chondrocytes and chondrons determined by micromanipulation and finite-element modelling. *J. R. Soc. Interface* **2010**, *7*, 1723–1733. [CrossRef]
30. Chen, M.; Zhang, Z.; Bott, T. Direct measurement of the adhesive strength of biofilms in pipes by micromanipulation. *Biotechnol. Tech.* **1998**, *12*, 875–880. [CrossRef]
31. Chen, M.; Zhang, Z.; Bott, T. Effects of operating conditions on the adhesive strength of Pseudomonas fluorescens biofilms in tubes. *Colloids Surf. B Biointerfaces* **2005**, *43*, 61–71. [CrossRef]
32. Liu, W.; Christian, G.; Zhang, Z.; Fryer, P. Development and Use of a Micromanipulation Technique for Measuring the Force Required to Disrupt and Remove Fouling Deposits. *Food Bioprod. Process.* **2002**, *80*, 286–291. [CrossRef]
33. Liu, W.; Christian, G.; Zhang, Z.; Fryer, P. Direct measurement of the force required to disrupt and remove fouling deposits of whey protein concentrate. *Int. Dairy J.* **2006**, *16*, 164–172. [CrossRef]
34. Liu, W.; Fryer, P.; Zhang, Z.; Zhao, Q.; Liu, Y. Identification of cohesive and adhesive effects in the cleaning of food fouling deposits. *Innov. Food Sci. Emerg. Technol.* **2006**, *7*, 263–269. [CrossRef]
35. Liu, W.; Zhang, Z.; Fryer, P. Identification and modelling of different removal modes in the cleaning of a model food deposit. *Chem. Eng. Sci.* **2006**, *61*, 7528–7534. [CrossRef]
36. Du, G.; Zhang, Z.; He, P.; Zhang, Z.; Sun, X. Determination of the mechanical properties of polymeric microneedles by micromanipulation. *J. Mech. Behav. Biomed. Mater.* **2021**, *117*, 104384. [CrossRef] [PubMed]
37. Zhang, Z. Novel Micromanipulation Studies of Biological and Non-Biological Materials, in School of Chemical Engineering. DSc Thesis, University of Birmingham, Birmingham, UK, 2016.
38. Roduit, C.; Saha, B.; Alonso-Sarduy, L.; Volterra, A.; Dietler, G.; Kasas, S. OpenFovea: Open-source AFM data processing software. *Nat. Methods* **2012**, *9*, 774–775. [CrossRef]
39. Dinarelli, S.; Girasole, M.; Longo, G. FC_analysis: A tool for investigating atomic force microscopy maps of force curves. *BMC Bioinform.* **2018**, *19*, 258. [CrossRef] [PubMed]
40. Monclus, M.A.; Young, T.J.; Di Maio, D. AFM indentation method used for elastic modulus characterization of interfaces and thin layers. *J. Mater. Sci.* **2010**, *45*, 3190–3197. [CrossRef]
41. Benítez, R.; Moreno-Flores, S.; Bolós, V.J.; Toca-Herrera, J.L. A new automatic contact point detection algorithm for AFM force curves. *Microsc. Res. Tech.* **2013**, *76*, 870–876. [CrossRef]
42. Chang, Y.-R.; Raghunathan, V.K.; Garland, S.P.; Morgan, J.T.; Russell, P.; Murphy, C.J. Automated AFM force curve analysis for determining elastic modulus of biomaterials and biological samples. *J. Mech. Behav. Biomed. Mater.* **2014**, *37*, 209–218. [CrossRef]
43. Li, G.; Yu, Y.; Han, W.; Zhu, L.; Si, T.; Wang, H.; Li, K.; Sun, Y.; He, Y. Solvent evaporation self-motivated continual synthesis of versatile porous polymer microspheres via foaming-transfer. *Colloids Surfaces A Physicochem. Eng. Asp.* **2021**, *615*, 126239. [CrossRef]
44. Smith, S.W. *The Scientist and Engineer's Guide to Digital Signal Processing*; California Technical Publishing: San Diego, CA, USA, 1997.
45. Pukelsheim, F. The three sigma rule. *Am. Stat.* **1994**, *48*, 88–91.
46. Montgomery, D.C.; Peck, E.A.; Vining, G.G. *Introduction to Linear Regression Analysis*, 5th ed.; Wiley Series in Probability and Statistics; Peck, E.A., Vining, G.G., Eds.; Wiley-Blackwell: Hoboken, NJ, USA, 2012.
47. Song, Y.; Chen, K.-F.; Wang, J.-J.; Liu, Y.; Qi, T.; Li, G.L. Synthesis of Polyurethane/Poly(urea-formaldehyde) Double-shelled Microcapsules for Self-healing Anticorrosion Coatings. *Chin. J. Polym. Sci.* **2019**, *38*, 45–52. [CrossRef]

48. Hu, J.; Chen, H.-Q.; Zhang, Z. Mechanical properties of melamine formaldehyde microcapsules for self-healing materials. *Mater. Chem. Phys.* **2009**, *118*, 63–70. [CrossRef]
49. Gergely, C.; Senger, B.; Voegel, J.C.; Hörber, J.K.H.; Schaaf, P.; Hemmerlé, J. Semi-automatized processing of AFM force-spectroscopy data. *Ultramicroscopy* **2001**, *87*, 67–78. [CrossRef]

 *micromachines*

Article

Omnidirectional Manipulation of Microparticles on a Platform Subjected to Circular Motion Applying Dynamic Dry Friction Control

Sigitas Kilikevičius *, Kristina Liutkauskienė, Ernestas Uldinskas, Ribal El Banna and Algimantas Fedaravičius

Department of Transport Engineering, Kaunas University of Technology, Studentų St. 56, 51424 Kaunas, Lithuania; kristina.liutkauskiene@ktu.lt (K.L.); ernestas.uldinskas@ktu.lt (E.U.); ribal.el@ktu.edu (R.E.B.); algimantas.fedaravicius@ktu.lt (A.F.)

* Correspondence: sigitas.kilikevicius@ktu.lt

Abstract: Currently used planar manipulation methods that utilize oscillating surfaces are usually based on asymmetries of time, kinematic, wave, or power types. This paper proposes a method for omnidirectional manipulation of microparticles on a platform subjected to circular motion, where the motion of the particle is achieved and controlled through the asymmetry created by dynamic friction control. The range of angles at which microparticles can be directed, and the average velocity were considered figures of merit. To determine the intrinsic parameters of the system that define the direction and velocity of the particles, a nondimensional mathematical model of the proposed method was developed, and modeling of the manipulation process was carried out. The modeling has shown that it is possible to direct the particle omnidirectionally at any angle over the full 2π range by changing the phase shift between the function governing the circular motion and the dry friction control function. The shape of the trajectory and the average velocity of the particle depend mainly on the width of the dry friction control function. An experimental investigation of omnidirectional manipulation was carried out by implementing the method of dynamic dry friction control. The experiments verified that the asymmetry created by dynamic dry friction control is technically feasible and can be applied for the omnidirectional manipulation of microparticles. The experimental results were consistent with the modeling results and qualitatively confirmed the influence of the control parameters on the motion characteristics predicted by the modeling. The study enriches the classical theories of particle motion on oscillating rigid plates, and it is relevant for the industries that implement various tasks related to assembling, handling, feeding, transporting, or manipulating microparticles.

Keywords: micromanipulation; microparticles; motion control; vibrations; dry friction; control; oscillating platform

Citation: Kilikevičius, S.; Liutkauskienė, K.; Uldinskas, E.; El Banna, R.; Fedaravičius, A. Omnidirectional Manipulation of Microparticles on a Platform Subjected to Circular Motion Applying Dynamic Dry Friction Control. *Micromachines* **2022**, *13*, 711. https://doi.org/10.3390/mi13050711

Academic Editor: Alessandro Cammarata

Received: 16 April 2022
Accepted: 28 April 2022
Published: 30 April 2022

Publisher's Note: MDPI stays neutral with regard to jurisdictional claims in published maps and institutional affiliations.

1. Introduction

The manipulation of microparticles is very important for many disciplines and sectors, such as micromachine technology, biotechnology, cell biology, material processing, semiconductor industries, and neuroscience [1–7]. Handling, transportation, and manipulation of microparticles or bulk and granular materials can be implemented by various methods and approaches that can generally be divided into two types: prehensile and non-prehensile. Prehensile methods usually involve some sort of force or form closure that is associated with grasping by microgrippers [8–12]. However, prehensile methods are most suited for the manipulation of individual objects, they always involve some mechanical effect on the object to be manipulated. They still struggle with precise force feedback at microscales, and the technological equipment used to perform micromanipulation tasks is usually very complex and expensive. During the processes of nonprehensile manipulation, the objects to be manipulated are subjected only to unilateral constraints. Therefore, the external mechanical effects acting on the object to be manipulated are reduced to a minimum, and the parts can be transported without damage. In addition, nonprehensile manipulation methods can typically offer lower equipment costs, larger workspaces, and shorter operational times.

Various non-prehensile manipulation methods are being used in practice and studied in the scientific literature. For example, nonprehensile manipulation operations can be performed by employing the devices that carry the objects to be moved [13], by pushing with robot end-effectors [14,15], and by controlling actuator arrays mounted under a flexible surface [16], etc. Micropositioning platforms take a significant share among nonprehensile manipulation methods and have attracted a lot of attention in recent years. Ablay [17] studied a magnetic micromanipulator with a model-free controller and a linear controller for the manipulation of microparticles in a fluid. Li et al. [18] presented a nanopositioning system composed of flexible beams mounted with magnetorheological elastomers. The properties of the beams, such as stiffness and damping, were able to be tuned under the influence of magnetic fields. Ferrara-Bello et al. [19] applied a micropositioning system actuated by three piezoelectric stacks to control the position and displacement in the three dimensions along the XYZaxis. The main disadvantage of micropositioning platforms is the limited size of their workspaces.

Microparticle manipulation implemented through vibration-assisted and acoustic techniques is widely used and investigated [20] as it is suitable for the manipulation of large numbers of microparticles without causing unwanted mechanical stress or other adverse effects such as contamination [21].

One of the acoustic methods that is suitable for microparticles is called acoustic levitation. This method exploits the acoustic radiation force in order to move microparticles that are sustained in the air [22,23]. Acoustic manipulation in microfluidic systems has recently gained significant attention in the field of biomedicine due to the potential to control individual particles, cells, or cellular clusters [24–27].

Vibration-assisted techniques utilize mechanical vibrations of the manipulation device to transport particles along a certain direction. Large workspaces and low operational times can be achieved using this approach. Asymmetry is an essential condition to achieve the motion of an object placed on a vibrating platform. It results in friction forces that are not canceled out over one cycle of vibrations. Several types of asymmetries are being employed for nonprehensile manipulation, such as time-asymmetries, kinematic asymmetries, wave asymmetries, or power asymmetries.

A time-asymmetry can be achieved through an asymmetric excitation of the platform when the forward motion takes a longer time compared to the backward motion in every cycle of the excitation [28]. This asymmetry can also be called a temporal or vibrational asymmetry. The dynamics of a body moving along a straight-line trajectory on a plate subjected to this kind of excitation were studied by Reznik et al. [28]. The dynamics of stick-slip motion under time-asymmetry were studied by Mayyas [29,30]. The time-asymmetry was created by mounting a platform on a nonlinear leaf spring that exhibits direction-dependent elasticity. Due to this peculiarity, the forward and backward accelerations of the plate are not equal when the plate is subjected to vibrational excitation.

Another type of asymmetry is kinematic asymmetry. It is implemented through an asymmetry of the vibration path or an asymmetry of the law of motion along this path. For example, this kind of asymmetry can be created when the direction of the harmonic oscillations is inclined with respect to the manipulation surface, i.e., the direction of the motion of particles to be manipulated. This asymmetry is usually applied to various vibratory conveyors and feeders [31–33]. Frei et al. [34] proposed a method for the manipulation of objects in individual paths by employing an array of multiple cells excited in two directions that caused a kinematic asymmetry. Vrublevskyi [35] studied the process of vibrational conveying with a kinematic asymmetry where an inclined surface was subjected to harmonic longitudinal and polyharmonic normal oscillations.

Wave asymmetries are also widely used to achieve the directional motion of particles. This kind of asymmetry is achieved by exciting traveling waves [36–38]. Various types of waves are used to implement this type of asymmetry. The mechanism of transport of dielectric particles on a conveyor employing a traveling electric field wave was investigated by Zouaghi et al. [39]. A wave asymmetry was employed by Kumar and DasGupta [40] to manipulate particles on a plate subjected to traveling circumferential harmonic waves.

The directional motion of an object placed on an oscillating surface can also be achieved through asymmetries classified as power asymmetries. These asymmetries can be subcategorized into types such as geometric and force asymmetries. One way to create a power asymmetry is to incline the system with respect to the horizontal plane. In this case, such an asymmetry can also be called a geometric asymmetry. Viswarupachari et al. [41] employed a geometric asymmetry along with a time-asymmetry to transport particles placed on a platform that was subjected to asymmetric vibrations. A power asymmetry can also be achieved when a constant force is applied to the object or when the resistance forces during the forward motions are not equal to the resistance forces during the backward motion. Asymmetries created in such a way can be designated as force asymmetries. Mitani et al. [42–44] applied an oscillating platform with a textured surface for the feeding of microparts. The motion of the microparts was achieved through a force asymmetry created by the anisotropic friction properties of the textured surface. Chen et al. proposed an oscillating trough with fin-like asperities for the transportation of particles. The fin-like asperities created a force asymmetry, which resulted in the directional motion of the particles.

Recently, a method of bidirectional vibrational transportation was demonstrated that was achieved through an asymmetry created by dynamically controlling the frictional conditions between the object being manipulated and the platform subjected to harmonic (sinusoidal) oscillations along the horizontal direction [45–47].

Unlike the omnidirectional manipulation systems with oscillating surfaces that are usually based on asymmetries of time, kinematic, wave, or power types, the presented work examines the case where the effective coefficient of friction is being dynamically controlled during each rotation cycle of the platform in such a way as to achieve the asymmetry of frictional conditions. This ensures the ability to manipulate various small particles on the platform in a complex trajectory. The objectives of the research are to determine the intrinsic parameters of the system that define the direction and velocity of the particles and to experimentally verify that the asymmetry created by dynamic dry friction control is technically feasible and can be applied for the omnidirectional manipulation of microparticles. The work addresses the scientific problem of manipulation at small scales, and the novelty of the research is that the motion of microparticles on a platform subjected to circular motion is achieved and controlled through the asymmetry created by dynamic friction control.

2. Methodology

2.1. Mathematical Model

Figure 1 shows a scheme of the dynamic model of omnidirectional manipulation of particles on a platform subjected to circular motion.

In this scheme, the stationary coordinate system is $O_1\chi\eta\zeta$, and the coordinate system $Oxyz$ is moving along with the rotating platform. Then, the coordinates of point (χ_i, η_i) on the platform in the stationary coordinate system are as follows:

$$\begin{cases} \chi(t) = \chi_i + R\cos\omega t \\ \eta(t) = \eta_i + R\sin\omega t \end{cases} \tag{1}$$

where t is the time, R is the radius of the circular motion, and ω is the angular frequency of the circular motion.

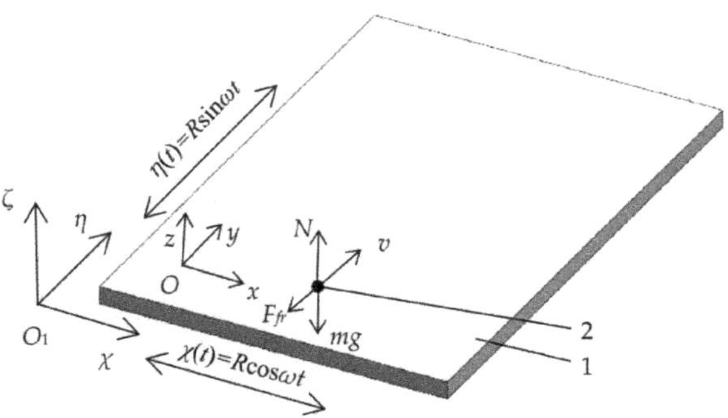

Figure 1. Scheme of the dynamic model of omnidirectional manipulation of particles on a platform subjected to circular motion: (1) platform; (2) particle.

In the Oxy coordinate system, the relative motion of the particle on the platform subjected to circular motion is described by the following differential equations:

$$\begin{cases} \ddot{x} + \frac{g\mu(t)\dot{x}}{\sqrt{\dot{x}^2+\dot{y}^2}} = R\omega^2 \cos \omega t \\ \ddot{y} + \frac{g\mu(t)\dot{y}}{\sqrt{\dot{x}^2+\dot{y}^2}} = R\omega^2 \sin \omega t \end{cases} \quad (2)$$

where $\dot{x}^2 + \dot{y}^2 \neq 0$, $\mu(t)$ is the coefficient of dry friction, which is being controlled with respect to the period of the circular motion of the platform in order to achieve the asymmetry of frictional conditions.

Nondimensionalization was applied to determine the intrinsic parameters of the system that define the direction and velocity of particles. In order to nondimensionalize Equation (2), the following nondimensional parameters were introduced:

$$\xi' = \frac{\dot{x}}{R\omega};\ \psi' = \frac{\dot{y}}{R\omega};\gamma = \frac{g}{R\omega^2};\tau = \omega t \quad (3)$$

Differentiation with respect to nondimensional time τ is denoted using the Lagrange notation. In this case, the nondimensionalized equations of motion can be written as follows:

$$\begin{cases} \xi'' + \frac{\mu(\tau)\gamma\xi'}{\sqrt{\xi'^2+\psi'^2}} = \cos \tau \\ \psi'' + \frac{\mu(\tau)\gamma\psi'}{\sqrt{\xi'^2+\psi'^2}} = \sin \tau \end{cases} \quad (4)$$

where $\xi'^2 + \psi'^2 \neq 0$, ξ and ψ are the horizontal and vertical components of the nondimensional displacement, respectively.

The dry friction coefficient $\mu(\tau)$ is controlled with respect to the circular motion of the platform by the following function:

$$\mu(\tau) = \begin{cases} \langle \mu_m \rangle, \text{ when } 2\pi n + \phi < \tau < 2\pi n + \phi + \lambda, \\ \mu_0, \text{ otherwise,} \end{cases} \quad (5)$$

where $n = (0, 1, 2, \dots)$, μ_0 is the nominal dry friction coefficient between the particle and the platform's surface, $\langle \mu_m \rangle$ is the dynamically modified time-averaged effective dry friction coefficient between the particle and the platform's surface, ϕ is the phase shift between the function governing the circular motion and the dry friction control function, and λ is the width of the dry friction control function. The principle of dynamic dry friction control is shown in Figure 2.

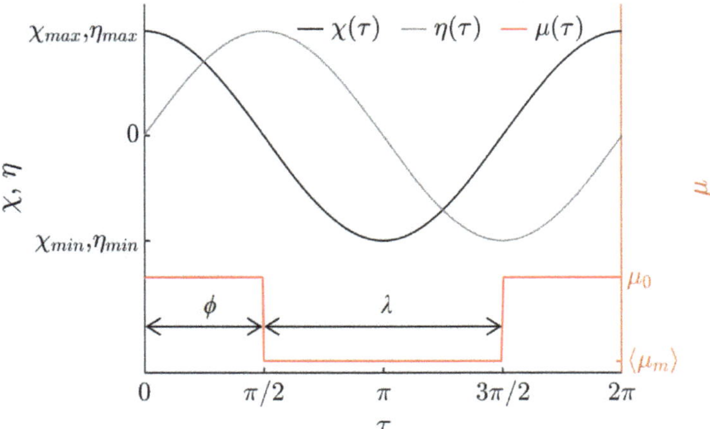

Figure 2. Principle of dynamic dry friction control.

The average nondimensional velocity of the particle can be found by the following:

$$\langle\vartheta\rangle = \frac{1}{2\pi}\int_0^{2\pi}\sqrt{\xi'^2 + \psi'^2}d\tau \tag{6}$$

Equation (4) can also be expressed in the polar coordinate system by defining the horizontal and vertical components of the nondimensional velocity as follows:

$$\xi' = \rho\cos\theta \tag{7}$$

$$\psi\prime = \rho\sin\theta \tag{8}$$

Then, Equation (4) in the polar coordinate system can be written as follows:

$$\begin{cases} \rho' + \mu(\tau)\gamma = \cos(\tau - \theta) \\ \rho\theta' = \sin(\tau - \theta) \end{cases} \tag{9}$$

where $\rho \neq 0$, ρ and θ are the magnitude and phase of the angular velocity vector, respectively.

2.2. Methodology of Experimental Investigation

Figure 3a shows a general view of the experimental setup built for omnidirectional manipulation of microparticles on a platform subjected to circular motion applying dynamic dry friction control. Figure 3a displays a schematic of the experimental setup that was used for the investigation. The setup consists of a platform (1) sustained by four elastic rods (2) (Figure 3a). The platform is subjected to circular motion by an electric motor (3) with an eccentric mechanism (4). A direct current power supply (HY3002-2, Mastech, Shenzhen, China) (5) supplies power to the electric motor. Four rectangular piezoelectric actuators (6) are mounted on the platform to excite a manipulation plate (7) in the vertical direction. The upper surface of the manipulation plate is polished to an average surface roughness of about 0.44 mm. The manipulation of microparticles (8) takes place on this surface. The phase of the circular motion of the platform is monitored by an optical reference sensor (P–95, Brüel & Kjær, Nærum, Denmark) (9). The sensor signal is processed by a vibration analyzer (Vibrotest 60, Brüel & Kjær, Nærum, Denmark) (10).

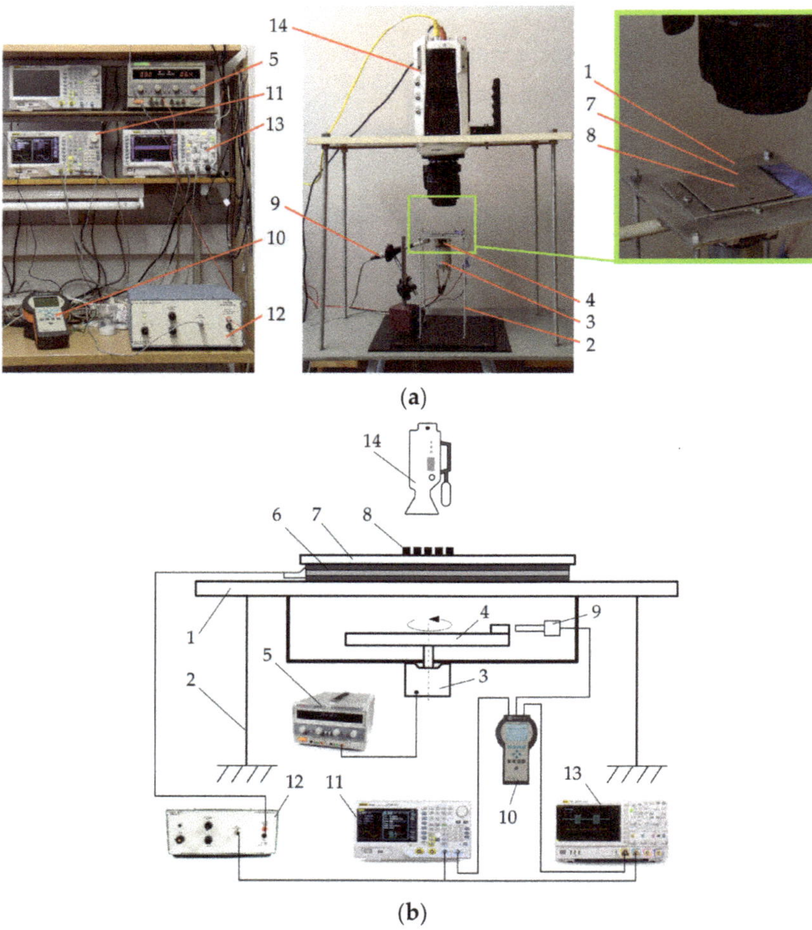

Figure 3. Experimental setup for omnidirectional manipulation applying dynamic dry friction control: (**a**) General view; (**b**) Scheme where the following components are shown: (1) platform; (2) elastic rods; (3) piezoelectric actuator; (4) eccentric mechanism; (5) direct current power supply; (6) piezoelectric actuators; (7) manipulation plate; (8) microparticles; (9) optical reference sensor; (10) vibration analyzer; (11) arbitrary waveform generator; (12) piezo linear amplifier; (13) digital oscilloscope; (14) high-speed camera.

The signal for dry friction control is composed of high-frequency pulses in burst mode that are generated by an arbitrary waveform generator (DG4202, RIGOL, Beijing, China) (11) and amplified by a piezo linear amplifier (EPA-104, Piezo Systems Inc., Cambridge, MA, USA) (12) and fed to the piezoelectric actuators. This signal is synchronized with respect to the phase of the circular excitation. Based on the principle of dry friction control shown in Figure 2, the piezoelectric actuators are being excited by a frequency of 2094 Hz for a fraction equal to λ and shifted by ϕ in each period of the circular motion of the manipulation plate. It has been demonstrated numerous times before that high-frequency vibrations introduced between sliding objects cause a reduction in the effective dry friction force between these objects as a result of the dynamic processes that occur in the contact region [48–50]. Therefore, in this fraction of the period when the manipulation plate is subjected to high-frequency vibrations by the piezoelectric actuators, the effective friction force between the manipulation plate's surface and the particle is reduced. In this way, the dry friction force can be controlled in a predefined manner with respect to the period of the circular motion of the manipulation plate. The dry friction control signal and the optical reference sensor readout are monitored and displayed by a digital oscilloscope (DS1054, RIGOL, Beijing, China) (13).

A high-speed camera (Phantom v711, 1280 × 800 CMOS sensor, 1 Mpx, 20 μm pixel size, Vision Research, Wayne, NJ, USA) (14) equipped with a macro lens (MP-E 65 mm f/2.8 1–5x, Canon Inc., Ōta, Tokyo, Japan) was used to record the motion of the particles. A video processing program based on the normalized cross-correlation approach was developed using MATLAB to digitize and analyze the motion of the particles.

3. Results

3.1. Modeling Results

Numerical modeling of omnidirectional manipulation on a platform subjected to a circular motion under dynamic friction control was carried out. For this purpose, software was developed in the MATLAB programming language (MathWorks, Natick, MA, USA). The equations were solved using the ode45 solver based on the Runge–Kutta (4, 5) formula and the Dormand–Prince pair.

In this study, the straight line between the starting and endpoints (Figure 4a, where P_0 is the starting point and P_1 is the endpoint) was assumed to be the distance covered by the particle. The angle between the distance covered by the particle and the horizontal axis was assumed to be the displacement angle α.

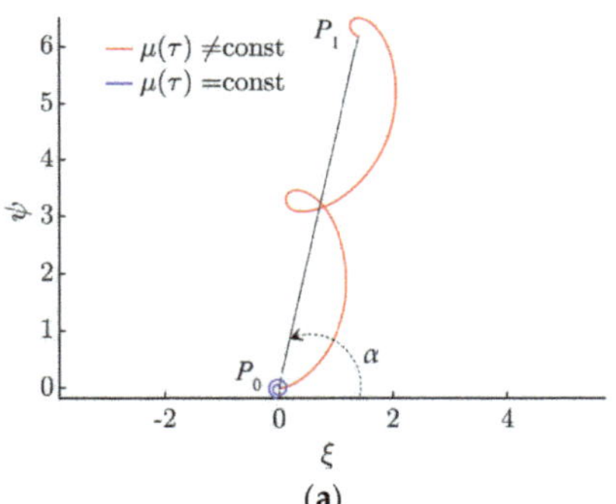

(a)

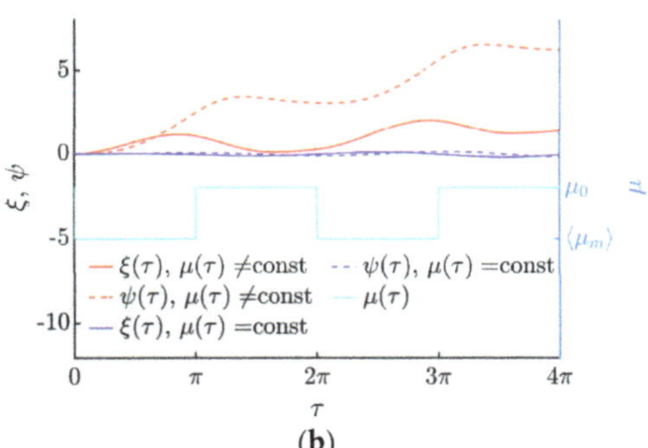

(b)

Figure 4. Nondimensional displacement of the particle when $\gamma = 4.9$, $\mu_0 = 0.2$, $\langle\mu_m\rangle/\mu_0 = 0.25$, $\lambda = \pi$, $\phi = 0$: (**a**) motion trajectories of the particle and the angle of displacement α. (**b**) horizontal ξ and vertical ψ components of the nondimensional displacement vs. the nondimensional time and the variation of the effective dry friction coefficient.

The modeling has revealed that the motion of a particle can be achieved through the asymmetry created by dynamic friction control (Figure 4a). Figure 4a shows the trajectories of the particles after two cycles of the circular motion of the platform. The blue trajectory represents the motion of the particle on the platform when the dry friction between the part and the platform is not being controlled. In this case, the particle is moving around the position of equilibrium (Figure 4a when μ = const), as shown in Figure 4b, where the horizontal ξ and vertical ψ components of the nondimensional displacement oscillate around a constant value (Figure 4b when μ = const). The red trajectory represents the motion of the particle when the dry friction coefficient between the part and the platform is being controlled, i.e., it is being periodically modified for some fraction of the period. In this case, the system's symmetry is eliminated, and the particle is constantly moving in a certain direction (Figure 4a when $\mu \neq$ const). Figure 4b shows how the variation of the effective friction coefficient influences the horizontal ξ and vertical ψ components of the nondimensional displacement. An increase in the magnitudes of ξ and ψ is starting to take place periodically in the intervals of τ where the friction coefficient is dynamically modified to be equal to $\langle\mu_m\rangle$.

The shape of the trajectory of the particle depends mainly on λ (Figure 5a). In a symmetric system (when $\lambda = 0$), the particle can travel some distance from the starting point in an undulating or spiral trajectory, but then it starts to circle around the point of equilibrium [51,52]. As a result of the asymmetry created by dynamic dry friction control, the particle can be moved in a preferred direction. Figure 5a shows that the shape of the trajectory becomes less circular and undulating, and the particle travels a greater distance from the starting point at higher values of λ since an increase in λ results in a greater asymmetry of the system. The displacement angle depends mainly on the phase shift ϕ (Figure 5b).

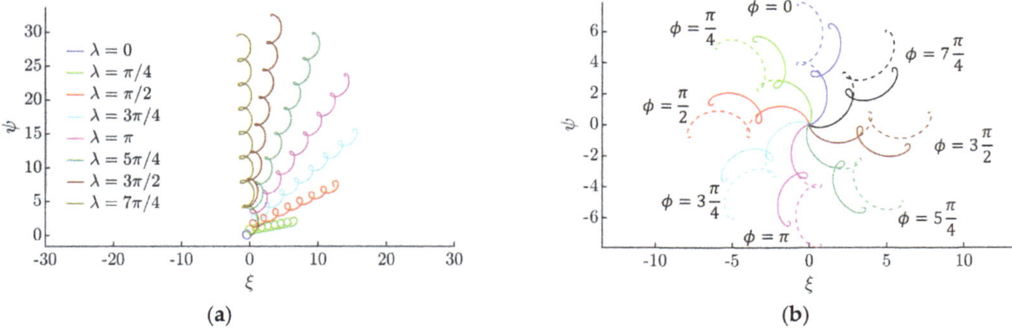

Figure 5. Trajectories of the particle: (**a**) under different values of λ after 8 cycles of the circular motion of the platform when $\phi = 0$, $\mu_0 = 0.2$, $\langle\mu_m\rangle/\mu_0 = 0.25$, $\gamma = 4$; (**b**) under different values of the phase shift ϕ after two cycles when $\mu_0 = 0.2$, $\langle\mu_m\rangle/\mu_0 = 0.25$, $\gamma = 4.9$, $\lambda = \pi$ (solid line), $\lambda = 13\pi/9$ (dashed line).

Figure 6 shows how the vector of the angular velocity of the particle varies during a stable period of the circular motion of the platform. In a symmetric system (when $\lambda = 0$), the magnitude of the vector of the angular velocity is constant (the polar plots are represented by the dashed circles in magenta). This indicates that the particle is moving around the position of equilibrium; therefore, it is not gaining displacement. The asymmetry created by dynamic dry friction control results in the asymmetric shapes of the polar plots that indicate that the particle is gaining displacement. An increase in λ results in an increase in the magnitude of the angular velocity vector ρ. The phase shift ϕ defines the phase of the vector of angular velocity θ at which the maximum value of ρ is reached.

The variation of the angular velocity vector during a stable period of the circular motion is shown in Figure 7a. The magnitude of the angular velocity vector ρ oscillates over some positive value under the influence of dynamic dry friction control. The amplitude of these oscillations depends on λ. These oscillations indicate that the particle is gaining displacement. An increase in λ results in a higher value of this amplitude. Figure 7b shows how the phase of the vector of angular velocity varies over nondimensional time.

The influence of the intrinsic system parameters (λ, ϕ, μ_0, γ, $\langle\mu_m\rangle/\mu_0$) on the nondimensional average velocity of the particle $\langle\vartheta\rangle$ was determined. The nondimensional average velocity $\langle\vartheta\rangle$ was found by dividing the nondimensional displacement gained by the particle after 17 cycles by the nondimensional time of travel.

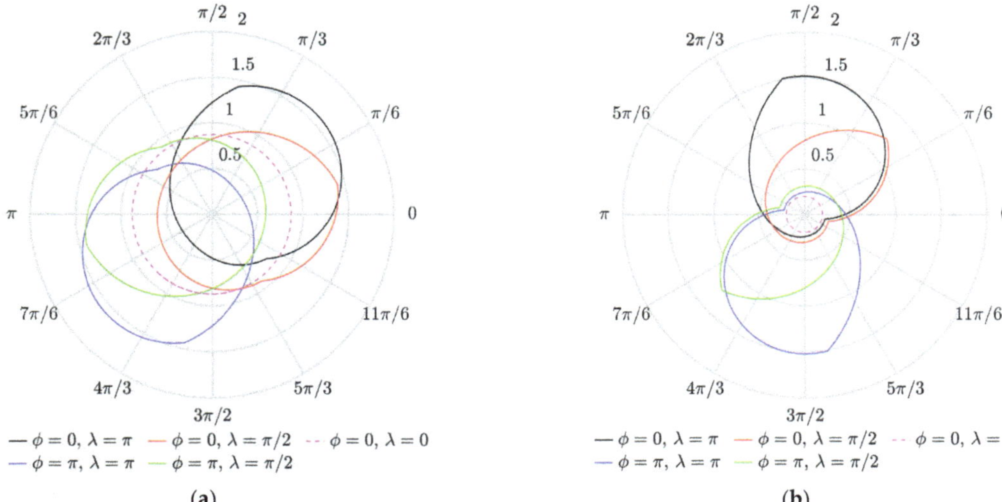

Figure 6. Variation of the angular velocity vector during the period of the eighth cycle of circular motion in polar coordinates when $\langle\mu_m\rangle/\mu_0 = 0.25$, $\gamma = 4.9$: (**a**) $\mu_0 = 0.1$; (**b**) $\mu_0 = 0.2$.

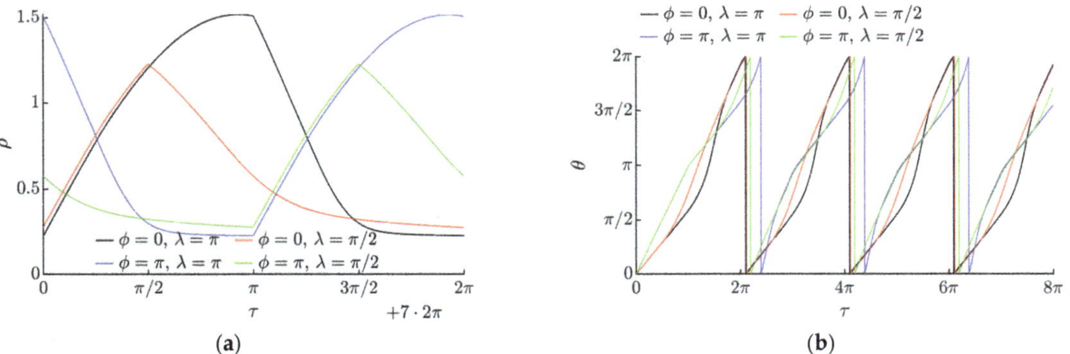

Figure 7. Angular velocity vector vs. the nondimensional time when $\mu_0 = 0.2$, $\langle\mu_m\rangle/\mu_0 = 0.25$, $\gamma = 4.9$: (**a**) magnitude ρ during the period of the 8th cycle; (**b**) phase θ during the first four cycles.

In a range of λ up to approximately $3\pi/2$, the average nondimensional velocity $\langle\vartheta\rangle$ increases when λ increases due to the increasing asymmetry of frictional conditions (Figure 8a). When the maximum of $\langle\vartheta\rangle$ is reached, a further increase in λ results in a decrease in the average nondimensional velocity $\langle\vartheta\rangle$ due to the decreasing asymmetry of fictional conditions in this range of such high values of λ.

Figure 8b shows the influence of γ on $\langle\vartheta\rangle$. An increase in γ results in a slight decrease in the average nondimensional velocity $\langle\vartheta\rangle$.

The ratio $\langle\mu_m\rangle/\mu_0$ represents how much the dry friction coefficient is dynamically modified with respect to the nominal dry friction coefficient. The modeling showed that this ratio has a significant influence on the average nondimensional velocity $\langle\vartheta\rangle$ (Figure 8c). As $\langle\mu_m\rangle/\mu_0$ approaches 1, the average nondimensional velocity $\langle\vartheta\rangle$ decreases until it becomes equal to 0 at $\langle\mu_m\rangle/\mu_0 = 1$. This is due to the fact that $\langle\mu_m\rangle/\mu_0$ values further from 1 result in a greater asymmetry of the system, and the system is in a symmetric state when $\langle\mu_m\rangle/\mu_0 = 1$. When $\langle\mu_m\rangle/\mu_0$ is less than 1, the friction is being periodically reduced, and when it is higher than 1, it is being periodically increased. These results show that the dry friction force can be periodically increased or decreased for some fraction of the period of the circular motion in order to achieve the asymmetry of the system by dynamic dry friction control. The red curve in Figure 8c shows that the control approach implemented by periodically decreasing the dry friction force ($\langle\mu_m\rangle/\mu_0 > 1$) is more efficient when the nominal coefficient μ_0 and λ are very low.

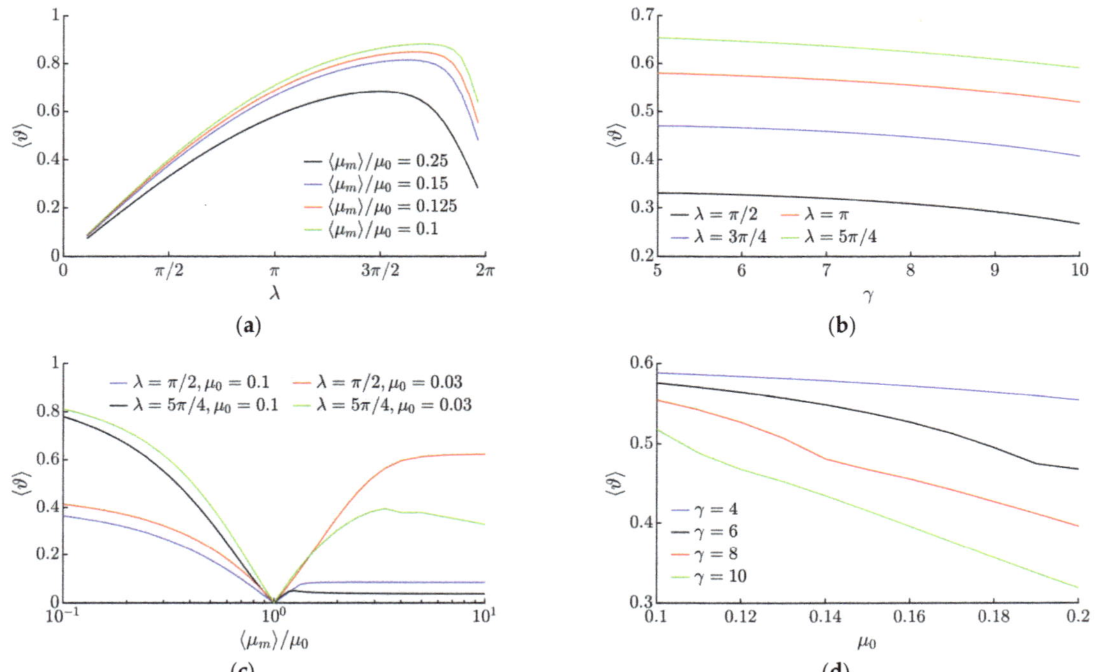

Figure 8. Average nondimensional velocity depending on the following: (**a**) λ when $\mu_0 = 0.1$, $\gamma = 5$, $\phi = 0$; (**b**) γ when $\mu_0 = 0.1$, $\phi = 0$, $\langle\mu_m\rangle/\mu_0 = 0.25$; (**c**) $\langle\mu_m\rangle/\mu_0$ when $\gamma = 9$, $\phi = \pi/2$; (**d**) μ_0 when $\langle\mu_m\rangle/\mu_0 = 0.25$, $\phi = 0$, $\lambda = \pi/2$.

Figure 8d shows the influence of the nominal dry friction coefficient μ_0 on the average nondimensional velocity at a value of $\langle\mu_m\rangle/\mu_0$ that is lower than 1. In this case, when μ_0 increases, the average nondimensional velocity tends to decrease. The decrease is more pronounced at higher values of γ (Figure 8d).

The modeling has shown that the average nondimensional velocity $\langle\vartheta\rangle$ does not depend on the phase shift ϕ between the function governing the circular motion and the dry friction control function.

Figure 9a shows a three-dimensional diagram of the average nondimensional velocity $\langle\vartheta\rangle$ as a function of λ and γ. Under the analyzed conditions, the average nondimensional velocity reaches its maximum value when λ is near $3\pi/2$.

The combined influence of the ratio $\langle\mu_m\rangle/\mu_0$ and the nominal dry friction coefficient μ_0 on the average nondimensional velocity is shown in Figure 9b. In the case where the friction coefficient is being periodically decreased ($\langle\mu_m\rangle/\mu_0 < 1$), the average nondimensional velocity mainly depends on the ratio $\langle\mu_m\rangle/\mu_0$, and it is less sensitive to the nominal dry friction coefficient μ_0. In the case where the friction coefficient is being periodically increased ($\langle\mu_m\rangle/\mu_0 > 1$), both μ_0 and $\langle\mu_m\rangle/\mu_0$ have a noticeable influence on the average nondimensional velocity. Higher values are observed at lower values of the nominal dry friction coefficient μ_0.

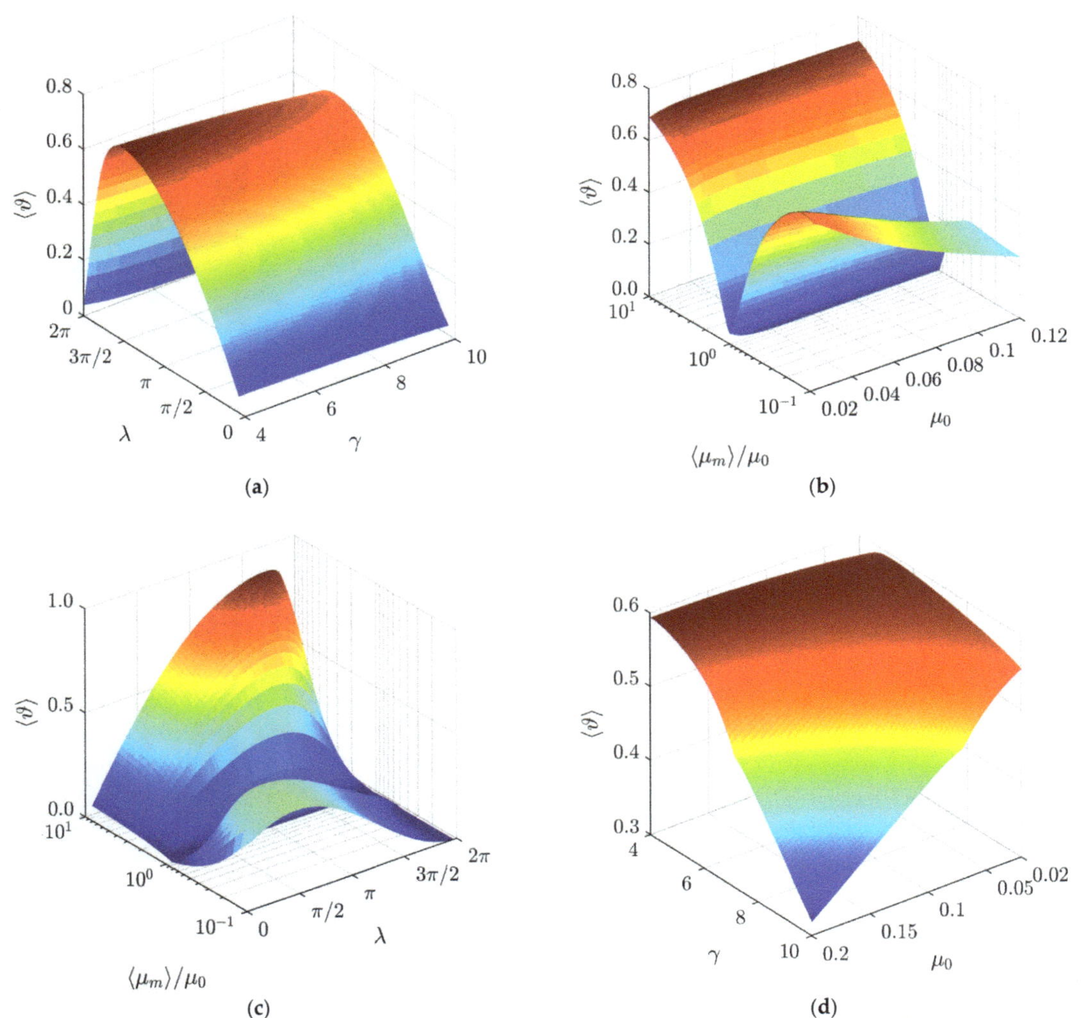

Figure 9. Average nondimensional velocity depending on the following: (**a**) λ and γ when $\mu_0 = 0.1$, $\langle\mu_m\rangle/\mu_0 = 0.25$, $\phi = 0$; (**b**) $\langle\mu_m\rangle/\mu_0$ and μ_0 when $\gamma = 5$, $\phi = 0$, $\lambda = \pi$; (**c**) $\langle\mu_m\rangle/\mu_0$ and λ when $\mu_0 = 0.1$, $\gamma = 5$, $\phi = 0$; (**d**) γ and μ_0 when $\langle\mu_m\rangle/\mu_0 = 4$, $\phi = 0$, $\lambda = \pi$.

Figure 9c shows a three-dimensional diagram of the average nondimensional velocity $\langle\vartheta\rangle$ as a function of $\langle\mu_m\rangle/\mu_0$ and λ. In the case where the friction coefficient is being periodically decreased ($\langle\mu_m\rangle/\mu_0 < 1$), the maximum values of $\langle\vartheta\rangle$ are obtained in a range of λ from $3\pi/2$ to $7\pi/4$ approximately. The value of λ, at which the maximum of $\langle\vartheta\rangle$ is reached, depends on $\langle\mu_m\rangle/\mu_0$. Under values of $\langle\mu_m\rangle/\mu_0$ that are further from 1, the maximum of $\langle\vartheta\rangle$ is reached at higher values of λ. In the case where the friction coefficient is being periodically increased ($\langle\mu_m\rangle/\mu_0 > 1$), the asymmetry of frictional conditions is shifted with respect to the function governing the circular motion differently compared to the case where $\langle\mu_m\rangle/\mu_0 < 1$. Therefore, the maximum values of $\langle\vartheta\rangle$ are obtained in a range of λ from approximately $\pi/2$ to $3\pi/4$. In this case, under values of $\langle\mu_m\rangle/\mu_0$ that are further from 1, the maximum of $\langle\vartheta\rangle$ is reached at lower values of λ.

Figure 9d shows a three-dimensional diagram of the average nondimensional velocity $\langle\vartheta\rangle$ as a function of γ and μ_0. A decrease in $\langle\vartheta\rangle$ is observed with an increase in both γ and μ_0.

The modeling has shown that the displacement angle α linearly depends on the phase shift ϕ (Figure 10a). This implies that it is possible to direct the particle omnidirectionally at any angle by changing ϕ.

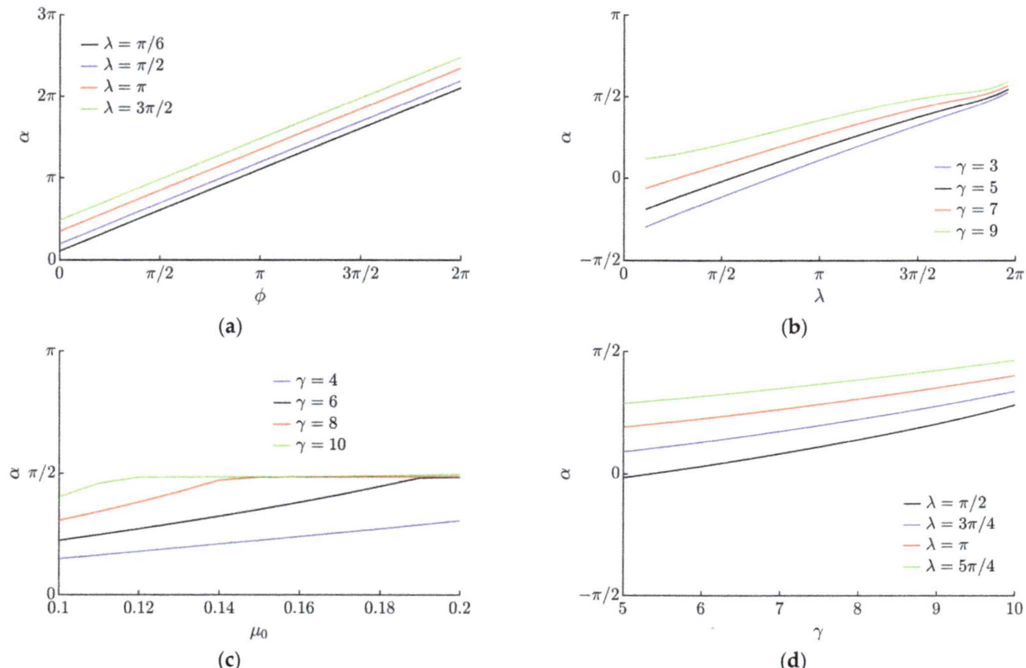

Figure 10. Displacement angle α depending on: (**a**) ϕ when $\mu_0 = 0.1$, $\gamma = 8.8$, $\langle\mu_m\rangle/\mu_0 = 0.25$; (**b**) λ when $\mu_0 = 0.1$, $\phi = 0$, $\langle\mu_m\rangle/\mu_0 = 0.25$; (**c**) μ_0 when $\langle\mu_m\rangle/\mu_0 = 0.25$, $\phi = 0$, $\lambda = \pi/2$; (**d**) γ when $\mu_0 = 0.1$, $\phi = 0$, $\langle\mu_m\rangle/\mu_0 = 0.25$.

Figure 10b shows the influence of λ on the displacement angle α. An increase in λ results in an increase in α to some extent.

An increase in the nominal dry friction coefficient μ_0 has a small effect on the displacement angle α (Figure 10c). As the nominal dry friction coefficient μ_0 increases, the displacement angle slightly increases when γ values are low.

The influence of the parameter γ on the displacement angle *a* is presented in Figure 10d. The modeling results demonstrate that when γ increases, the displacement angle slightly increases as well.

Figure 11a shows a three-dimensional diagram of the displacement angle α as a function of λ and γ. An increase in both λ and γ results in an increase in α.

Figure 11b shows a three-dimensional diagram of the displacement angle α as a function of the ratio $\langle\mu_m\rangle/\mu_0$ and the nominal dry friction coefficient μ_0. In the case where the friction coefficient is being periodically decreased ($\langle\mu_m\rangle/\mu_0 < 1$), the ratio $\langle\mu_m\rangle/\mu_0$ does not have a significant influence on the displacement angle α. However, an increase in μ_0 results in a slight increase in α in this case. In the case where the friction coefficient is being periodically increased ($\langle\mu_m\rangle/\mu_0 > 1$), both $\langle\mu_m\rangle/\mu_0$ and μ_0 have some influence on α. When $\langle\mu_m\rangle/\mu_0$ increases, the displacement angle α slightly increases until it reaches a critical value and then starts to decrease slightly again. The value of $\langle\mu_m\rangle/\mu_0$, at which this critical value is reached, depends on μ_0. Under higher values of μ_0, the critical value is reached at lower values of $\langle\mu_m\rangle/\mu_0$. The displacement angle α in the case where $\langle\mu_m\rangle/\mu_0 > 1$ is shifted compared to the case when $\langle\mu_m\rangle/\mu_0 < 1$. This can be explained by the fact that the asymmetry of frictional conditions is shifted differently with respect to the function governing the circular motion in these two cases.

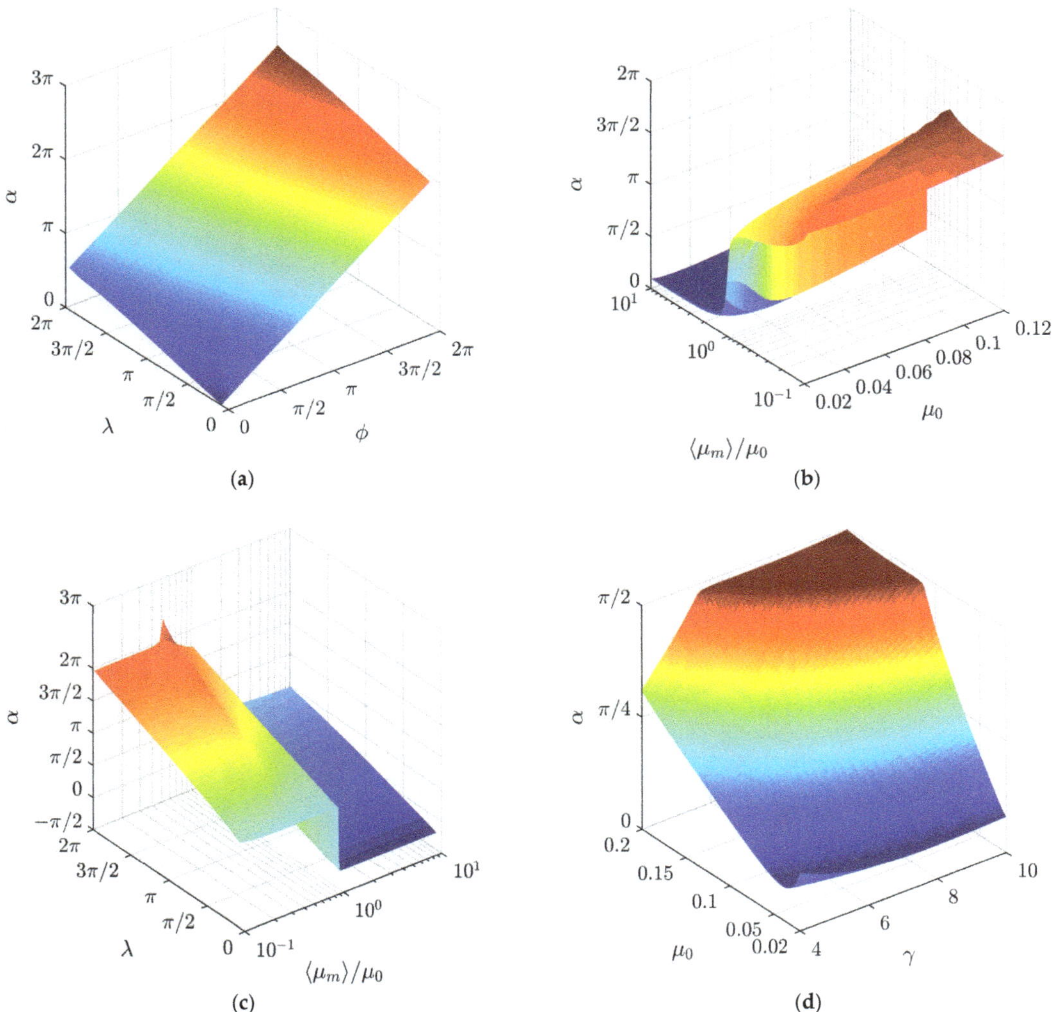

Figure 11. Displacement angle α depending on: (**a**) λ and ϕ when $\mu_0 = 0.1$, $\gamma = 8$, $\langle\mu_m\rangle/\mu_0 = 0.25$; (**b**) $\langle\mu_m\rangle/\mu_0$ and μ_0 when $\gamma = 5$, $\lambda = \pi$, $\phi = 0$; (**c**) λ and $\langle\mu_m\rangle/\mu_0$ when $\mu_0 = 0.1$, $\gamma = 5$, $\phi = 0$; (**d**) μ_0 and γ when $\langle\mu_m\rangle/\mu_0 = 0.25$, $\phi = 0$, $\lambda = \pi$.

A similar shift is observed in a three-dimensional diagram of the displacement angle as a function of λ and $\langle\mu_m\rangle/\mu_0$ (Figure 11c). Figure 11c shows a similar influence of $\langle\mu_m\rangle/\mu_0$ on α, as it was discussed previously. It also shows that λ has a more significant influence on α. The displacement angle α increases with an increase in λ.

Figure 11d shows a three-dimensional diagram of the displacement angle α as a function of μ_0 and γ. At higher values of both μ_0 and γ, α does not change much under the influence of these parameters. A further decrease in μ_0 results in a decrease in α as well, with an exception at very low μ_0 and γ values.

3.2. Experimental Results

The proposed method of omnidirectional manipulation was experimentally tested with multilayer ceramic capacitors (MLCC) as such particles are widely used in microelectronics [42–44,53,54]. The developed setup was tested using 0402- and 0603-type MLCCs (Mouser Electronics, Mansfield, MA, USA).

The experiments verified that the asymmetry created by dynamic dry friction control is technically feasible and can be applied for the omnidirectional manipulation of microparticles. In the experiments, the particles were moving on the manipulation surface, and the characteristics of the motion depended on the friction control parameters ϕ and λ.

In the experiments, the average velocity of the particle $\langle v \rangle$ was calculated by dividing the distance between the starting and end points of the displacement by the time of travel.

Figure 12 shows the experimental results for the average velocity of a 0603-type MLCC. An increase in λ resulted in an increase in the average velocity $\langle v \rangle$ (Figure 12a). This trend was consistent with the trend obtained by the modeling (Figure 8a). An increase in the average velocity $\langle v \rangle$ was observed with an increase in the radius of the circular motion of the platform (Figure 12b).

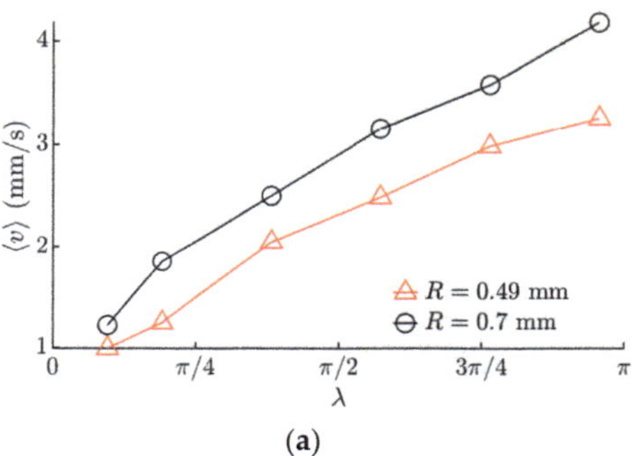

(**a**)

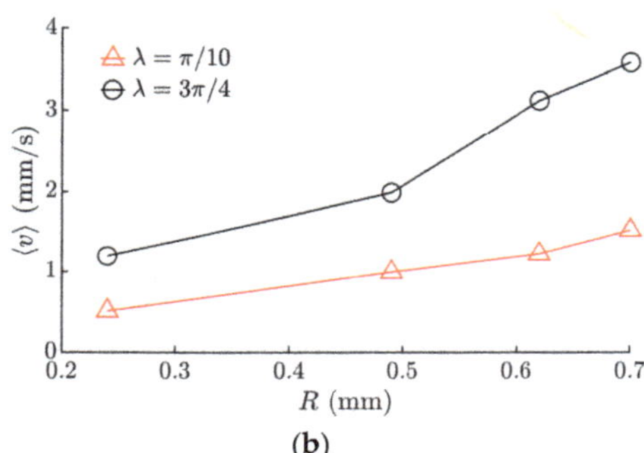

(**b**)

Figure 12. Experimental results of the average velocity depending on the following: (**a**) λ when $\omega = 62.83$ rad/s, $\phi = 0$; (**b**) the radius of the circular motion of the platform R when $\omega = 62.83$ rad/s, $\phi = 0$.

The experiments showed that the displacement angle mainly depends on the phase shift ϕ. An increase in ϕ resulted in a proportional increase in the displacement angle α (Figure 13a). This finding was consistent with the relationship obtained by the modeling (Figure 10a). A slight increase in the displacement angle was observed with an increase in λ (Figure 13b). This trend was consistent with the trend obtained by the modeling (Figure 8b).

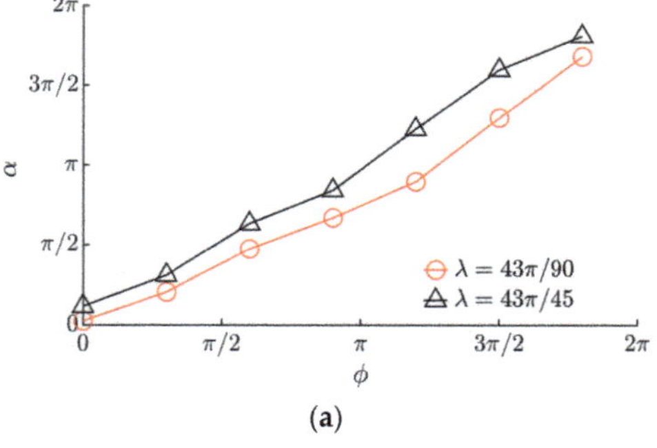

(**a**)

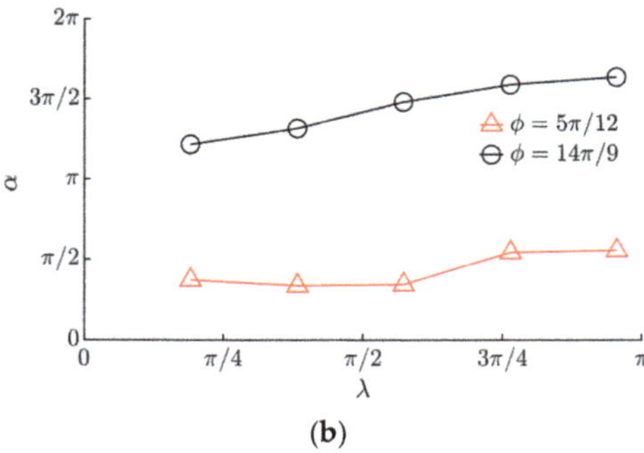

(**b**)

Figure 13. Experimental results of α depending on the following: (**a**) ϕ when $R = 0.49$ mm, $\omega = 62.83$ rad/s; (**b**) λ when $R = 0.49$ mm, $\omega = 62.83$ rad/s.

Figure 14a shows an image captured during the manipulation of a single 0603-type MLCC. In this image, the trajectory of motion is shown by the red curve. The captured trajectory is very similar to the trajectories obtained by the modeling (Figure 5b).

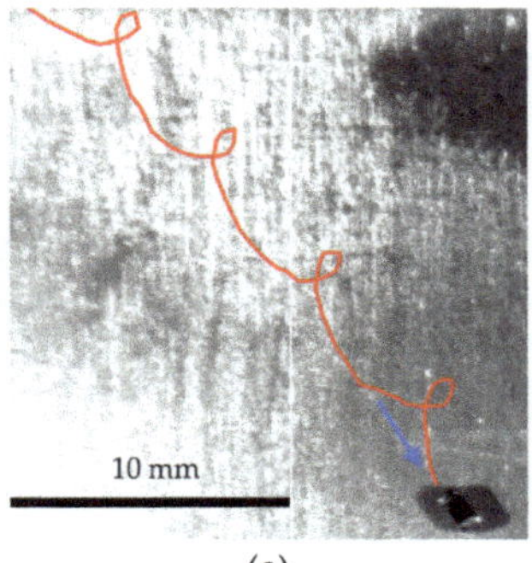

(**a**)

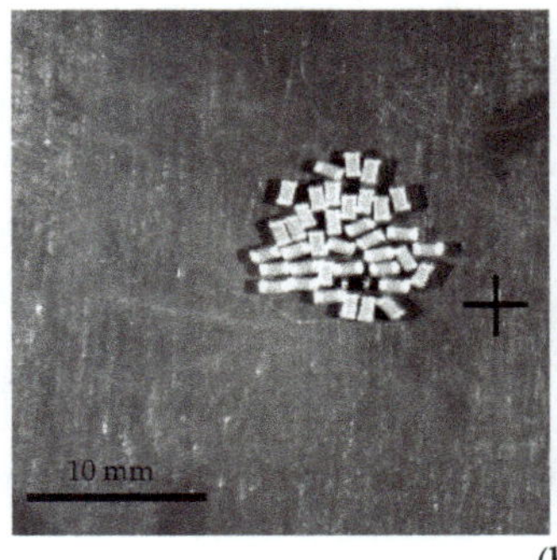

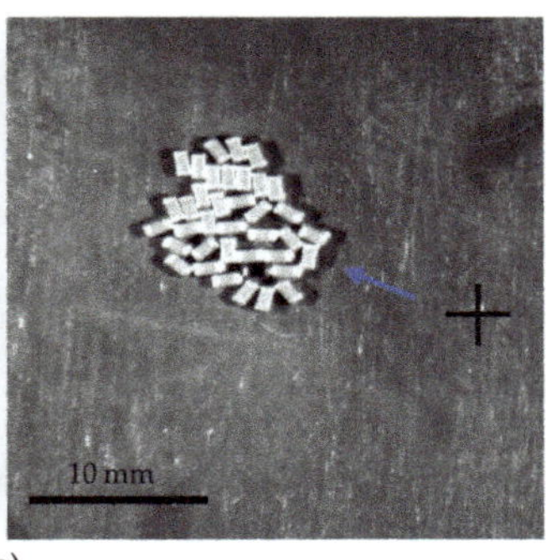

(**b**)

Figure 14. Images captured during the experiments: (**a**) captured trajectory of a single 0603-type MLCC when $R = 0.49$ mm, $\omega = 62.83$ rad/s, $\phi = 9\pi/5$, $\lambda = 43\pi/90$; (**b**) two frames separated by a time interval of 0.792 s that were captured during the manipulation of multiple 0603-type MLCC when $R = 0.49$ mm, $\omega = 62.83$ rad/s, $\phi = 9\pi/10$, $\lambda = 43\pi/45$.

The experiments have shown that the proposed method can be applied for the manipulation of an individual particle or multiple particles. Figure 14b shows two frames separated by a time interval of 0.792 s that were captured during the manipulation of multiple 0603-type MLCC. The motion of a single 0603-type MLCC on the manipulation plate is demonstrated in Supplementary Material Video S1. The motion of multiple MLCCs on the manipulation plate is demonstrated in Supplementary Material Video S2.

4. Conclusions

A method for omnidirectional manipulation of microparticles on a platform subjected to circular motion is proposed, which is achieved by applying dynamic dry friction control.

In order to determine the intrinsic parameters of the system that define the direction and velocity of particles, a nondimensional mathematical model of the proposed method was developed. Modeling of the manipulation process applying dynamic dry friction control was carried out. The modeling has revealed that the motion of a particle can be achieved and controlled through the asymmetry created by dynamic friction control. An asymmetry necessary to achieve the particle's motion can be created when the effective friction coefficient is either periodically increased or decreased for some fraction of the period of the circular motion. The modeling has shown that it is possible to direct the particle omnidirectionally at any angle by changing the phase shift ϕ between the function governing the circular motion and the dry friction control function. The average nondimensional velocity does not depend on the phase shift. The shape of the trajectory of the particle depends mainly on λ. The shape of the trajectory becomes less circular and undulating at higher values of λ since an increase in λ results in a greater asymmetry of the system. Due to this, the velocity of the particle can be controlled through this parameter λ. Furthermore, an increase in λ results in an increase in the displacement angle as well. The ratio between the dynamically modified dry friction coefficient and the nominal dry friction coefficient $\langle\mu_m\rangle/\mu_0$ determines the size of the asymmetry. Therefore, values of $\langle\mu_m\rangle/\mu_0$ that are further from one result in a higher velocity. The modeling has also shown that an increase in the parameter γ results in a slight decrease in both the average velocity and the displacement angle. When the nominal dry friction coefficient increases, the average velocity tends to decrease.

An experimental investigation of the omnidirectional manipulation of microparticles on a horizontal platform subjected to the circular motion was carried out by implementing the method of dynamic dry friction control. The experiments verified that the asymmetry created by dynamic dry friction control is technically feasible and can be applied for the omnidirectional manipulation of microparticles. In the experiments, the particles were moving on the manipulation plate, and the characteristics of motion depended on the friction control parameters. The experimental results were consistent with the modeling results and showed that it is possible to direct the particle omnidirectionally at any angle by changing the phase shift ϕ. The experiments also showed that the velocity of the particle can be controlled by changing λ. The proposed method can be applied to the manipulation of an individual particle or multiple particles. The experiments qualitatively confirmed the influence of the control parameters on the motion characteristics predicted in the modeling, since the trends of the results observed in the experiments were in agreement with the trends of the modeling results.

Potential practical applications of the proposed method can be lab-on-a-chip applications, microassembly lines, feeders, and distributors of microparticles, handling and transportation systems of bulk materials, or other systems for manipulation of delicate components. Due to its relatively simple technical implementation, the proposed method can replace other more expensive and complex devices such as microgrippers.

The study enriches the classical theories of particle motion on oscillating rigid plates, and it is relevant for the industries that implement various tasks related to assembling, handling, feeding, transporting, or manipulating microparticles.

Supplementary Materials: The following supporting information can be downloaded at: https://www.mdpi.com/article/10.3390/mi13050711/s1, Video S1: motion of a single particle; Video S2: motion of multiple particles on the platform subjected to circular motion applying dynamic dry friction control.

Author Contributions: Conceptualization, S.K. and A.F.; methodology, S.K., K.L., E.U. and A.F.; software, S.K. and K.L.; validation, S.K., K.L., E.U., R.E.B. and A.F.; formal analysis, S.K., K.L., E.U. and A.F.; investigation, S.K., K.L., E.U., R.E.B. and A.F.; resources, S.K.; writing—original draft preparation, S.K.; writing—review and editing, S.K., K.L. and A.F., visualization, S.K. and K.L.; supervision, S.K.; project administration, S.K.; funding acquisition, S.K. All authors have read and agreed to the published version of the manuscript.

Funding: This research was funded by a grant (No. S-MIP-19-64) from the Research Council of Lithuania.

Conflicts of Interest: The authors declare no conflict of interest.

References

1. Warnat, S.; King, H.; Wasay, A.; Sameoto, D.; Hubbard, T. Direct integration of MEMS, dielectric pumping and cell manipulation with reversibly bonded gecko adhesive microfluidics. *J. Micromech. Microeng.* **2016**, *26*, 097001. [CrossRef]
2. Wu, Z.; Xu, Q. Survey on recent designs of compliant micro-/nano-positioning stages. *Actuators* **2018**, *7*, 5. [CrossRef]
3. Janusas, T.; Urbaite, S.; Palevicius, A.; Nasiri, S.; Janusas, G. Biologically Compatible Lead-Free Piezoelectric Composite for Acoustophoresis Based Particle Manipulation Techniques. *Sensors* **2021**, *21*, 483. [CrossRef] [PubMed]
4. Wang, G.; Ding, Y.; Long, H.; Guan, Y.; Lu, X.; Wang, Y.; Yang, L. Simulation of Optical Nano-Manipulation with Metallic Single and Dual Probe Irradiated by Polarized Near-Field Laser. *Appl. Sci.* **2022**, *12*, 815. [CrossRef]
5. Yang, Y.; Ma, T.; Zhang, Q.; Huang, J.; Hu, Q.; Li, Y.; Wang, C.; Zheng, H. 3D Acoustic Manipulation of Living Cells and Organisms Based on 2D Array. *IEEE Trans. Biomed. Eng.* **2022**. [CrossRef]
6. Chircov, C.; Grumezescu, A.M. Microelectromechanical Systems (MEMS) for Biomedical Applications. *Micromachines* **2022**, *13*, 164. [CrossRef]
7. Rashid, N.F.A.; Deivasigamani, R.; Wee, M.F.; Hamzah, A.A.; Buyong, M.R. Integration of a Dielectrophoretic Tapered Aluminum Microelectrode Array with a Flow Focusing Technique. *Sensors* **2021**, *21*, 4957. [CrossRef]
8. Ahmad, B.; Chambon, H.; Tissier, P.; Bolopion, A. Laser Actuated Microgripper Using Optimized Chevron-Shaped Actuator. *Micromachines* **2021**, *12*, 1487. [CrossRef]
9. Fontana, G.; Ruggeri, S.; Legnani, G.; Fassi, I. Performance assessment of a modular device for micro-sphere singularization. *Precis. Eng.* **2021**, *71*, 29–35. [CrossRef]

10. Potekhina, A.; Voicu, R.; Muller, R.; Al-Zandi, M.H.; Wang, C. Design and characterization of a polymer electrothermal microgripper with a polynomial flexure for efficient operation and studies of moisture effect on negative deflection. *Microsyst. Technol.* **2021**, *27*, 2723–2731. [CrossRef]
11. Zore, A.; Čerin, R.; Munih, M. Impact of a Robot Manipulation on the Dimensional Measurements in an SPC-Based Robot Cell. *Appl. Sci.* **2021**, *11*, 6397. [CrossRef]
12. Tiwari, B.; Billot, M.; Clévy, C.; Agnus, J.; Piat, E.; Lutz, P. A Two-Axis Piezoresistive Force Sensing Tool for Microgripping. *Sensors* **2021**, *21*, 6059. [CrossRef] [PubMed]
13. Ragulskis, K.; Spruogis, B.; Bogdevičius, M.; Matuliauskas, A.; Mištinas, V.; Ragulskis, L. Mechanical systems of precise robots with vibrodrives, in which the direction of the exciting force coincides with the line of relative motion of the system. *Mechanika* **2021**, *27*, 408–414. [CrossRef]
14. Fan, C.; Shirafuji, S.; Ota, J. Modal Planning for Cooperative Non-Prehensile Manipulation by Mobile Robots. *Appl. Sci.* **2019**, *9*, 462. [CrossRef]
15. Feemster, M.; Piepmeier, J.A.; Biggs, H.; Yee, S.; ElBidweihy, H.; Firebaugh, S.L. Autonomous microrobotic manipulation using visual servo control. *Micromachines* **2020**, *11*, 132. [CrossRef]
16. Rizwan, M.; Shiakolas, P.S. On the optimum synthesis of a microconveyor platform for micropart translocation using differential evolution. *Inverse Probl. Sci. Eng.* **2013**, *21*, 1335–1351. [CrossRef]
17. Ablay, G. Model-Free Controller Designs for a Magnetic Micromanipulator. *J. Dyn. Syst. Meas. Control* **2021**, *143*, 031003. [CrossRef]
18. Li, Z.; Liu, P.; Yan, P. Design and Analysis of a Novel Flexure-Based Dynamically Tunable Nanopositioner. *Micromachines* **2021**, *12*, 212. [CrossRef]
19. Ferrara-Bello, A.; Vargas-Chable, P.; Vera-Dimas, G.; Vargas-Bernal, R.; Tecpoyotl-Torres, M. XYZ Micropositioning System Based on Compliance Mechanisms Fabricated by Additive Manufacturing. *Actuators* **2021**, *10*, 68. [CrossRef]
20. Chen, Z.; Liu, X.; Kojima, M.; Huang, Q.; Arai, T. Advances in Micromanipulation Actuated by Vibration-Induced Acoustic Waves and Streaming Flow. *Appl. Sci.* **2020**, *10*, 1260. [CrossRef]
21. Röthlisberger, M.; Schuck, M.; Kulmer, L.; Kolar, J.W. Contactless Picking of Objects Using an Acoustic Gripper. *Actuators* **2021**, *10*, 70. [CrossRef]
22. Wijaya, H.; Latifi, K.; Zhou, Q. Two-dimensional manipulation in mid-air using a single transducer acoustic levitator. *Micromachines* **2019**, *10*, 257. [CrossRef] [PubMed]
23. Wang, Y.; Wu, L.; Wang, Y. Study on Particle Manipulation in a Metal Internal Channel under Acoustic Levitation. *Micromachines* **2022**, *13*, 18. [CrossRef] [PubMed]
24. Guex, A.G.; Di Marzio, N.; Eglin, D.; Alini, M.; Serra, T. The waves that make the pattern: A review on acoustic manipulation in biomedical research. *Mater. Today Bio* **2021**, *10*, 100110. [CrossRef]
25. Akkoyun, F.; Gucluer, S.; Ozcelik, A. Potential of the acoustic micromanipulation technologies for biomedical research. *Biomicrofluidics* **2021**, *15*, 061301. [CrossRef]
26. Li, J.; Shen, C.; Tony, J.H.; Cummer, S.A. Acoustic tweezer with complex boundary-free trapping and transport channel controlled by shadow waveguides. *Sci. Adv.* **2021**, *7*, eabi5502. [CrossRef]
27. Devendran, C.; Collins, D.J.; Neild, A. The role of channel height and actuation method on particle manipulation in surface acoustic wave (SAW)-driven microfluidic devices. *Microfluid. Nanofluidics* **2022**, *26*, 9. [CrossRef]
28. Reznik, D.; Canny, J.; Goldberg, K. Analysis of part motion on a longitudinally vibrating plate. In Proceedings of the 1997 IEEE/RSJ International Conference on Intelligent Robot and Systems, Grenoble, France, 11 September 1997; IEEE: Piscataway, NJ, USA, 1997; pp. 421–427. [CrossRef]
29. Mayyas, M. Parallel Manipulation Based on Stick-Slip Motion of Vibrating Platform. *Robotics* **2020**, *9*, 86. [CrossRef]
30. Mayyas, M. Modeling and analysis of vibratory feeder system based on robust stick–slip motion. *J. Vib. Control* **2021**, 10775463211009633. [CrossRef]
31. Klemiato, M.; Czubak, P. Control of the transport direction and velocity of the two-way reversible vibratory conveyor. *Arch. Appl. Mech.* **2019**, *89*, 1359–1373. [CrossRef]
32. Buzzoni, M.; Battarra, M.; Mucchi, E.; Dalpiaz, G. Motion analysis of a linear vibratory feeder: Dynamic modeling and experimental verification. *Mech. Mach. Theory* **2017**, *114*, 98–110. [CrossRef]
33. Gursky, V.; Krot, P.; Korendiy, V.; Zimroz, R. Dynamic Analysis of an Enhanced Multi-Frequency Inertial Exciter for Industrial Vibrating Machines. *Machines* **2022**, *10*, 130. [CrossRef]
34. Frei, P.U.; Wiesendanger, M.; Büchi, R.; Ruf, L. Simultaneous planar transport of multiple objects on individual trajectories using friction forces. In *Distributed Manipulation*; Böhringer, K.F., Choset, H., Eds.; Springer: Boston, MA, USA, 2000; pp. 49–64. [CrossRef]
35. Ihor, V. Vibratory conveying by harmonic longitudinal and polyharmonic normal vibrations of inclined conveying track. *J. Vib. Acoust.* **2022**, *144*, 011004. [CrossRef]
36. Kharkongor, B.; Pohlong, S.S.; Mahato, M.C. Net transport in a periodically driven potential-free system. *Physica A Stat. Mech. Appl.* **2021**, *562*, 125341. [CrossRef]
37. Palencia, J.L.D. Travelling Waves Approach in a Parabolic Coupled System for Modelling the Behaviour of Substances in a Fuel Tank. *Appl. Sci.* **2021**, *11*, 5846. [CrossRef]

38. Cao, H.X.; Jung, D.; Lee, H.; Go, G.; Nan, M.; Choi, E.; Kim, C.; Park, J.; Kang, B. Micromotor Manipulation Using Ultrasonic Active Traveling Waves. *Micromachines* **2021**, *12*, 192. [CrossRef]
39. Zouaghi, A.; Zouzou, N.; Dascalescu, L. Assessment of forces acting on fine particles on a traveling-wave electric field conveyor: Application to powder manipulation. *Powder Technol.* **2019**, *343*, 375–382. [CrossRef]
40. Kumar, A.; DasGupta, A. Generation of circumferential harmonic travelling waves on thin circular plates. *J. Sound Vib.* **2020**, *478*, 115343. [CrossRef]
41. Viswarupachari, C.; DasGupta, A.; Pratik Khastgir, S. Vibration induced directed transport of particles. *J. Vib. Acoust.* **2012**, *134*, 051005. [CrossRef]
42. Mitani, A.; Matsuo, Y. Feeding of microparts along an asymmetric surface using horizontal and symmetric vibrations—Development of asymmetric surfaces using anisotropic etching process of single-crystal silicon. In Proceedings of the 2011 IEEE International Conference on Robotics and Biomimetics, Karon Beach, Thailand, 7–11 December 2011; IEEE: Piscataway, NJ, USA, 2011; pp. 795–800. [CrossRef]
43. Le, P.H.; Dinh, T.X.; Mitani, A.; Hirai, S. A study on the motion of micro-parts on a saw-tooth surface by the PTV method. *J. Dyn. Control Syst.* **2012**, *6*, 73–80. [CrossRef]
44. Le, P.H.; Mitani, A.; Xuan, T.; Hirai, S. Feed and align microparts on symmetrically vibrating saw-tooth surface. In Proceedings of the 11th World Congress on Intelligent Control and Automation, Shenyang, China, 29 June–4 July 2014; IEEE: Piscataway, NJ, USA, 2014; pp. 5282–5286. [CrossRef]
45. Kilikevičius, S.; Fedaravičius, A.; Daukantienė, V.; Liutkauskienė, K.; Paukštaitis, L. Manipulation of Miniature and Micro-miniature Bodies on a Harmonically Oscillating Platform by Controlling Dry Friction. *Micromachines* **2021**, *12*, 1087. [CrossRef] [PubMed]
46. Kilikevičius, S.; Fedaravičius, A.; Daukantienė, V.; Liutkauskienė, K. Analysis on Conveying of Miniature and Microparts on a Platform Subjected to Sinusoidal Displacement Cycles with Controlled Dry Friction. *Mechanika* **2022**, *28*, 38–44. [CrossRef]
47. Kilikevičius, S.; Fedaravičius, A. Vibrational Transportation on a Platform Subjected to Sinusoidal Displacement Cycles Employing Dry Friction Control. *Sensors* **2021**, *21*, 7280. [CrossRef] [PubMed]
48. Benad, J.; Nakano, K.; Popov, V.L.; Popov, M. Active control of friction by transverse oscillations. *Friction* **2019**, *7*, 74–85. [CrossRef]
49. Kapelke, S.; Seemann, W. On the effect of longitudinal vibrations on dry friction: Modelling aspects and experimental investigations. *Tribol. Lett.* **2018**, *66*, 79. [CrossRef]
50. Menga, N.; Bottiglione, F.; Carbone, G. Dynamically induced friction reduction in micro-structured interfaces. *Sci. Rep.* **2021**, *11*, 1–12. [CrossRef]
51. Bakšys, B.; Ramanauskytė, K. Vibratory positioning and search of automatically assembled parts on a horizontally vibrating plane. *J. Vibroeng.* **2009**, *11*, 56–67.
52. Kilikevičius, S.; Liutkauskienė, K.; Fedaravičius, A. Nonprehensile Manipulation of Parts on a Horizontal Circularly Oscillating Platform with Dynamic Dry Friction Control. *Sensors* **2021**, *21*, 5581. [CrossRef]
53. Yu, D.; Dai, K.; Zhang, J.; Yang, B.; Zhang, H.; Ma, S. Failure Mechanism of Multilayer Ceramic Capacitors under Transient High Impact. *Appl. Sci.* **2020**, *10*, 8435. [CrossRef]
54. Yau, Y.; Hwu, K.; Shieh, J. Applying FPGA Control with ADC-Free Sampling to Multi-Output Forward Converter. *Electronics* **2021**, *10*, 1010. [CrossRef]

micromachines

MDPI

Article

Design and Modeling of a Novel Tripteron-Inspired Triaxial Parallel Compliant Manipulator with Compact Structure

Yanlin Xie, Yangmin Li * and Chifai Cheung

State Key Laboratory of Ultra-precision Machining Technology, Department of Industrial and Systems Engineering, The Hong Kong Polytechnic University, Hung Hom, Kowloon, Hong Kong; yanlin.xie@connect.polyu.hk (Y.X.); benny.cheung@polyu.edu.hk (C.C.)

* Correspondence: yangmin.li@polyu.edu.hk

Abstract: Compliant mechanisms are popular to the applications of micro/nanoscale manipulations. This paper proposes a novel triaxial parallel-kinematic compliant manipulator inspired by the Tripteron mechanism. Compared to most conventional triaxial compliant mechanisms, the proposed manipulator has the merits of structure compactness and being free of assembly error due to its unique configuration and the utilize of 3D printing technology. The compliance matrix modeling method is employed to determine the input stiffness of the compliant manipulator, and it is verified by finite-element analysis (FEA). Results show that the deviations between simulation works and the derived analytical models are in an acceptable range. The simulation results also reveal that the compliant manipulator can achieve a 16 μm × 16 μm × 16 μm cubic workspace. In this motion range, the observed maximum stress is much lower than the yield strength of the material. Moreover, the dynamic characteristics of the manipulator are investigated via the simulations as well.

Keywords: tripteron; triaxial; parallel-kinematic; compliant manipulator; compact structure

Citation: Xie, Y.; Li, Y.; Cheung, C. Design and Modeling of a Novel Tripteron-Inspired Triaxial Parallel Compliant Manipulator with Compact Structure. *Micromachines* **2022**, *13*, 678. https://doi.org/10.3390/mi13050678

Academic Editor: Alessandro Cammarata

Received: 18 March 2022
Accepted: 24 April 2022
Published: 27 April 2022

1. Introduction

Piezo-driven compliant mechanisms have received increasing attention in both academic and industrial communities, due to their merits of micro/nanoscale resolution and fast response. They have been widely applied to a variety of fields such as atomic force microscopy [1], precise assembly [2], cell micro-injection [3], micro/nanoscratching [4] and micro electromechanical systems [5]. The compliant mechanisms transmit motion based on the deformation of flexure hinges, which are different from conventional mechanisms on the basis of rigid links and gears. They are capable of overcoming the disadvantages of traditional mechanisms including the backlash and assembly error [6,7]. The piezoelectric actuator (PEA) is a normally-used driving component in compliant mechanisms owing to its high resolution, rapid response and ease of compact size [8–10]. With all outstanding superiorities as mentioned, the piezo-driven compliant mechanism has been one of the promising choices for precision manipulators [11–13].

The translational XYZ compliant mechanism is a significant manipulator in some applications such as the atomic force microscopy and micro-injection. They can be commonly divided into serial mechanisms and parallel mechanisms based on the kinematic principle. For the serial mechanisms, each of the axes works independently and is easy to control, since they do not have coupling motions between axes. Wadikhaye et al. [14] designed a serial-kinematic XYZ positioner with rapid response and compact structure for the atomic force microscopy. The positioner has a reachable motion range of 8 μm × 6 μm × 2 μm with the frequencies of 10, 7.5 and 64 kHz along the x, y and z-axis, respectively. Kenton et al. [15] proposed a serial-kinematic triaxial compliant mechanism. A positioning range of 9 μm × 9 μm × 1 μm is achieved with the fabricated prototype. The experimental results also demonstrate that the mechanism has the resonant frequencies of 24.2, 6 and 70 kHz along the x, y and z-axis, respectively. A three-axis serial-kinematic positioning device driven by piezoelectric actuators was developed for fast tool servo application [16]. The natural frequencies of the positioner along the x, y and z-axis were measured at 1.06, 0.65 and 0.54 kHz, respectively. However, the accumulation of error is inevitable for the serial structure, and it is at the sacrifice of the natural frequency because the serial connection increases the mass of motion parts of other axes, resulting in different resonant frequencies among axes [17,18].

On the contrary, the motion platform is directly connected with all axes in the parallel mechanisms, so that the cumulative error and motion mass increase can be avoided. Many parallel XYZ compliant mechanisms have been developed due to their virtues of high precision and high resonant frequency. Li et al. [19] proposed a parallel XYZ compliant mechanism with good motion decoupling properties in both the input end and the output platform. In the design, the compound parallelogram flexures were adopted to realize the total decoupling. The finite element analysis (FEA) simulation results show that it has identical dynamic performance in all axes, and experimental results demonstrate that the developed mechanism can achieve submicron accuracy. By introducing the biaxial right circular flexure hinges, Zhu et al. [20] developed a parallel triaxial translational monolithic compliant mechanism, which aims to achieve high bandwidth, high stiffness and high compactness. For the sake of enlarging the mechanism workspace, a 3-DoF XYZ precision positioner based on the modified differential lever amplifier was proposed in the literature of [21].Nonetheless, the base frames in the aforementioned works were not clearly displayed and the overall body sizes were difficult to evaluate. In the literature of [22], a complete parallel XYZ micromanipulator with the base frame was presented. Tang et al. [23] also conceived and designed a decoupled XYZ flexure parallel mechanism mounted on the base frame. Hao et al. [24] developed a 3 DoF translational compliant manipulator with three XY orthogonally-placed flexure mechanisms fixed on the base frame. With the same research group, spatial double four-beam modules were employed in the design of a parallel modular XYZ compliant mechanism referring to a 3-PPPR (P: prismatic, R: revolute) kinematic principle. The developed mechanism also took into account the base frame in the design [25]. In addition, Awtar et al. [26,27] proposed a parallel triaxial translational flexure mechanism mounted on the base with a travel range of 10 mm $\times$ 10 mm $\times$ 10 mm. Good decoupled translational motions among axes and high stiffness in the rotational motions were observed. However, the body sizes of some of the above-mentioned manipulators are bulky, while the base frame structures are considered. Moreover, the assembly errors between amplifier mechanisms/guidance mechanisms and base frames in these studies are difficult to be avoided.

The kinematics-based approach, the building blocks approach and the topology optimization approach [28–30] are three classical compliant mechanisms designing methods. On the basis of the kinematics-based approach, a novel mechanism inspired by the Tripteron [31] is proposed in this paper to solve the aforementioned problems, which may provide an alternative way to construct the parallel XYZ compliant manipulator. 3D printing technology is expected to monolithically fabricate the proposed manipulator for eliminating assembly error and realizing structure compactness. The rest of this paper is structured as follows. Section 2 illustrates the overall configuration and working principle of the parallel XYZ manipulator. In Section 3, the compliance matrix modeling method is utilized to analyse the stiffness/compliance of the manipulator. FEA simulation works are conducted to verify the theoretical analyses in Section 4. Section 5 draws a brief conclusion of this work.

2. Mechanism Design

As shown in Figure 1, the Tripteron is a triaxial translational parallel mechanism [31,32]. It is composed of an orthogonally arranged base frame, three kinematically identical legs and an end-effector. Referring to the kinematics model of the Tripteron, the leg has three revolute joints on its own and a prismatic joint connected with the base frame. When the prismatic joint is driven by the linear motor, the end-effector can achieve consistent translational motion along the driving axis. As a result, the Tripteron has three translational motions in the Cartesian coordinate. Investigation into compliant Triteron is still scarce, although a compliant Tripteron has been presented previously, and the kinematostatic model has been established in the literature of [33]. In the reported literature, cruciform hinges were employed, whereas the right circular flexure hinges and beam flexure hinges will be adopted in the current paper. Furthermore, the reported compliant Tripteron is not a monolithic mechanism, which is different from the proposed mechanism in this paper.

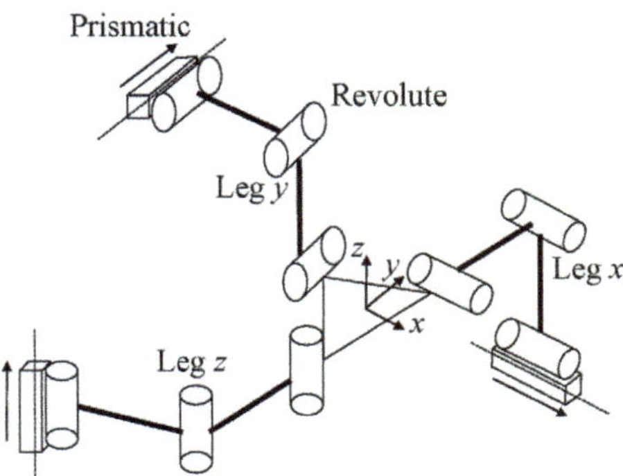

Figure 1. The kinematic diagram of Tripteron .

As enumerated above, a new translational XYZ parallel compliant mechanism based on the Tripteron is devised as shown in Figure 2. It consists of a base, three kinematically identical legs and a motion platform. The base structure is a guidance mechanism (GM) embedded with a PEA along each axis of the Cartesian coordinates. The leg is a rectangular-shaped compliant mechanism. The leg is worked as a driving unit while it moves toward the elongation direction of the directly connected PEA, otherwise it works as a passive guidance unit along the other two axes. For the convenience of analyses in the following sections, each leg is further subdivided into leg-p which is directly connected with the platform and leg-b which is directly connected with the base frame.

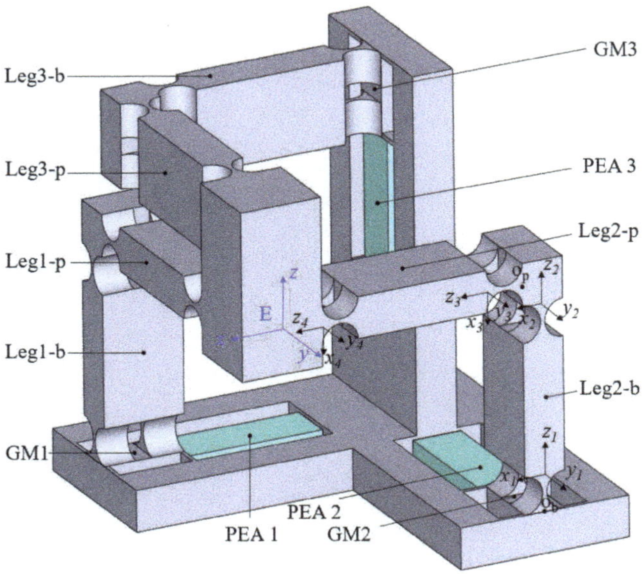

Figure 2. Mechanical structure of the proposed triaxial compliant manipulator.

The right circular flexure hinge and beam flexure hinge are commonly used flexure hinges. The right circular flexure hinge has the merits of little axis drift, high-precision and high transversal stiffness. The beam flexure hinge can distribute the deformation in the whole beam and avoid the concentration of stress, which provides a larger displacement of deformation without failure. As a result, the GM 1 is constructed with beam flexure hinges, which mainly aims to provide large output displacement. On the other hand, the guidance mechanism has four parallel connected beam flexure hinges, which can avoid the coupled motion caused by other driving units including the transverse motion and twisted deformation. This is significant for protecting the PEAs from being damaged, since they are sensitive to the tangential force and bending deformation. In contrast, the right circular flexure hinges are used in the design of the legs, so that the high lateral stiffness can alleviate the deformation when they are working as driving units, and the compliance can reduce the constraint stiffness when they are working as passive guidance units.

The working principle of the proposed 3-DoF compliant mechanism can be illustrated as follows. The PEA 1 elongates when the voltage is exerted on it, driving the GM 1 and corresponding connected leg toward the positive x-axis, and thus the motion platform moves toward positive x-axis accordingly. Based on the same principle, the motions along y-axis and z-axis can be achieved when the voltages are applied at PEA 2 and PEA 3. As a result, a compliant mechanism with translational 3-DoF is developed.

3. Modeling and Analysis

3.1. Modeling Method

There are many modeling methods for analyzing compliant mechanisms as reviewed in the literature of [34–36], including Castigliano's second theorem, elastic beam theory, compliance matrix modeling (CMM) method, finite element method, pseudo-rigid-body (PRB) method, the chained beam constraint model and the 3-D dynamic stiffness model. Both merits and shortcomings were comprehensively reported. In this paper, the PRB method [37] and CMM method [38,39] are mainly discussed. The PRB method only takes account of the compliance of flexure hinges along the working direction, so that it is not able to achieve a complete compliance analysis. Moreover, it involves loop closure theory, virtual work principle and Lagrange's equation, which is complex for complicated mechanisms. On the contrary, the CMM method on the basis of Hooke's law can establish 6-DoF compliance in the space of flexure hinges with high accuracy. Furthermore, it has high calculation efficiency due to the effective operation of matrix with the help of a computer. As a result, the compliance/stiffness of the proposed compliant mechanism in this paper is analyzed based on the compliance matrix modeling method.

As shown in Figure 3, the compliance matrix in its local coordinate can be expressed as:

$$C_f = \begin{bmatrix} c_1 & 0 & 0 & 0 & c_3 & 0 \\ 0 & c_2 & 0 & -c_4 & 0 & 0 \\ 0 & 0 & c_5 & 0 & 0 & 0 \\ 0 & -c_4 & 0 & c_6 & 0 & 0 \\ c_3 & 0 & 0 & 0 & c_7 & 0 \\ 0 & 0 & 0 & 0 & 0 & c_8 \end{bmatrix}, \tag{1}$$

where the specific values of compliance factors of different flexure hinges are listed in the literature [40]. Equation (2) shows the conversion of the compliance matrix from its local coordinate to a new coordinate frame.

$$C_2 = T_1^2 C_1 (T_1^2)^T, \tag{2}$$

where $C_1 = C_f$ is the compliance matrix of the flexure hinge with respect to the fixed-end. The coordinate transformation matrices T_1^2 include the rotational matrix and the translational matrix. $\overline{R}_x, \overline{R}_y, \overline{R}_z$ are the rotational matrices around the x, y and z-axis, and they can be written as

$$\overline{R}_x(\gamma) = \begin{pmatrix} R_x(\gamma) & 0 \\ 0 & R_x(\gamma) \end{pmatrix}, \tag{3}$$

$$\overline{R}_y(\beta) = \begin{pmatrix} R_y(\beta) & 0 \\ 0 & R_y(\beta) \end{pmatrix}, \tag{4}$$

$$\overline{R}_z(\alpha) = \begin{pmatrix} R_z(\alpha) & 0 \\ 0 & R_z(\alpha) \end{pmatrix}, \tag{5}$$

where $R_x(\gamma)$, $R_y(\beta)$ and $R_z(\alpha)$ denote the rotation around the corresponding axis. The translational matrix $\overline{q} = (q_x, q_y, q_z)$ can realize the translation of the compliance matrix, and it is given by

$$\overline{q}(q_x, q_y, q_z) = \begin{pmatrix} I & \hat{q} \\ 0 & I \end{pmatrix}, \tag{6}$$

where $\hat{q}$ represents the outer product for a vector $q = (q_x, q_y, q_z)$, and it can be derived as Equation (7), in which q is the coordinate of the new coordinate system relative to the transferred coordinate frame. The identity matrix I is described by Equation (8).

$$\hat{q} = \begin{pmatrix} 0 & -q_z & q_y \\ q_z & 0 & -q_x \\ -q_y & q_x & 0 \end{pmatrix}, \tag{7}$$

$$I = \begin{pmatrix} 1 & 0 & 0 \\ 0 & 1 & 0 \\ 0 & 0 & 1 \end{pmatrix}. \tag{8}$$

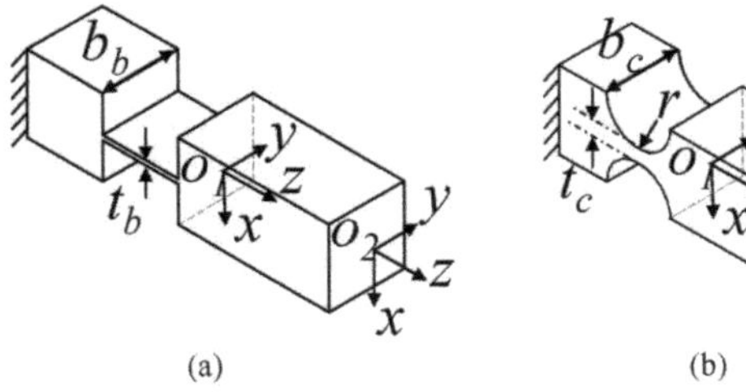

Figure 3. The beam flexure hinge (**a**) and right circular flexure hinge (**b**).

3.2. Input Stiffness of the Proposed Mechanism

For the purpose of determining the compliance/stiffness of the compliant mechanism, the mechanism has been divided into two parts for easy analysis. The compliances/stiffnesses of the guidance mechanisms on the base and legs connected motion platform and guidance mechanisms are separately analyzed first, and are then added together based on the principle of parallel and series connection.

As shown in Figure 4, the GM 3 on the base along the z-axis is picked out for analysis. On the basis of the compliance matrix modeling method, the compliance of hinge a can be generated as

$$C_a^G = T_a^G C_a (T_a^G)^T. \tag{9}$$

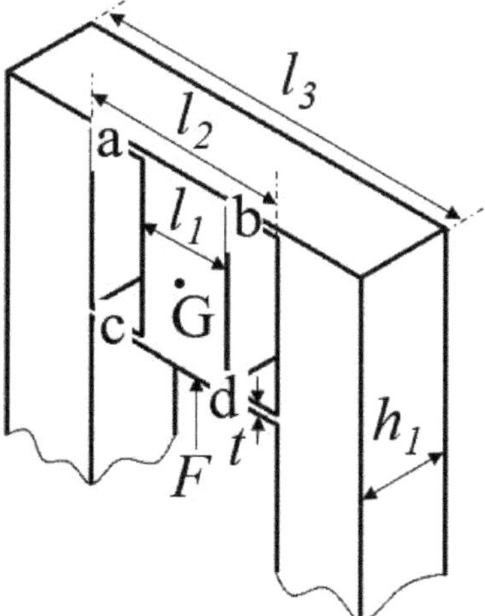

Figure 4. The guidance mechanism in the base frame.

Due to flexure hinges b and a are symmetric around the x-axis, the compliance of hinge b can be derived as

$$C_b^G = \overline{R}_x(\pi)C_a^G(\overline{R}_x(\pi))^T. \tag{10}$$

The above mentioned principle is also applicable to flexure hinges c and d, and thus their compliance matrices can be obtained as

$$C_c^G = T_c^G C_c (T_c^G)^T, \tag{11}$$

$$C_d^G = \overline{R}_x(\pi)C_c^G(\overline{R}_x(\pi))^T. \tag{12}$$

Owing to the four flexure hinges being parallel connected with the motion platform G, the vertical compliance C_v^G and vertical stiffness K_v^G can be determined by

$$C_v^G = ((C_a^G)^{-1} + (C_b^G)^{-1} + (C_c^G)^{-1} + (C_d^G)^{-1})^{-1}, \tag{13}$$

$$K_v^G = (C_v^G)^{-1}. \tag{14}$$

Following the previous assumption that the leg connected with the z-axis is the driving unit as shown in Figure 5, and the leg1-p and leg2-p are worked as guidance mechanisms, referring to Figure 2, leg2 is selected for the analysis. Considering that two layers of flexure hinges parallel connect the motion platform and leg2-b, the compliance of point E with respect to the point Op can be derived as

$$C_E^p = ((T_3^E C_3 (T_3^E)^T + T_4^E C_4 (T_4^3)^T)^{-1} + (T_3^E C_3 (T_3^E)^T + T_4^E C_4 (T_4^3)^T)^{-1}) \cdot \tag{15}$$

Due to leg1-p and leg2-p being parallel connected with the motion platform, and having circular symmetry of 90° around the z-axis, the compliance C_G^{1p2p} and the stiffness K_G^{1p2p} can be derived as

$$C_G^{1p2p} = (K_G^{1p2p})^{-1} = ((C_E^p)^{-1} + (T_{1p}^{2p} C_E^p (T_{1p}^{2p})^T)^{-1})^{-1}, \tag{16}$$

where T_{1p}^{2p} is the transformation matrix transferring from leg1-p to leg2-p, and it can be written as

$$T_{1p}^{2p} = \begin{pmatrix} R_z(-\pi/2) & 0 \\ 0 & R_z(-\pi/2) \end{pmatrix}. \tag{17}$$

As a result, the total input stiffness along the z-axis of the compliant mechanism K_z can be obtained as the following equation since the GM 3 and guidance mechanism $1p2p$ are parallel connected with the motion platform

$$K_z = K_v^G + K_G^{1p2p}. \tag{18}$$

According to the parallel kinematics of the GM 3 and guidance mechanism $1p2p$, the output stiffness should be equal to the input stiffness. However, the distortion of the cantilever structure of the driving leg will reduce the output stiffness to some extent. The stiffnesses/compliances of the compliant mechanisms along other two axes can also be determined based on the same principle.

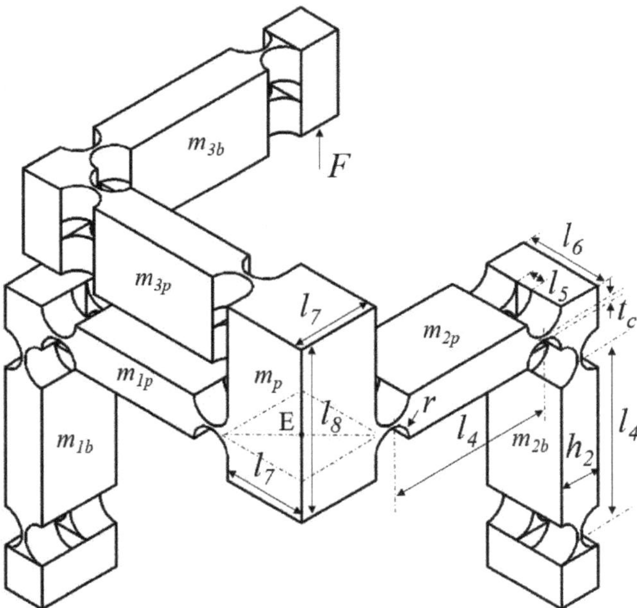

Figure 5. Schematic of the motion platform and three legs.

4. Model Validation and Performance Evaluation via FEA

Table 1 shows the structural parameters of the developed parallel triaxial compliant mechanism. 3D printing technology is expected to fabricate the mechanism, due to its capability of manufacturing monolithic complex structure with high dimension accuracy on the basis of the photosensitive resin material [41–43]. The detail physical properties of resin are: Young's modulus $E = 2.2$ GPa, Poisson's ratio $\mu = 0.394$, density $\rho = 1.18 \times 10^3$ kg/m^3, and yield strength $\delta_s = 50$ MPa. According to the above mentioned parameters, FEA simulations are conducted to validate the established theoretical model and to evaluate the performance of the proposed compliant mechanism with the help of commercial software ANSYS.

The details of the FEA model are shown in Figure 6, including the constraints, loads, etc. In addition, a flowchart of conducting the FEA simulations is also given in Figure 7. Starting from building the 3-D model of the compliant manipulator via the CAD modelling software, the model is then imported to the ANSYS workbench to perform the simulation analysis. During the simulation, the material of the mechanism was defined according to the aforementioned details of the material properties. Following is the mesh generation. The mesh has 152,295 elements. The skewness criteria mesh metrics is adopted to evaluate the mesh quality and the average value is 0.327, which shows that it has a very good quality.

After that, the constraints and loads are defined in the model, and the simulation results can be obtained after the calculation.

Table 1. Dimensional parameters of the proposed triaxial compliant manipulator.

GM 3		Legs and the Motion Platform	
Symbol	**Value (mm)**	**Symbol**	**Value (mm)**
l_1	10	l_4	41
l_2	22	l_5	4
l_3	42	l_6	20
h_1	10	h_2	10
t	0.6	t_c	1
		r	4.5
		l_7	20
		l_8	40

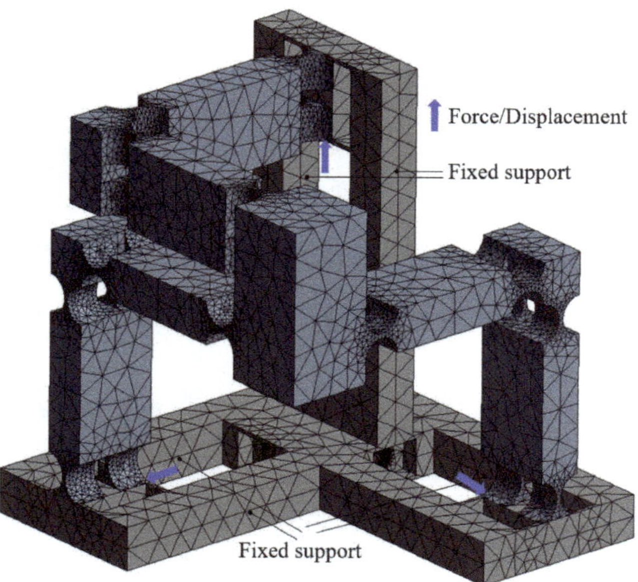

Figure 6. Details of the FEA model.

The correlation between the output displacement of the moving platform and the input force is shown in Figure 8. In the figure, three almost coinciding straight lines can be seen, which means that the three axes have a nearly identical linear relationship. The input stiffness is derived by dividing the input force with the output displacement of the moving platform, which is the reciprocal of the slope of the line in Figure 8. The comparison between analytical models and simulation results are given in Table 2. The deviations are mainly induced by the deformation of the driving legs due to the cantilever structure, which can be further explained in the next discussion of motion loss.

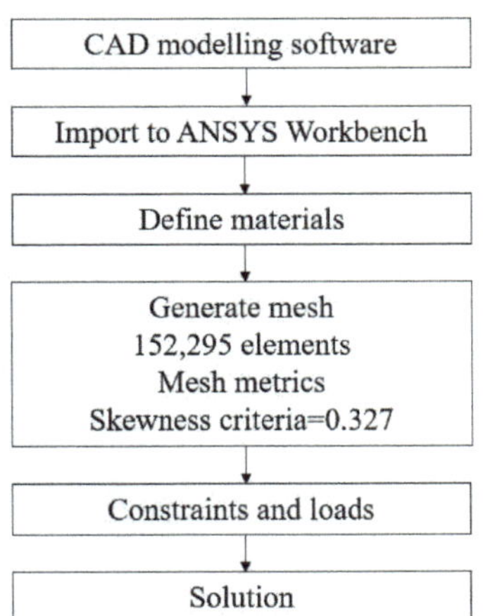

Figure 7. Flowchart of the FEA simulation.

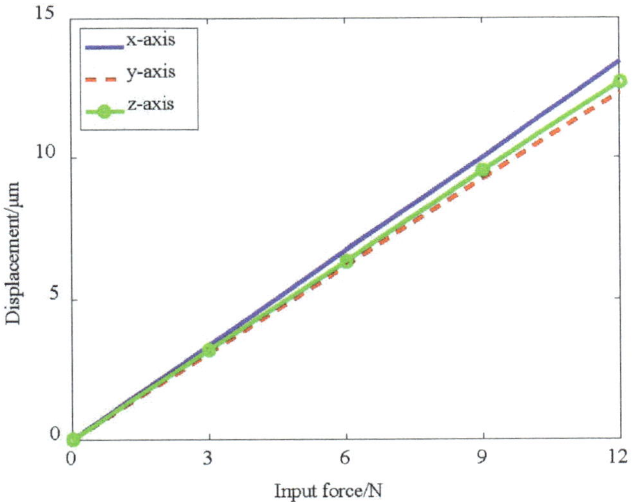

Figure 8. The correlation between input force and output displacement.

Table 2. Comparison results of input stiffness.

Methods	*x*-Axis Input/Output Stiffness (N/μm)	*y*-Axis Input/Output Stiffness (N/μm)	*z*-Axis Input/Output Stiffness (N/μm)
Analytical model	0.82	0.82	0.82
FEA	0.89/0.69	0.97/0.70	0.94/0.72
Error	−18%/19%	−15%/17%	−13%/14%

Figure 9 illustrates the output displacement of the motion platform with a 20 μm input displacement on each axis. One can observe that the corresponding output displacement of the motion platform is 16 μm with the same direction as the driving axis, which has a motion loss of 20%. The motion loss is mainly caused by the constraint of the passive legs and the deformation of the driving leg which is a cantilever structure. In the case of 20 μm input, the deflection of the cantilever beam can be determined by $\omega = Fl^3/3EI$ and it is equal to 1 μm, which is still smaller than the motion loss of 4 μm. This deviation may be further linked to the following two reasons. Firstly, the right circular flexure hinge on the leg may increase the distortion of the structure. For the next reason is that the leg is fixed on the GM, and the GM is connected with the base frame using beam flexure hinges, which may also induce the distortion.

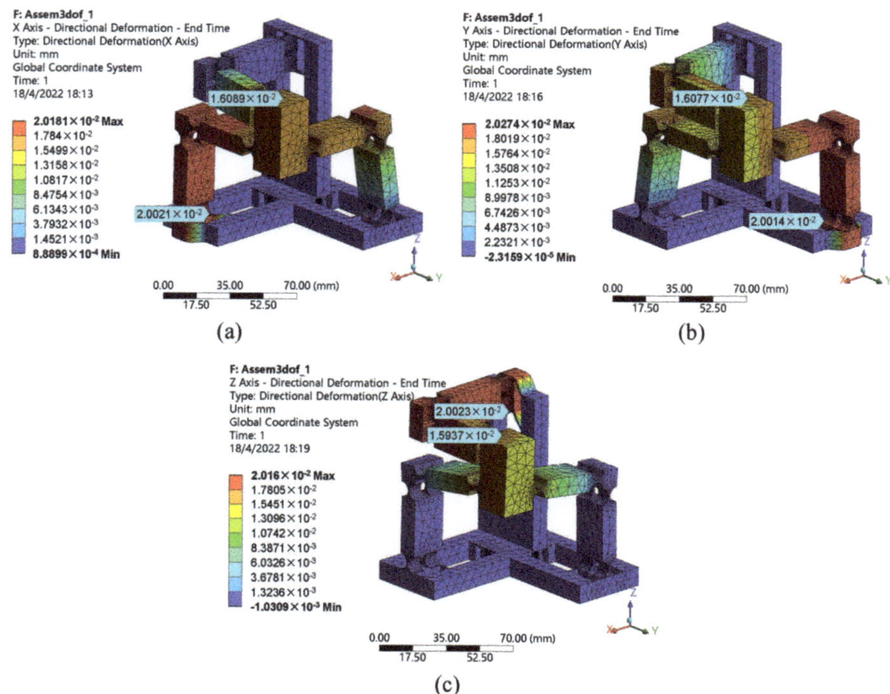

Figure 9. The corresponding output displacement of the motion platform with a 20 μm input on different axes: (**a**) x-axis, (**b**) y-axis, (**c**) z-axis.

The decoupling property of the compliant mechanism is depicted in Table 3. When the motion platform is driven toward x-axis with a displacement of 16, the cross-axis motions along the y and z axes are 1.4 and −0.84 μm, respectively. A major motion along the y-axis induces the parasitic motions of 2.82 and −0.62 μm along the x and z axes. Similarly, the couple motions along the x and y axes caused by the output displacement of 16 μm in the z-axis are −0.96 and 1.41 μm. The results reveal that the decoupling properties among all axes of the proposed mechanism are not as good as they should be, and it may be linked to the distortion of the driving legs.

Table 3. Performance of the proposed triaxial compliant manipulator.

Input Displacement (μm)			Output Displacement (μm)		
x	y	z	x	y	z
20	-	-	16	1.40	−0.84
-	20	-	2.82	16	−0.62
-	-	20	−0.96	1.41	16

With the consideration of the maximum stroke of the selected PEA, the workspace of the mechanism can be obtained as shown in Figure 10. The reachable workspace is a cube with 16 μm along each axis. In addition, the maximal stress of 6.44 MPa occurred in the beam flexure hinges can be observed from Figure 11 when the maximum stroke of 20 μm of the PEA is exerted on the input end. It is much lower than the yield strength 50 MPa, indicating that the proposed mechanism is capable of working normally without failure.

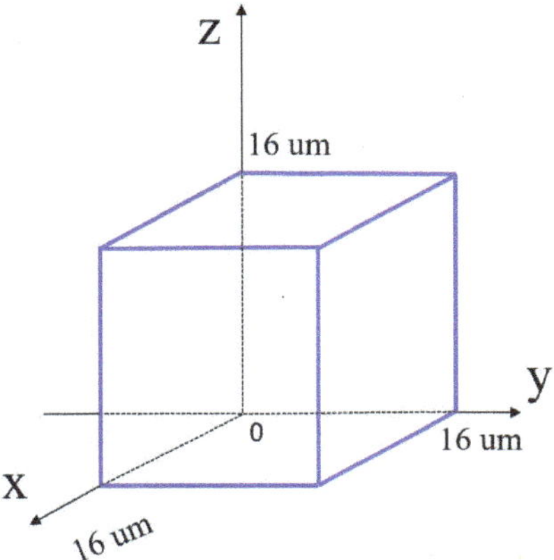

Figure 10. The reachable workspace of the proposed triaxial compliant manipulator.

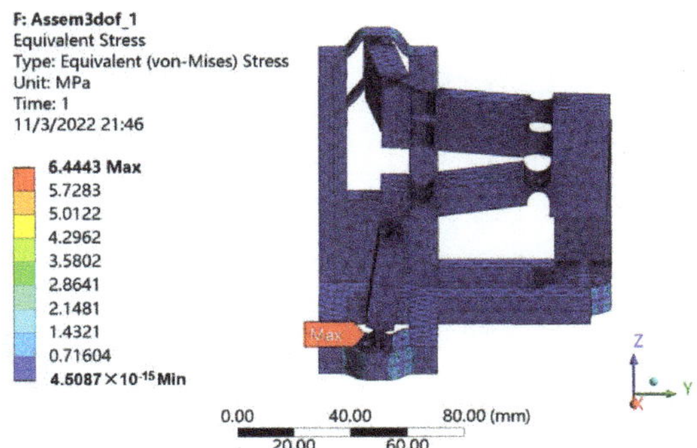

Figure 11. Stress distribution.

The dynamic performances are also studied with the FEA methodology. The first six mode shapes are shown in Figure 12. The first mode has a frequency of 50.58 Hz. The second and third modes of the proposed compliant mechanism have the frequencies of 72.87 Hz and 119.9 Hz, respectively. The corresponding value of the fourth, fifth, sixth modes are 288.39 Hz, 354.49 Hz and 390.46 Hz.

It can be observed that the first three resonance frequencies are different and the reasons are as follows. According to the working principle of the manipulator, when the motion platform moves toward x-axis, m_{1b}, m_{1p} and m_{2p} share the same displacement of the motion platform. However, m_{2b} and m_{3p} only have half of the displacement. When the motion platform moves toward y-axis, m_{1p}, m_{3p}, m_{2b} and m_{2p} share the same displacement of the motion platform. However, m_{1b} and m_{3b} only have half of the displacement. When the motion platform moves toward the z-axis, m_{3b}, m_{3p} share the same displacement of the motion platform. However, m_{1p} and m_{2p} only have half of the displacement. As a result, the equivalent masses in different directions are different, with the consideration of the same equivalent stiffness as given in Table 2, and thus the vibration modes have different resonance frequency values.

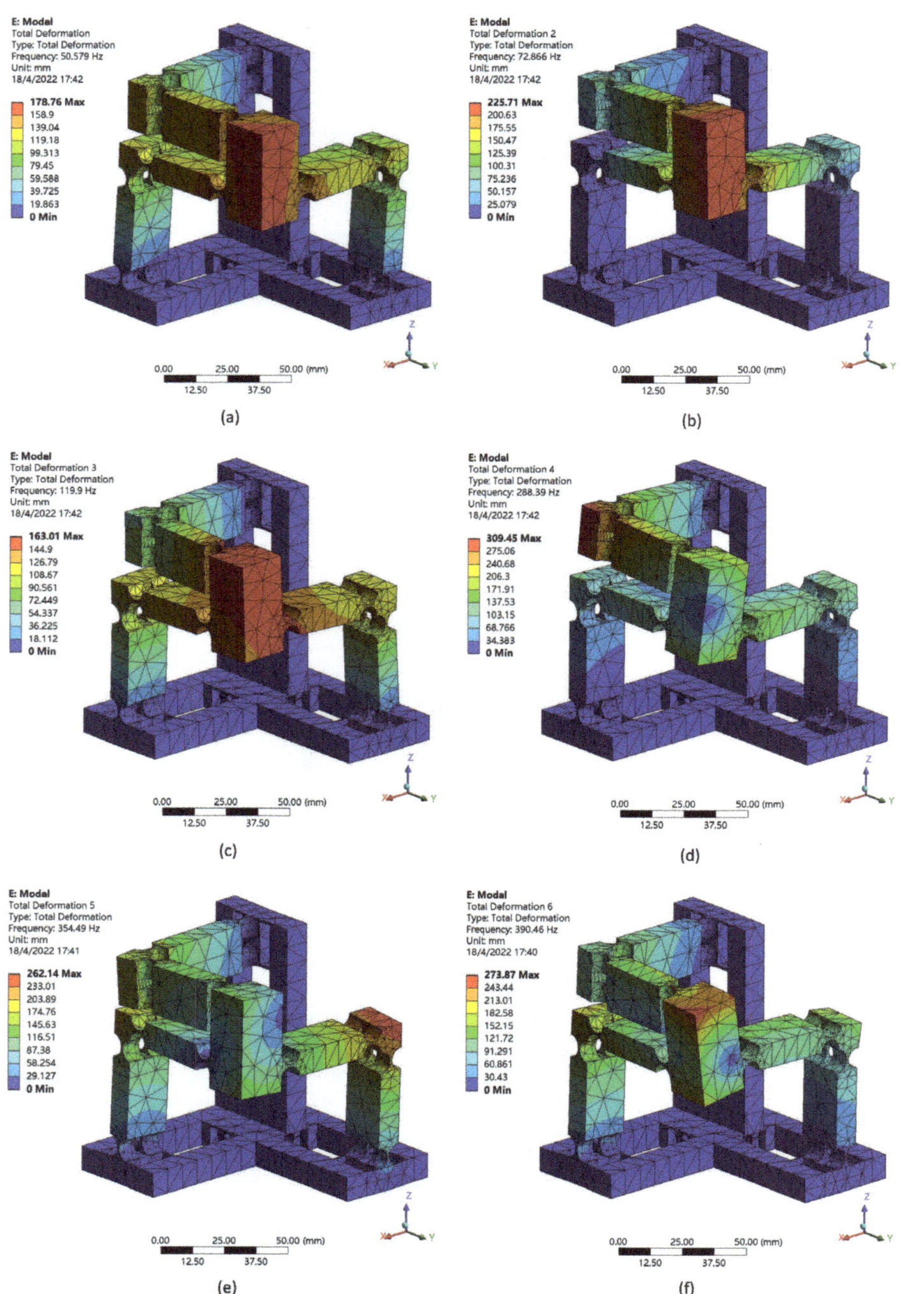

Figure 12. First six mode shapes: (**a**) mode 1, (**b**) mode 2, (**c**) mode 3, (**d**) mode 4, (**e**) mode 5, (**f**) mode 6.

5. Conclusions

A novel triaxial parallel-kinematic compliant manipulator inspired by the Tripteron mechanism is proposed and analyzed in this paper. The manipulator is composed of a base structure, three legs and a motion platform. Unlike the design of most conventional triaxial compliant mechanisms, the base frames of the mechanisms were ignored. The proposed compliant manipulator has taken it into consideration in the design. Due to the unique configuration of the proposed manipulator, it is able to realize the triaxial translational motions with a compact structure. The assembly error can also be avoided by fabricating the mechanism monolithically with the help of 3D printing technology. The stiffness/compliance model of the mechanism is established based on the compliance matrix modeling method and verified by FEA simulations. The deviations between them are in a reasonable range. A cubic workspace with 16 μm along each axis is observed with a maximum stress much lower than the yield strength, which means the manipulator is in safe working condition, although the decoupling performance is not as good as expected. In addition, the dynamic performances of the manipulator are also indicated via the simulations. In the future research, 3D printing technology is expected to fabricate the prototype to experimentally study the comprehensive performances of the proposed manipulator.

Author Contributions: Y.X. established the model of the compliant mechanism and performed simulations to verify it; Y.X. also wrote the original draft; Y.L. provided the guidance of the mechanism design, and contributed to reviewing as well as editing the manuscript; C.C. conducted careful reviewing and revision of the manuscript. All authors have read and agreed to the published version of the manuscript.

Funding: This research was funded by the National Natural Science Foundation of China (Grant No. 51575544).

Institutional Review Board Statement: Not applicable.

Informed Consent Statement: Not applicable.

Data Availability Statement: Not applicable.

Acknowledgments: The authors would like to express their sincere thanks for the financial support from the State Key Laboratory of Ultra-precision Machining Technology of The Hong Kong Polytechnic University(BBXG) and a grant from the Research Committee of The Hong Kong Polytechnic University under the student account code RK34.

Conflicts of Interest: The authors declare no conflict of interest.

References

1. Fleming, A.J.; Aphale, S.S.; Moheimani, S.R. A new method for robust damping and tracking control of scanning probe microscope positioning stages. *IEEE Trans. Nanotechnol.* **2009**, *9*, 438–448. [CrossRef]
2. Komati, B.; Clévy, C.; Lutz, P. High bandwidth microgripper with integrated force sensors and position estimation for the grasp of multistiffness microcomponents. *IEEE/ASME Trans. Mechatron.* **2016**, *21*, 2039–2049. [CrossRef]
3. Wang, G.; Xu, Q. Design and precision position/force control of a piezo-driven microinjection system. *IEEE/ASME Trans. Mechatron.* **2017**, *22*, 1744–1754. [CrossRef]
4. Wang, F.; Huo, Z.; Liang, C.; Shi, B.; Tian, Y.; Zhao, X.; Zhang, D. A novel actuator-internal micro/nano positioning stage with an arch-shape bridge-type amplifier. *IEEE Trans. Ind. Electron.* **2018**, *66*, 9161–9172. [CrossRef]
5. Wang, D.; Yang, Q.; Dong, H. A monolithic compliant piezoelectric-driven microgripper: Design, modeling, and testing. *IEEE/ASmE Trans. Mechatron.* **2011**, *18*, 138–147. [CrossRef]
6. Wu, S.; Shao, Z.; Su, H.; Fu, H. An energy-based approach for kinetostatic modeling of general compliant mechanisms. *Mech. Mach. Theory* **2019**, *142*, 103588. [CrossRef]
7. Yang, X.; Zhu, L.; Li, S.; Zhu, W.; Ji, C. Development of a Novel pile-up Structure based nanopositioning mechanism driven by piezoelectric actuator. *IEEE/ASME Trans. Mechatron.* **2020**, *25*, 502–512. [CrossRef]
8. Gao, X.; Liu, Y.; Zhang, S.; Deng, J.; Liu, J. Development of a novel flexure-based XY platform using single bending hybrid piezoelectric actuator. *IEEE/ASME Trans. Mechatron.* **2022**, 1–11. [CrossRef]
9. Liang, C.; Wang, F.; Huo, Z.; Shi, B.; Tian, Y.; Zhao, X.; Zhang, D. A 2-DOF monolithic compliant rotation platform driven by piezoelectric actuators. *IEEE Trans. Ind. Electron.* **2019**, *67*, 6963–6974. [CrossRef]

10. Gao, X.; Zhang, S.; Deng, J.; Liu, Y. Development of a small two-dimensional robotic spherical joint using a bonded-type piezoelectric actuator. *IEEE Trans. Ind. Electron.* **2019**, *68*, 724–733. [CrossRef]
11. Yang, X.; Zhu, W.L.; Zhu, Z.; Zhu, L.M. Design, assessment, and trajectory control of a novel decoupled robotic nanomanipulator. *IEEE/ASME Trans. Mechatron.* **2022**, 1–12. [CrossRef]
12. Mitrovic, A.; Nagel, W.S.; Leang, K.K.; Clayton, G.M. Closed-loop range-based control of dual-stage nanopositioning systems. *IEEE/ASME Trans. Mechatron.* **2020**, *26*, 1412–1421. [CrossRef]
13. Hao, G.; Zhu, J. Design of a monolithic double-slider based compliant gripper with large displacement and anti-buckling ability. *Micromachines* **2019**, *10*, 665. [CrossRef]
14. Wadikhaye, S.; Yong, Y.; Moheimani, S.R. Design of a compact serial-kinematic scanner for high-speed atomic force microscopy: An analytical approach. *Micro Nano Lett.* **2012**, *7*, 309–313. [CrossRef]
15. Kenton, B.J.; Leang, K.K. Design and control of a three-axis serial-kinematic high-bandwidth nanopositioner. *IEEE/ASME Trans. Mechatron.* **2011**, *17*, 356–369. [CrossRef]
16. Liu, Y.T.; Li, B.J. A 3-axis precision positioning device using PZT actuators with low interference motions. *Precis. Eng.* **2016**, *46*, 118–128. [CrossRef]
17. Xiao, R.; Shao, S.; Xu, M.; Jing, Z. Design and analysis of a novel piezo-actuated XYθz micropositioning mechanism with large travel and kinematic decoupling. *Adv. Mater. Sci. Eng.* **2019**. [CrossRef]
18. Muraoka, M.; Sanada, S. Displacement amplifier for piezoelectric actuator based on honeycomb link mechanism. *Sens. Actuators A Phys.* **2010**, *157*, 84–90. [CrossRef]
19. Li, Y.; Xu, Q. A totally decoupled piezo-driven XYZ flexure parallel micropositioning stage for micro/nanomanipulation. *IEEE Trans. Autom. Sci. Eng.* **2010**, *8*, 265–279. [CrossRef]
20. Zhu, Z.; To, S.; Zhu, W.L.; Li, Y.; Huang, P. Optimum design of a piezo-actuated triaxial compliant mechanism for nanocutting. *IEEE Trans. Ind. Electron.* **2017**, *65*, 6362–6371. [CrossRef]
21. Gao, J.; Zeng, Z.; Tang, H.; Chen, X.; Qiu, Q.; He, S.; He, Y.; Yang, Z. Design and assessment of a piezo-actuated 3-DOF flexible nanopositioner with large stroke. In Proceedings of the 2016 IEEE International Conference on Manipulation, Manufacturing and Measurement on the Nanoscale (3M-NANO), Chongqing, China, 18–22 July 2016; pp. 19–24.
22. Bacher, J.P.; Bottinelli, S.; Breguet, J.M.; Clavel, R. Delta3: Design and control of a flexure hinge mechanism. In *Microrobotics and Microassembly III, Proceedings of the Intelligent Systems and Advanced Manufacturing, Boston, MA, USA, 28–31 October 2001*; SPIE: Bellingham, WA, USA, 2001; Volume 4568, pp. 135–142.
23. Tang, X.; Chen, I.M.; Li, Q. Design and nonlinear modeling of a large-displacement XYZ flexure parallel mechanism with decoupled kinematic structure. *Rev. Sci. Instruments* **2006**, *77*, 115101. [CrossRef]
24. Hao, G.; Kong, X. A 3-DOF translational compliant parallel manipulator based on flexure motion. In Proceedings of the ASME International Design Engineering Technical Conferences and Computers and Information in Engineering Conference, San Diego, CA, USA, 30 August–2 September 2009; Volume 49040, pp. 101–110.
25. Hao, G.; Kong, X. Design and modeling of a large-range modular XYZ compliant parallel manipulator using identical spatial modules. *ASME J. Mech. Robot.* **2012**, *4*, 021009. [CrossRef]
26. Awtar, S.; Ustick, J.; Sen, S. An XYZ parallel-kinematic flexure mechanism with geometrically decoupled degrees of freedom. *ASME J. Mech. Robot.* **2013**, *5*, 015001. [CrossRef]
27. Awtar, S.; Quint, J.; Ustick, J. Experimental characterization of a large-range parallel kinematic XYZ flexure mechanism. *ASME J. Mech. Robot.* **2021**, *13*, 015001. [CrossRef]
28. Gallego, J.A.; Herder, J. Synthesis methods in compliant mechanisms: An overview. In Proceedings of the ASME International Design Engineering Technical Conferences and Computers and Information in Engineering Conference, San Diego, CA, USA, 30 August–2 September 2009; Volume 49040, pp. 193–214.
29. Sigmund, O. On the design of compliant mechanisms using topology optimization. *J. Struct. Mech.* **1997**, *25*, 493–524. [CrossRef]
30. Kumar, P.; Schmidleithner, C.; Larsen, N.; Sigmund, O. Topology optimization and 3D printing of large deformation compliant mechanisms for straining biological tissues. *Struct. Multidiscip. Optim.* **2021**, *63*, 1351–1366. [CrossRef]
31. Kong, X.; Gosselin, C.M. Kinematics and singularity analysis of a novel type of 3-CRR 3-DOF translational parallel manipulator. *Int. J. Robot. Res.* **2002**, *21*, 791–798. [CrossRef]
32. Gosselin, C.; Kong, X.; Foucault, S.; Bonev, I. A fully decoupled 3-dof translational parallel mechanism. In Proceedings of the 4th Chemnitz Parallel Kinematics Seminar, Chemnitz, Germany, 20–21 April 2004; pp. 595–610.
33. Quennouelle, C.; Gosselin, C. Kinematostatic modeling of compliant parallel mechanisms. *Meccanica* **2011**, *46*, 155–169. [CrossRef]
34. Ling, M.; Howell, L.L.; Cao, J.; Chen, G. Kinetostatic and dynamic modeling of flexure-based compliant mechanisms: A survey. *Appl. Mech. Rev.* **2020**, *72*, 030802. [CrossRef]
35. Bilancia, P.; Berselli, G. An overview of procedures and tools for designing nonstandard beam-based compliant mechanisms. *Comput.-Aided Des.* **2021**, *134*, 103001. [CrossRef]
36. Ling, M.; Song, D.; Zhang, X.; He, X.; Li, H.; Wu, M.; Cao, L.; Lu, S. Analysis and design of spatial compliant mechanisms using a 3-D dynamic stiffness model. *Mech. Mach. Theory* **2022**, *168*, 104581. [CrossRef]
37. Howell, L.L.; Magleby, S.P.; Olsen, B.M. *Handbook of Compliant Mechanisms*; John Wiley & Sons Incorporated: Hoboken, NJ, USA, 2013; p. 311.

38. Xie, Y.; Li, Y.; Cheung, C.F.; Zhu, Z.; Chen, X. Design and analysis of a novel compact XYZ parallel precision positioning stage. *Microsyst. Technol.* **2021**, *27*, 1925–1932. [CrossRef]
39. Chen, X.; Li, Y. Design and analysis of a new high precision decoupled XY compact parallel micromanipulator. *Micromachines* **2017**, *8*, 82. [CrossRef]
40. Koseki, Y.; Tanikawa, T.; Koyachi, N.; Arai, T. Kinematic analysis of a translational 3-dof micro-parallel mechanism using the matrix method. *Adv. Robot.* **2002**, *16*, 251–264. [CrossRef]
41. Moritoki, Y.; Furukawa, T.; Sun, J.; Yokoyama, M.; Shimono, T.; Yamada, T.; Nishiwaki, S.; Kageyama, T.; Fukuda, J.; Mukai, M.; et al. 3D-printed micro-tweezers with a compliant mechanism designed using topology optimization. *Micromachines* **2021**, *12*, 579. [CrossRef] [PubMed]
42. Li, X.; Yu, D.; Cao, T.; Wen, Z.; Lu, C.; Liu, W.; Zhu, C.; Wu, D. 3D printing and dynamic modeling of a polymer-based bimodal piezoelectric motor. *Smart Mater. Struct.* **2020**, *30*, 025003. [CrossRef]
43. Clark, L.; Shirinzadeh, B.; Bhagat, U.; Smith, J.; Zhong, Y. Development and control of a two DOF linear–angular precision positioning stage. *Mechatronics* **2015**, *32*, 34–43. [CrossRef]

Article

Development of an Electromagnetic Micromanipulator Levitation System for Metal Additive Manufacturing Applications

Parichit Kumar, Saksham Malik, Ehsan Toyserkani and Mir Behrad Khamesee *

Department of Mechanical and Mechatronics Engineering, University of Waterloo, Waterloo, ON N2L 3G1, Canada; parichit.kumar@uwaterloo.ca (P.K.); saksham.malik@uwaterloo.ca (S.M.); ehsan.toyserkani@uwaterloo.ca (E.T.)

* Correspondence: khamesee@uwaterloo.ca

Abstract: Magnetism and magnetic levitation has found significant interest within the field of micromanipulation of objects. Additive manufacturing (AM), which is the computer-controlled process for creating 3D objects through the deposition of materials, has also been relevant within the academic environment. Despite the research conducted individually within the two fields, there has been minimal overlapping research. The non-contact nature of magnetic micromanipulator levitation systems makes it a prime candidate within AM environments. The feasibility of integrating magnetic micromanipulator levitation system, which includes two concentric coils embedded within a high permeability material and carrying currents in opposite directions, for additive manufacturing applications is presented in this article. The working principle, the optimization and relevant design decisions pertaining to the micromanipulator levitation system are discussed. The optimized dimensions of the system allow for 920 turns in the inner coil and 800 turns in the outer coil resulting in a $N_{inner\ coil}$:$N_{outer\ coil}$ ratio of 1.15. Use of principles of free levitation, which is production of levitation and restoration forces with the coils, to levitate non-magnetic conductive materials with compatibility and applications within the AM environment are discussed. The Magnetomotive Force (MMF) ratio of the coils are adjusted by incorporation of an resistor in parallel to the outer coil to facilitate sufficient levitation forces in the axial axis while producing satisfactory restoration forces in the lateral axes resulting in the levitation of an aluminum disc with a levitation height of 4.5 mm. An additional payload of up to 15.2 g (59% of mass of levitated disc) was added to a levitated aluminum disk of 26 g showing the system capability coping with payload variations, which is crucial in AM process to gradually deploy masses. The final envisioned system is expected to have positional stability within the tolerance range of a few µm. The system performance is verified through the use of simulations (ANSYS Maxwell) and experimental analyses. A novel method of using the ratio of conductivity (σ) of the material to density (ρ) of the material to determine the compatibility of the levitation ability of non-magnetic materials with magnetic levitation application is also formulated. The key advantage of this method is that it does not rely on experimental analyses to determine the levitation ability of materials.

Keywords: magnetic levitation; additive manufacturing; eddy current levitation; direct energy deposition

Citation: Kumar, P.; Malik, S.; Toyserkani, E.; Khamesee, M.B. Development of an Electromagnetic micromanipulator levitation system for Metal Additive Manufacturing Applications. *Micromachines* **2022**, *13*, 585. https://doi.org/10.3390/mi13040585

Academic Editor: Alessandro Cammarata

Received: 9 March 2022
Accepted: 4 April 2022
Published: 9 April 2022

1. Introduction

In recent times, magnetism has garnered significant interest within academic spheres. Magnetism has found applications in various fields, such as the damping capability [1], levitation and manipulation of a magnetized object [2], energy harvesting [3], sensing applications [4] and manipulation of magnetorheological fluid (MRF) [5] among several others, owing to its compatibility and adaptability to these applications.

Magnetic levitation is a promising field that brings forth an alternative contact-free method of moving objects. The absence of the physical aspect of actuation opens avenues for its applicability within both the macro and micro manipulation applications. Within the macro-scale, magnetic levitation has found applications in transportation [6] and wireless charging [7] among several others.

There has been significant research associated with the development of magnetic systems within the micromanipulation sphere. Ref. [8] discussed the development and actuation of bio-inspired robots, which mimic the motion style of different micro-organisms and are actuated using the principles of magnetism.

Ref. [9] studied the use of a magnetic micromanipulator levitation system to provide a wireless mode of navigation for surgical devices and drug containers within the human body using the principles of magnetism. Ref. [10] studied the development of micro-robots working on the principle of magnetic levitation. Ref. [11] studied the characteristics of a floating magnet freely levitated between two diamagnetic places without an external energy input.

Ref. [12] highlighted the use of magnetic fields for the actuation and micromanipulation of dynamically self-assembled magnetic droplets. Ref. [13] discussed the design, fabrication, actuation and control of versatile micro and nano robots with significant applications within the biomedical industry, while also discussing the bio-safety aspects of the system. Despite the extensive research in the field, a manipulator capable of supporting levitation with stability in the micrometer (μm) range and compatible with additive manufacturing environment has not been developed.

Additive manufacturing (AM) has also been a primary point of emphasis for research within the academic environment. The prevalence of AM in fields, including tooling, repair and reconditioning [14]; the aerospace industry [15]; medical implants [16] with a special focus on the development of metallic implants [17]; dental devices [18]; and the micro-fabrication of microrobots [19], has significantly strengthened the importance of the field. However, despite the vastness of and the emphasis placed on the two fields, there is minimal to no overlap between magnetism and AM.

Metal AM processes are reliant on the delivery of feedstock materials, such as sheets, powder or wire on a substrate/object, melting the feedstock materials using a reliable energy source, such as lasers, electron beam/arc among others, and the subsequent solidification of the material in a layer-by-layer approach [20]. The ability to make complex parts using metals makes it a prime candidate for further research.

The two most commonly used techniques for metal additive manufacturing are laser powder fed additive manufacturing (LPF-AM) (also known as the direct energy deposition (DED) technique) and the laser powder bed additive manufacturing (LPB-AM) technique (also known as powder bed fusion (PBF) technique). The primary emphasis is placed on LPF-AM technique due to its compatibility with the electromagnetic micromanipulator levitation system discussed in this research. The critical components of interest within the LPF-AM system are typically a high energy laser system, a powder feeding component/nozzle and a movable worktable [21] or robotic arm [22].

The LPF-AM technique generates tool paths for deposition using sliced cross-sectional dimensions obtained from CAD models. The material nozzle head carries the material (usually powder form) along with an inner gas (e.g., argon) to deliver the feedstock material onto the build surface. Through the use of a power source (e.g., laser), the material deposited is melted and solidified to build parts using the layer-by-layer approach as described in [23].

According to [24], the critical process parameters associated with the implementation of LPF-AM technique arise from two critical steps: The delivery of heat and mass onto the build surface and the interaction of the heat and mass at the build surface. The parameters of relevance include the laser source ignition, gas-powder delivery, powder stream and laser beam interaction and melt pool formation by heat conduction, among several others. These parameters aid in determining the quality of parts built with LPF-AM. As discussed in Section 6, the key process parameters to determine the compatibility of the electromagnetic micromanipulator levitation system are the mass deposition rate, the velocity of powder and the nozzle angle.

The design for the electromagnetic micromanipulator levitation system discussed in this paper was developed with the objective of creating a compact arrangement of electric coils to be used in an AM environment. The specific application of this system correlates with a LPF-AM type of metal 3D printing where a computer-controlled robotic arm melts a metal filament with a laser or an electron beam and create complex structures.

The use of magnetic micromanipulator levitation system for AM applications offers some significant advantages. First, building the part on a levitated geometry bypasses the need for a substrate as utilized in conventional AM applications [25]. Furthermore, since the levitated geometry is anticipated to be a portion of the built part, the amount of postmanufacturing operations required is reduced significantly. A proof of concept for the application of magnetism and acoustic levitation systems for AM application was developed [26]. However, the article presents a high-level view of the idea.

Designing an Electromagnetic (EM) micromanipulator levitation system is not quite straightforward as multiple factors need to be considered. The material selection, geometric constraints, input power and peak frequency capabilities, undesirable inductance and impedance related drawbacks are only a few of the major criteria that need to be evaluated. This article offers a comprehensive analysis of such a system, comparing simulation analyses conducted in ANSYS Maxwell, a world-renowned EM simulation software, with the experimental apparatus developed. Key emphasis is placed upon the constraints that led to the development of the system causing the above-mentioned drawbacks and the strategies employed to overcome them.

Material properties have been a crucial point of emphasis for magnetic levitation applications. Conventional magnetic levitation techniques focus on the material properties of the core [27] or the ferromagnetism of the levitated part [28]. The novel magnetic levitation technique presented in this article aims to use the principle of eddy current levitation to facilitate the levitation of nonmagnetic materials, such as aluminum.

While this technique has been developed previously [29,30], the emphasis has always been placed on the levitation of aluminum. However, the compatibility of other conducting materials for eddy current applications has not been explored. Ref. [31] presented the compatibility of various conductive metals for induction heating applications. They developed the levitation ability of materials as the ratio of the maximum levitation force (F_{lev}) to the weight of the object (W) as presented in Equation (1).

$$Lev\ Ability = \frac{F_{lev}}{W} \tag{1}$$

However, Ref. [31] relied heavily on the use of experimental analyses to measure the maximum levitation force to develop the levitation ability of the materials. The research presented in that article aimed to develop a parameter to ease the process of determining the levitation ability of materials and subsequently the compatibility of different conducting materials for eddy current levitation applications. The theory presented was verified through simulation analyses.

The objective of the research presented in this article is the development of a magnetic micromanipulator levitation system that is suitable for AM applications. The schematic for the envisioned system is shown in Figure 1a. The initial constraints, optimization and associated design decisions, implementation through simulation and experimental analyses and the compatibility of the developed system with AM applications are presented. The final envisioned manipulator is expected to provide a positional stability of a few micrometers in the axial and lateral axes. This positional stability is anticipated to be offered with the effects of powder deposition as well.

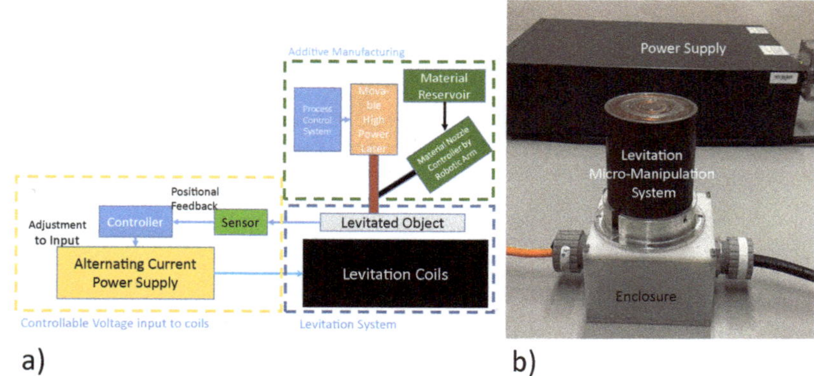

Figure 1. Micromanipulator levitation system. (**a**) Schematic for the envisioned system. (**b**) Experimental apparatus.

2. Theory

2.1. Working Principle

The working principle of the micromanipulator levitation system was described in our previous work [32] as follows.

- According to Ampere's Law, the time varying currents of the coil will produce time-varying magnetic fields
- According to Faraday's Law, the axial component of the time-varying magnetic fields will induce a voltage in conductors placed in close proximity to the micromanipulator levitation system. This will result in the induction of currents in the conductors. These induced currents are called eddy currents.
- According to Lorentz' Law, the induced eddy currents will interact with the radial component of the source magnetic field to generate repulsive force. This resulting force will be responsible for the levitation of the levitated object.

Additional details of the working principle are presented in Section 2.5.

2.2. Electromagnetic Micromanipulator Levitation System

The micromanipulator levitation system is the set of electromagnetic coils responsible for the generation of levitation force resulting in the free suspension of the levitated disc as shown in Figure 1b.

The two concentric coils carry current in the opposite directions. The inner coil is responsible for the production of levitation force in the axial (z) axis while the outer coil is responsible for the production of restoration forces in the lateral (x,y) axes to facilitate levitation at the equilibrium point. The two coils are placed within a highly magnetized iron core to facilitate magnetic focusing towards the levitated geometry.

2.3. Micromanipulator Levitation System as a Series RLC Circuit

Impedance is defined as the measure of overall opposition of the circuit elements to the current flowing through it. Impedance is different from conventional resistance because impedance is variable with frequency. Higher frequencies result in high impedance. Impedance is also significantly dependent on the inductance and capacitance of the circuit as well. The relationship of these components is shown in Equation (2). Inductive reactance (shown by X_L) is the opposition to the current offered by the inductor in the circuit. Capacitive reactance (given by X_C) is the opposition to the current offered by the capacitor in the circuit.

$$Z = \sqrt{R^2 + (X_L - X_C)^2} \quad (2)$$

$$X_L = 2\pi f L\,,\ X_C = \frac{1}{2\pi f C}$$

where R is the DC resistance of the circuit, X_L is the inductive reactance, X_C is the capacitive reactance, f is the frequency of supplied voltage, L is the inductance of the circuit, and C is the capacitance of the circuit.

Impedance becomes quite relevant in applications with alternating currents. Thus, a key point of emphasis of this research was to minimize the overall impedance of the system. The lower the impedance of the system, the higher the current across the coils with a given voltage input. Thus, minimizing the impedance results in an increase in the current across the levitator, which subsequently results in higher levitation forces.

2.4. Magnetomotive Force of Coils

The force of levitation is directly affected by the Magnetomotive Force (MMF) of the coils. The MMF is the product of the number of turns of the coils to the current through the coils [33]. According to [29], the average force of levitation is derived as:

$$F_{Z_{avg}} = \frac{(N1 - N2)^2 I_0^2 \mu_0 A_{airgap}}{4z^2} \quad (3)$$

where $N1$ and $N2$ are the number of turns of turns of coil 1 and coil 2, respectively, I_0 is the current through the coils, μ_0 is the permeability of free space, A_{airgap} is the area of the air gap under the disc, and z is the distance of the disc from the levitation coil.

$N \times I$ is the magnetomotive force, which is the line integral of the magnetic intensity around a closed line. Maximizing $N \times I$ is the objective function of the development of this system, while minimizing the inductance 'L', as well as the resulting impedance 'Z', which is the cost function. A higher resultant levitation force is a necessity to counteract the opposing force (weight) imposed by the deposition of material on the disc substrate in an AM environment. The inductance of a multi-coil, multi-core system is a non-linear property, and theoretical calculation of the same is not an objective for this research, it will be calculated and minimized through simulations using ANSYS Maxwell (Version: ANSYS electronics desktop 2020 R2. Pennsylvania, USA.

2.5. Levitation Ability

The working principle described in Section 2.1 are modelled analytically as shown in Equations (4) (Faraday's Law), (5) (Ohms Law) and (6) (Lorentz' Law):

$$E_{induced} = \frac{d\phi}{dt} \quad (4)$$

$$J = E_{induced} \frac{\sigma A_{conductor}}{l} \quad (5)$$

$$F_{lev} = \int J \times B dV \quad (6)$$

where J is the induced eddy currents, ϕ is the magnetic flux, B is the magnetic field, F_{Lev} is the levitation force, $E_{induced}$ is the induced emf, l is the length of conductor, $A_{conductor}$ is the area of the conductor, σ is the conductivity of the material and dV is the differential volume of the conductor. The gravitational force of the levitated object is modelled as in Equation (7).

$$F_{gravity} = mg = \rho V g \quad (7)$$

where m is the mass of the object, ρ is the density of the material of the conductor, V is the volume of the levitated object, and g is the gravitational constant. According to [31], the levitation ability of a material is the defined as the ratio of the maximum levitation force experienced by the levitated object to the gravitational force experienced by the levitated object. This is described analytically in Equation (5).

According to [31], the levitation ability of a material is the defined as the ratio of the maximum levitation force experienced by the levitated object to the gravitational force experienced by the levitated object. This is described analytically in Equation (8).

$$Lev\ Ability = \frac{F_{lev}}{F_{gravity}} = \frac{\int J \times BdV}{\rho V g} \quad (8)$$

As can be seen from Equations (4)–(6), the only material property that affects the levitation force of the conductor is the conductivity of the material. As evident from Equation (7), the only material property of the material that affects the gravitational force on the object is the density of the material. Thus, the levitation ability of materials can be represented as the ratio of the conductivity to the density of the material as shown in Equation (9).

$$Lev\ Ability = \frac{F_{lev}}{F_{gravity}} = \frac{\int J \times BdV}{\rho V g}\ \alpha\ \frac{\sigma}{\rho} \quad (9)$$

Several different materials were considered for the analysis. A strong emphasis was placed on materials used for additive manufacturing applications. The technique is critical to determine the compatibility of different materials employed for AM operations with the magnetic micromanipulator levitation system. The levitation ability of several conductors is plotted in Figure 2. According to Figure 2, aluminum and its alloys have a high levitation ability owing to the high conductivity and low density. Other materials, such as titanium and nickel and its alloys have a low levitation ability due to their high density.

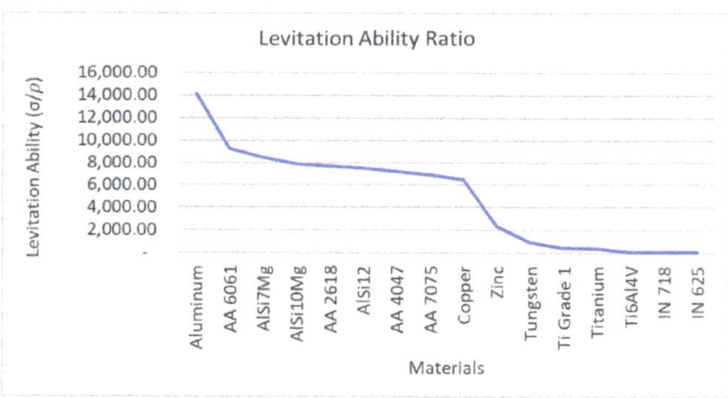

Figure 2. Levitation ability: the ratio of $\frac{\sigma}{\rho}$ for different materials.

3. Design and Optimization of the Micromanipulator Levitation System

The optimization technique used here is the trial-and-error direct substitution method. As shown in Figure 3, each design variable is selected individually, and the design variable is iterated within a range of values, keeping every other parameter constant. The value of the design variable with maximum levitation force output is selected as the optimized value. The optimization procedure entails the use of finite element analysis (FEA) software ANSYS Maxwell to compute the objective function (which is the levitation force in the axial axis).

The levitator setup (with system components as shown in Figure 4a) is supposed to be placed within an AM machine. Thus, there was a dimensional constraint that the outer diameter of the system could not exceed 90 mm in the radial axis. This constraint was developed using the DMD-IC106 AM machine as reference. The DMD-IC106 has a working envelop of 350 mm × 350 mm × 350 mm [34]. For the developed system to be minimally invasive, the volume occupied by the magnetic micromanipulator levitation system was minimized. By placing a 90 mm constraint on the system, the final levitation system, consisting of the coils and its enclosure, has a total volume 11,735,850 mm^3. Thus, the levitation system only occupies 28% of the available working envelope, therefore, leaving sufficient volume for the material deposition activities.

The levitated object in consideration is a disc of radius 25 mm and height 5 mm. Some of the parameters kept constant throughout the optimization process are shown in Figure 4b.

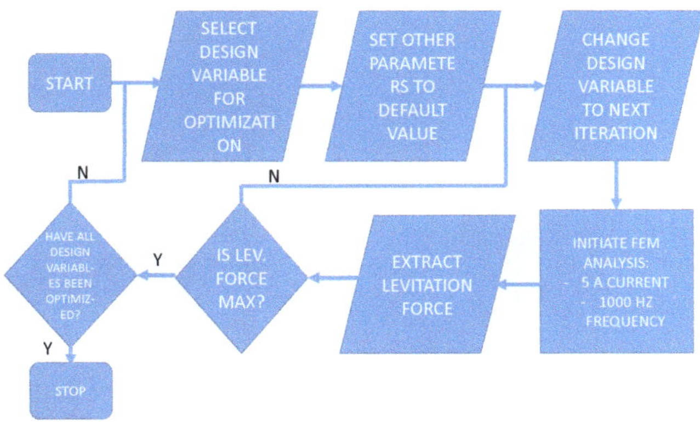

Figure 3. Flowchart of the optimization technique.

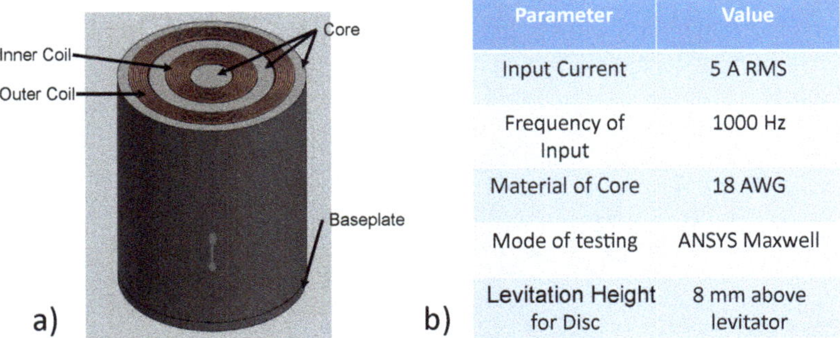

Parameter	Value
Input Current	5 A RMS
Frequency of Input	1000 Hz
Material of Core	18 AWG
Mode of testing	ANSYS Maxwell
Levitation Height for Disc	8 mm above levitator

Figure 4. System components and parameters of micromanipulator levitation system. (**a**) CAD Model of the envisioned System. (**b**) Definition of constant parameters for optimization.

3.1. *Optimization of Width of Coils*

The trend of levitation force as a function of the coil 1 width and coil 2 width is presented in Figure 5a and 5b, respectively. From the data obtained, the optimized width of the inner coil was 12 mm, and the optimized width of the outer coil was 9 mm.

3.2. *Optimization of the Radial Placement of teh Coil*

Following the optimization of the two coil widths, we determined the placement of these coils within the highly magnetized ferrite cores. The ratio of the mean radius of coil 2 (R2) to the mean radius of coil 1 (R1) was used as the optimization parameter. The trend of the levitation force as a function of R2/R1 was study to determine the optimum placement of the coils. The resulting plot is shown in Figure 6a. As shown in Figure 6a, the optimum R2/R1 is 2.1562 (R2 = 34.5 mm, R1 = 16 mm).

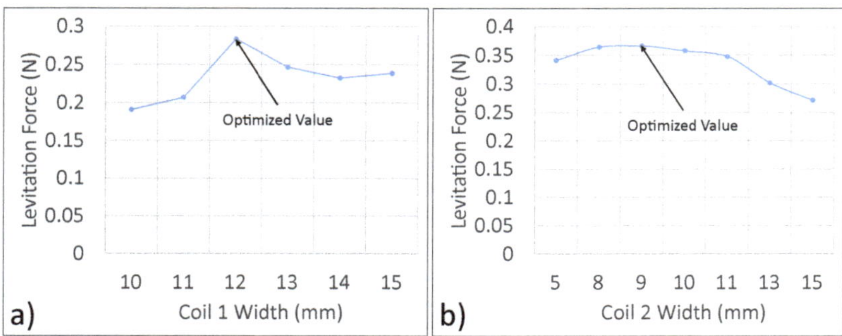

Figure 5. Optimization of the width of coils. (**a**) Optimization of the width of Coil 1. (**b**) Optimization of the width of Coil 2.

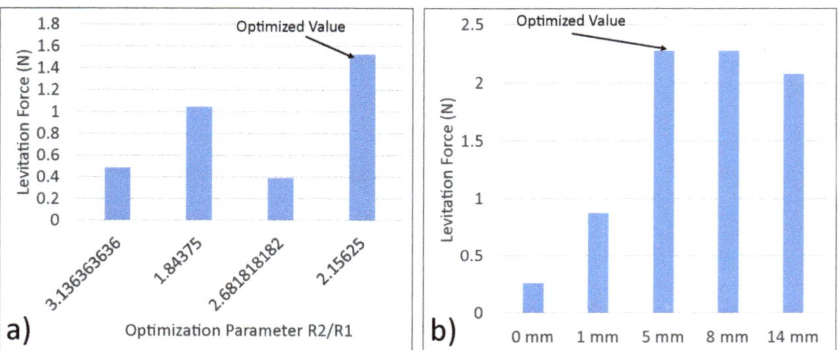

Figure 6. (**a**) Optimization of radial placement of coils; (**b**) Optimization of baseplate height.

3.3. Calculation of Coil Height

Having computed the width of the coils, the radial placement of the coils and the gauge of the wire to be used, the fill factor (*FF*) of the coil was used to determine the height of the coil. The fill factor is the ratio of the area of electrical conductor (cross-sectional area of wire) to area of the provided space (cross-sectional of coil) as shown in Equation (10).

$$FF = \frac{d^2 \cdot \frac{\pi}{4} \cdot n}{b \cdot h} \tag{10}$$

where d is the diameter of the wire, n is the number of conductors in the coil, b is the width and h is the height of the coil. For the sake of this analysis, a fill factor of 0.72 was used.

$$0.72 = \frac{\frac{\pi}{4}\,(1.1)^2\,1000}{12\,h_{coil1}} \qquad h_{coil1} = 110\text{ mm} \tag{11}$$

To keep the same height for coil 1 and coil 2, the number of turns for coil 2 were adjusted as shown in Equation (12).

$$N = \frac{110 \times (0.8) \times 9}{\frac{\pi}{4} \times (1.1)^2} = 750.05 \sim 750\ textturns \tag{12}$$

During the coil manufacturing process, with the incorporation of epoxy and potential variations in the assumed vs. real wire diameter values, the numbers of turns that could be incorporated in coil 1 and coil 2 were 920 and 800, respectively.

3.4. Optimization of Baseplate

A baseplate is a plate attached to the bottom of the micromanipulator levitation system. The key objective of the baseplate is to improve the magnetic field focusing capability of the system. Five different cases were considered for the analysis here. First, there was no baseplate added to the levitator. Next, 1 mm, 5 mm, 8 mm and 14 mm baseplates were added. The levitation forces for the cases were compared. The case with the maximum levitation force was pursued as the final optimized levitator setup. As shown in Figure 6b, the levitation force is maximum for the 5 mm baseplate.

3.5. Selection of Core Material

The next step was to find the ideal material for the core. Ferrite, electric steel and pure iron were shortlisted due to their high permeability. Solid cores were compared. Comparisons are based on levitation force and restoration force (with 5 mm displacement in x axis). The analysis was also conducted at 60 Hz to get a deeper understanding of the material performance.

As evident from Figure 7a, pure iron cores produce the highest levitation force. Figure 7b shows that pure iron also produces comparable restoration forces at both 1000 and 60 Hz. The final system parameters are listed in Table 1.

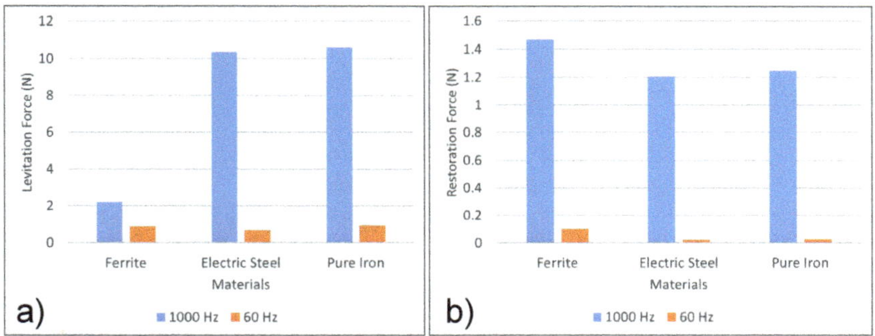

Figure 7. Comparison of performance of the materials. (**a**) Comparison of core materials: Levitation Force (N). (**b**) Comparison of core materials: Restoration Force (N).

Table 1. Specifications of the levitation system.

Parameter	**Value**
Outer Diameter of Levitation System	90 mm
Height of Levitation System	115 mm
Core Material	Pure Iron
No. of turns Coil 1 N1	920
No. of turns Coil 2 N2	800
Distance between disc and levitator	0 mm
Wire AWG	18 AWG
Disc Radius	25 mm

Pure iron (99.6% pure) was utilized to serve as the core of the micromanipulator levitation system. The BH curve, which is the graph plotted between magnetic flux density (B) and the magnetic field strength (H), of the material is from the specification sheet of the material [35]. For the range of magnetic field anticipated for the micromanipulator levitation system (<1000 G), the magnetic permeability of the core was found to be 700.

3.6. Frequency of Operation

Due to the dimensional constraints of the micromanipulator levitation system developed, the height of the micromanipulator levitation system needed to be high to incorporate the number of turns for coil 1 and coil 2 as explained in Section 3.3. This resulted in a significant increase in the inductance (and subsequently the impedance). This is highlighted through the variation of impedance vs. frequency for the micromanipulator levitation system obtained from ANSYS Maxwell as shown in Figure 8. To facilitate operation through a voltage controlled power supply, a frequency with a low impedance output was selected for operation.

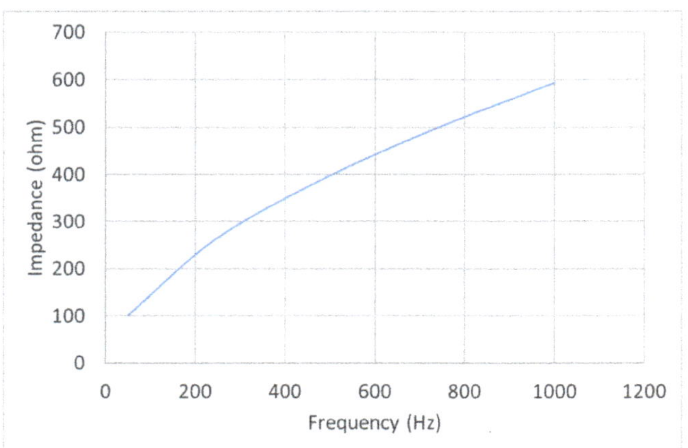

Figure 8. Impedance vs. frequency for the micromanipulator levitation system from ANSYS Maxwell.

4. Free Levitation Experiment

4.1. Initial Levitation Experiment

As discussed in Section 2.2, a set of two concentric coils carrying current in the opposite directions was employed for the levitation experiment. Some of the key characteristics of the system are shown in Table 1. ANSYS Maxwell was used for the simulation analyses. The simulation analysis of the proposed system is shown in Figure 9. According to the simulation data obtained, a current micromanipulator levitation system apparatus attempting to levitate a disc with a 25 mm radius will not be able to do so successfully. The simulation data was verified through experiments, where high-frequency vibration of the levitated disc was observed.

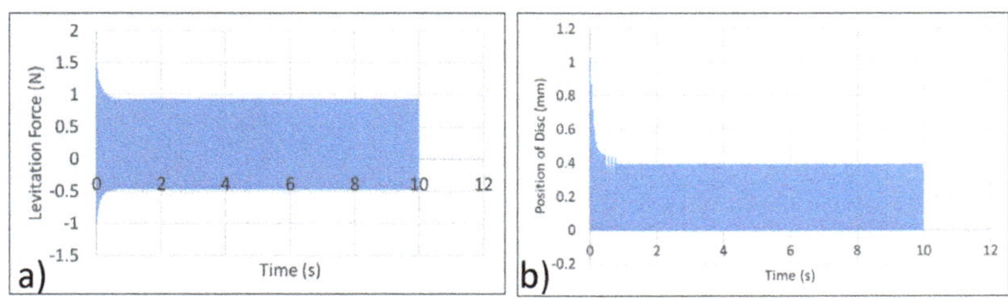

Figure 9. Simulation data from ANSYS Maxwell. (**a**) Levitation force vs. Time. (**b**) Position of disc vs. Time.

4.2. Strength of Coils

Since the two coils carry current in the opposite directions, through the principle of superposition, the two coils oppose each others magnetic fields. From theory, it is known that the inner coil is responsible for the production of axial levitation force while the outer coil is responsible for the production of restoration forces in the lateral axes as outlined in Section 2.1. Since sufficient force is not observed in the axial axis, there was a need to reduce the strength of the outer coil to facilitate a higher levitation force in the axial axis.

As explained in Section 2.4, the strength of the coils is dependent on two factors: The number of turns of the coil and the current through the coil as shown in Equation (13).

$$Coil\ Strength = \frac{N1\ I}{N2\ I} = \frac{N1}{N2} \tag{13}$$

The current ratio of the strength of coils of 1.15 produces significant restoration forces in the lateral axes. However, the system is unable to produce sufficient levitation forces in the axial axis to facilitate free levitation. Since the number of turns were established before-hand, the strength of coils was altered by changing the current through the coil.

From the initial experiment, it was clear that the strength of the outer coil need to be reduced to facilitate an increase the levitation forces in the axial axis. This was achieved through the addition of a resistor in parallel to coil 2. The circuit setup for the system is shown in Figure 10. As expected, the current through coil 2 would split between the coil and the resistor in parallel. Through simulation and experimental analysis, steady state levitation was with a 40 Ω resistor in parallel to coil 2 at an operating frequency of 85 Hz.

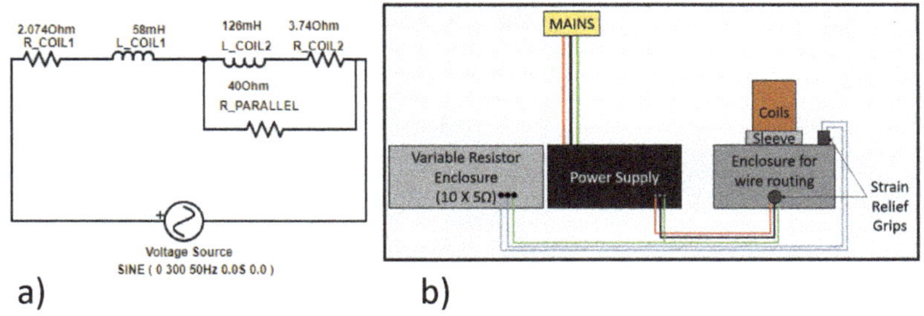

Figure 10. (**a**) Circuit representation of the coils + resistor in parallel setup. (**b**) Schematic of the experimental apparatus with adjusted strength of coils.

4.3. Simulation Analysis

ANSYS Maxwell was used for the analysis. A 300 V input at 85 Hz was supplied to the micromanipulator levitation system. With the 300 V at 85 Hz input, the current through coil 1 (I1) is 4.6582 A RMS and the current through coil 2 (I2) is 2.7844 A RMS. The levitation force in the axial axis and the position of disc as a function of time were extracted from the simulation results. The data are presented in Figure 11.

4.4. Experimental Analysis

To verify the simulation analysis conducted, a 300 V RMS at 85 Hz was supplied to the levitation coils. The power supply employed for the experiment was BK precision Model 9830B, which can support an RMS power output of 3000 VA (10 A RMS at 300 V RMS) with a response time of 1.5 ms and the frequency of operation between the range of 43–1000 Hz. The steady state position of the disc was then compared to the data obtained through simulations. The experiment shows the steady state position of the disc to be about 4.5 mm. This is in close agreement with the simulation data. Thus, the experiment was deemed a success as shown in Figure 12. Thus, steady state levitation of the disc was achieved.

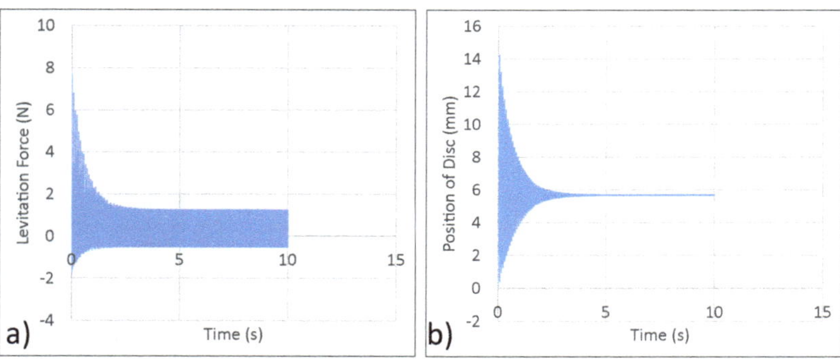

Figure 11. Simulation result with the resistor in parallel (**a**) Levitation force vs. time—with resistor in parallel. (**b**) Position of disc vs. time—with resistor in parallel.

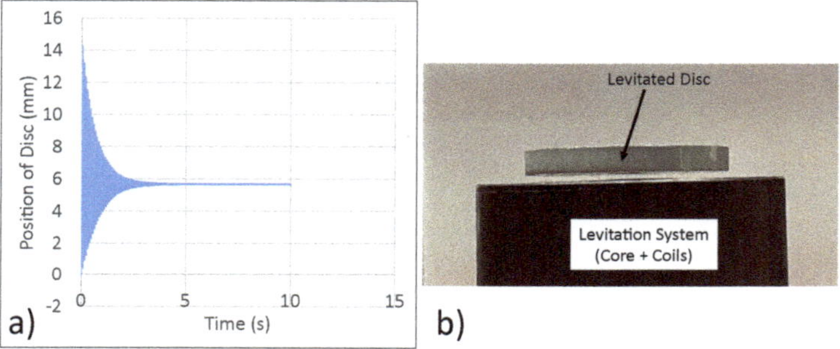

Figure 12. Experimental verification of analysis (**a**) Position of Disc vs. Time—ANSYS Maxwell. (**b**) Steady State Position of Disc—Experiment.

4.5. Experimental Analysis with Additional Payload

A critical aspect of AM is the addition of material on the substrate as a function of time. Thus, it was critical to depict the retention of stability of the levitated substrate with the addition of mass as a function of time. In order to highlight the systems ability to support mass added during the AM operation and its subsequent compatibility within the AM environment, an experiment was conducted to mock up AM environment where the mass of the levitated geometry will increase with time.

The following experiment was conducted to ensure the system can cope with payload variation in AM environment. Here, two additional distinct masses were added to the levitated disc and its behavior was observed. Two different payloads (15.2 and 4.4 g) were placed on top of the levitated substrate to observe the system's ability to retain stability. Following the addition of the payload, steady state levitation was observed for both payloads as shown in Figure 13.

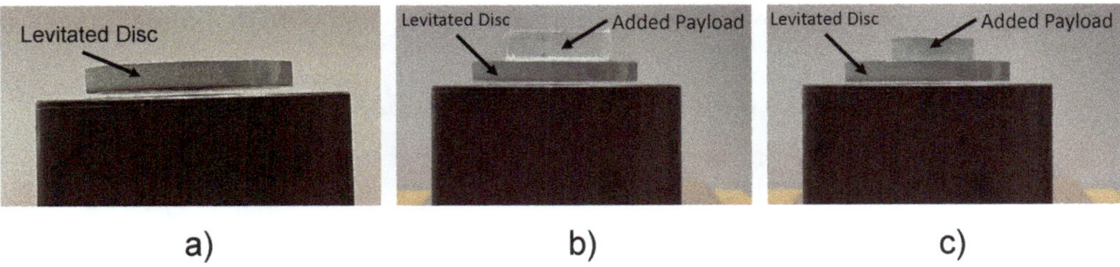

Figure 13. Experiment with additional payloads to highlight stable levitation with added mass as expected within the AM environment. (**a**) Free levitation of aluminum disc. (**b**) Added payload: 4.4 g (17% of initial disc). (**c**) Added payload: 15.2 g (59% of initial disc).

5. Semi-Levitation Experiment

5.1. Need for Semi-Levitation Experiment

With the successful implementation of free levitation of a suspended disc, the emphasis was shifted to a critical drawback of the system: the range of disc radii that can be supported for free levitation. With the increase in disc radii, the weight of the disc increases and will thus require a larger strength of coil for the inner coil.

In addition, the disc will lose its lateral stability as disc radius increases. This was verified through the employment of ANSYS Maxwell simulation analyses. Figure 14 shows the variation of restoration forces (obtained from ANSYS Maxwell) in the lateral (x) axis with a 3 mm displacement provided in the positive x-axis. As it can be seen, for the radii of disc over 40 mm, the forces in the x-axis are positive, therefore, contributing negatively to the stability of the system.

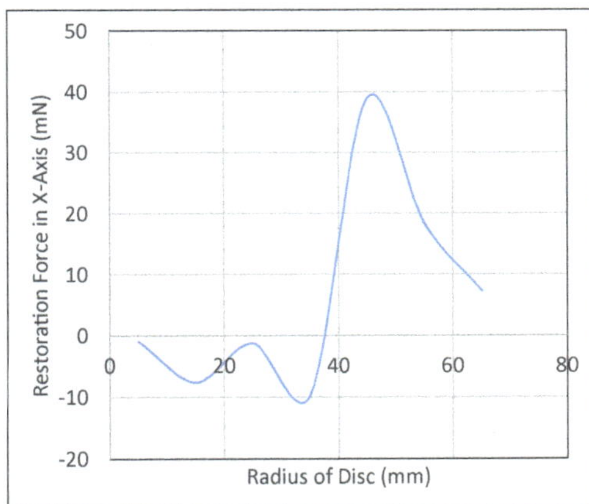

Figure 14. The need for a micromanipulator semi-levitation system.

5.2. Working Principle

The two coils carry currents in the same direction to maximize the levitation forces. Since the two coils carry current in the same direction, through the principle of superposition, the effect of the magnetic field of the system will increase. However, since the outer coil also carries current in the same direction as the inner coil, no restoration forces will be produced in the lateral axes. To counteract the loss of stability, the system uses 4 external stands that serve as supports to provide restoration forces in the lateral axes.

5.3. Simulation and Experimental Analyses

A 300 V RMS input at 50 Hz is supplied to the coils. First ANSYS Maxwell is used to extract the variation of position of disc as a function of time. Next, the levitation coils are connected in series carrying current in the same direction. The resulting data from ANSYS Maxwell is shown in Figure 15. From simulation, the steady state position of the disc was 7.8 mm. From experimental analysis, the steady state position of the disc was 8 mm. Thus, simulation and experimental analysis are in close agreement.

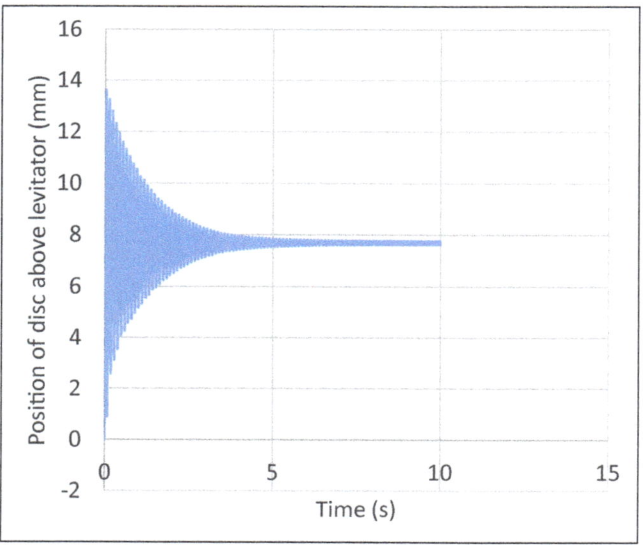

Figure 15. Performance of micromanipulator semi-levitation system.

6. Compatibility of Micromanipulator Levitation System with Additive Manufacturing Applications

The envisioned interaction of the magnetic micromanipulator levitation system within the AM enclosure is shown in Figure 16. Having successfully conducted the free levitation and semi-levitation experiment, it was integral to determine the compatibility of the magnetic micromanipulator levitation system with AM applications. Section 4.5 presents the experimental verification of the levitated substrate capable of supporting added payload as a function of time without losing stability. The compatibility of the micromanipulator levitation system is further tested by accounting for the effect of powder deposition.

6.1. Context

In Laser Powder-Fed AM, the powder stream is fed onto the substrate and fused through the use of lasers on the substrate. The levitated geometry is expected to be subjected to a similar process. Thus, the deposition of powder during the AM process can have a detrimental impact of the stability of the levitated geometry. According to [25], some of the critical AM process parameters are:

- Mass deposition rate: 5 g/min [25].
- Velocity of powder: 2 m/s [36].
- Nozzle angle: 60 degree [25].

Two analyses are conducted to verify the stability of the system with powder deposition.

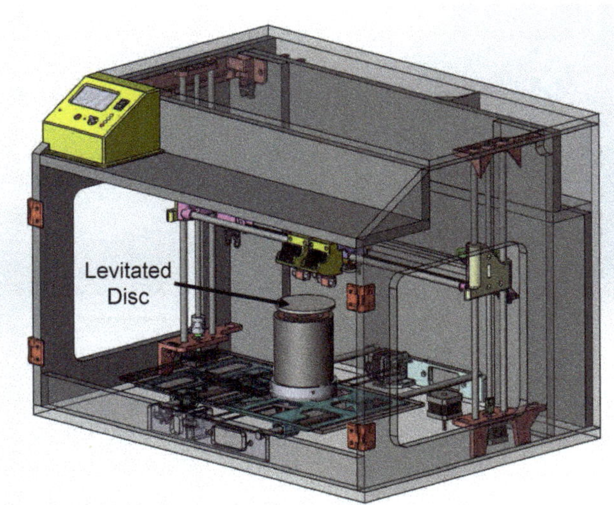

Figure 16. Envisioned MagLev system within the AM enclosure.

6.2. *Axial Stability*

The first step is to verify that sufficient levitation force is produced to overcome the effect of powder deposition in the axial axis. To verify the performance, a worse case scenario analysis is conducted to ensure reliable performance during real operation. Some of the critical assumptions for this analysis are:

- The analysis assumes steady flow of powders and the impact of air friction is negligible. Since AM operations occur in a vacuum, this assumption is fair.
- Since the size of the particles are very small and originating at the same source, the collisions between the particles can be ignored.
- The nozzle angle is 0 degrees. This ensures all the force of powder deposition only acts in the axial axis.
- The coefficient of restitution is 0. This implies that all the kinetic energy of the powder is transferred to the levitated geometry and the final velocity of the powder is 0 m/s.
- The mass deposition is assumed to be continuous at 5 g/s (higher than the expected mass deposition rate)

The principle of transfer of momentum is highlighted in Equation (14).

$$F_{impact} = m_{powder}(V_{powder,final} - V_{powder,initial}) \quad (14)$$

Inputting the data presented in Section 6.1, the impact force results are highlighted in Equation (15).

$$F_{impacts} = 0.005(0 - 2) = 0.01 \text{ N} \quad (15)$$

Adding the force of continuous deposition along with the force of gravity in ANSYS Maxwell, it can be clearly seen that the system retains its stability. The system stabilizes 3 mm above the levitator as shown in Figure 17. It should be noted that a worse case scenario analysis was conducted with a highly exaggerated mass deposition rate. Highlighting the system's ability to facilitate stable suspension in the worse-case-scenario analysis shows that the system will allow for stable suspension with real operation as well.

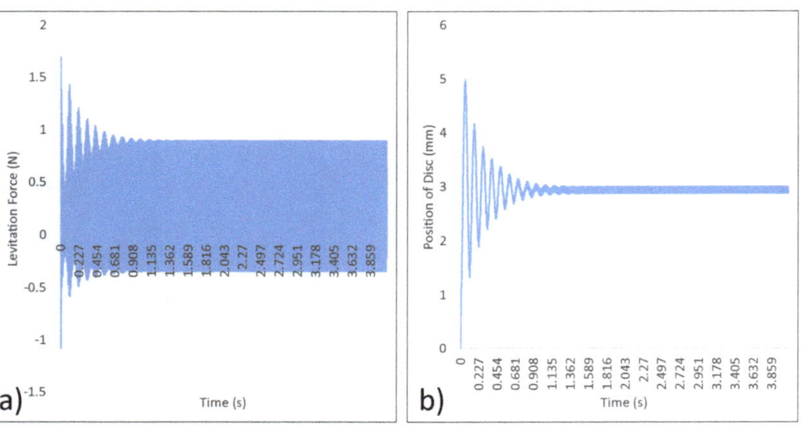

Figure 17. Stability of the disc in the axial axis (**a**) Levitation force vs. time (**b**) Position of disc vs. time.

6.3. Lateral Stability

Lateral stability of the system was verified experimentally. Here, the micromanipulator levitation systems ability to come back to its equilibrium point was tested by displacing the disc significantly (20 mm displacement) in the lateral axis. Upon turning the micromanipulator levitation system on, the disc moves back to its expected equilibrium position. This is highlighted in Figure 18, which shows the variation of position of disc vs. time, with the achievement of steady state levitation at the equilibrium point.

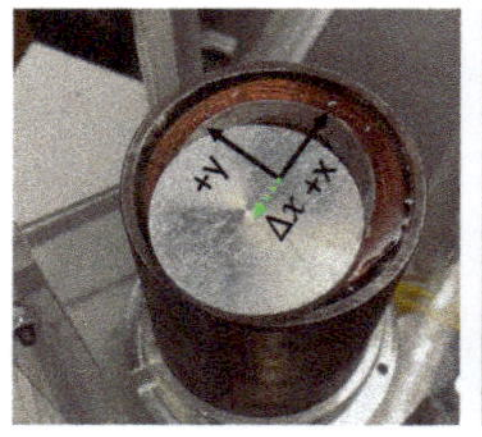

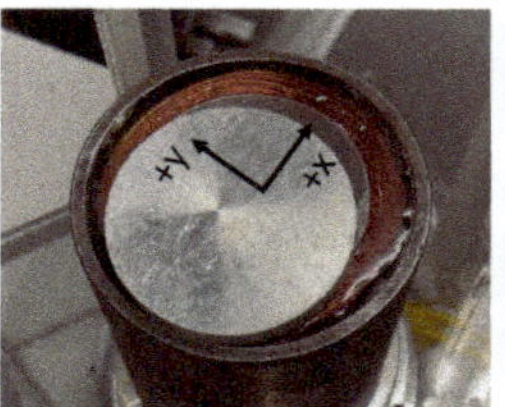

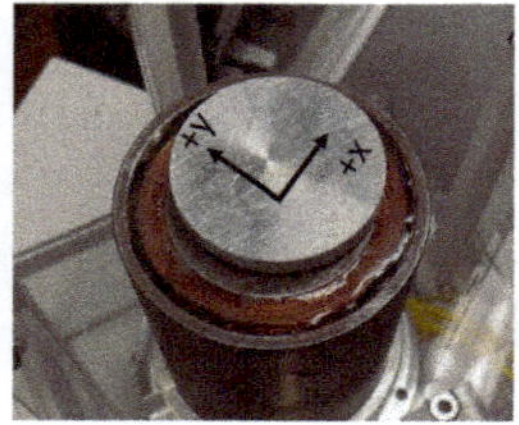

Time-step 1
Significant Displacement
(o⍼) in –x axis

Time-step 2
Restoration force in + x axis

Time-step 3
Intermediate timestep

Time-step 4
Stability at equilibrium point

Figure 18. Stability of disc in the lateral axis.

From the experiment, it can be clearly seen that the system is capable for maintaining stability even with large displacements in the lateral axes. The forces of impact of powder deposition will not cause displacements in the lateral axes larger than 20 mm. Thus, the stability of micromanipulator levitation system will be retained in the axial axis with powder deposition.

7. Verification of Levitation Ability

The semi-levitation experimental apparatus is utilized for the verification of the levitation ability of materials. Ref. [31] states that the maximum levitation force should be emphasized for the development and verification of levitation ability of materials. With the two coils in series, the magnetic field is maximized through the principle of superposition. Thus, the levitation force produced is maximized in the semi-levitation experimental apparatus. The levitation force on the disc is documented 2 mm above the levitation. A 300 V RMS input is supplied to the coils at a frequency of 50 Hz.

Simulation analyses are relied upon for the verification of the levitation ability of materials. Through simulations, a wide variety of materials with applications in the AM environment could be subject to testing. The use of several materials will bolster the validity of the principle of levitation ability. The reliability of simulation analyses has also been verified in Sections 4 and 5 by depicting the close agreement between simulation and experimental analyses.

The analytical model and simulation data for different materials have been normalized relative to the levitation ability of aluminum. The ratio of levitation force to gravitational force (weight) was obtained through ANSYS Maxwell. The data are documented in Figure 19. As it can be seen, the two plots follow the same trend. Thus, the viability of the principle of the levitation ability of materials was verified.

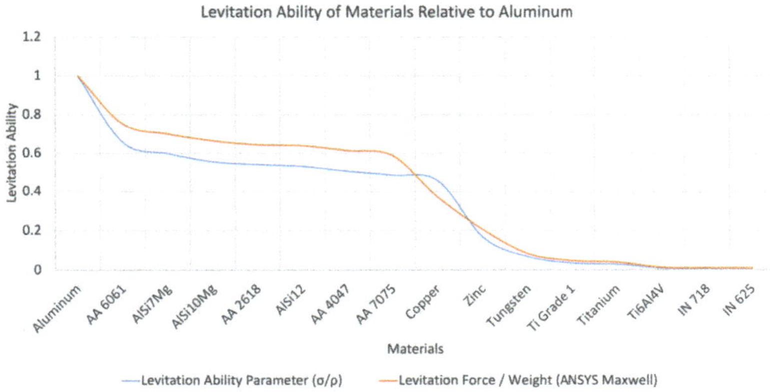

Figure 19. Levitation ability. Analytical model, simulation.

8. Conclusions and Future Work

The simulation and experimental work shown in this article highlights the initial viability of electromagnetic levitation of materials for additive manufacturing applications. The optimization of the micromanipulator levitation system and most relevant design decisions are discussed. To facilitate levitation using the magnetic micromanipulator levitation system, a resistor was added in parallel to the outer coil to reduce the MMF of the outer coil, therefore, facilitating levitation of the geometry that was to be suspended. With a 300 V input at 85 Hz input and a 40 Ω resistor in parallel to the outer coil, an aluminum disc of 25 mm radius and 5 mm height was levitated at a height of 4.5 mm, thereby, highlighting successful levitation with the system.

The compatibility of the micromanipulator levitation system developed with AM applications is also discussed in great detail. Three distinct analyses were conducted to verify the compatibility. First, the ability of the micromanipulator levitation system to support the suspension of an additional payload (to mimic the effect of added mass due to mass deposition in the AM processes) was tested experimentally. The system was able to support the suspension of the initial geometry plus up to 15.2 g (59% of the initial geometry mass).

Next, the effect of the impact of powder deposition, which is critical to the AM process, on the stability of the levitated geometry was reported. Through simulation analysis, the levitated geometry retained its stability in the axial axis despite having a powder deposition force of 0.01 N (which is much higher than the expected force of powder deposition) acting on the same. The system also produced sufficient restoration forces in the lateral axes, which was highlighted through experiments. The levitated geometry was given a significant 20 mm initial displacement (as a disturbance) in the x-axis.

The levitated geometry subsequently came back to the equilibrium point once the system was turned on. The novel methodology of using the ratio of conductivity of the levitated geometry's material (σ) to the density of the material (ρ) to determine the compatibility of different materials within the AM sphere with magnetic levitation applications was also discussed. The newly developed method relies solely on material properties for the determination of levitation ability of materials as opposed to experimental analyses. The testing of the micromanipulator levitation system within the AM enclosure to satisfactorily build a part using the technique LPF-AM is the critical next step of this research. The testing of the compatibility will be conducted using the DMD-IC106, an industrial metal 3D printer. Some important factors to consider include ensuring that no collision exists between the components of the AM machine and the micromanipulator levitation system, ensuring the micromanipulator levitation system is protected from the intrusive conductive dust present in the AM machine, and ensuring that no damage occurs to the laser due to laser-back reflection (LBR). LBR occurs when the light from the laser beam reflects back into the optic resulting in optical result loss and damage to the equipment.

Next, the positional stability offered by the micromanipulator levitation system will be verified with and without the effects of powder deposition. The final envisioned system is expected to provide a positional stability of a few micrometers (μm). The development a set of process parameters for deposition of the selected alloy on the levitated part to arrive at high-quality printed components is also critical. This includes parameters, such as the laser power, powder feed rate and nozzle angle, to facilitate optimum AM operations.

Author Contributions: Manufacturing, methodology, implementation, investigation, data collection, formal analyses and preparation of first draft: P.K. and S.M. Conceptualization, methodology, supervision, funding acquisition, project administration, review and editing: M.B.K. and E.T. All authors have read and agreed to the published version of the manuscript.

Funding: This work was by the umbrella of the Holistic Innovation in Additive Manufacturing (HI-AM) Network through the Natural Sciences and Engineering Research Council of Canada (NSERC) and Canada Foundation for Innovation (CFI).

Data Availability Statement: Data available from the authors upon request.

Conflicts of Interest: The authors declare no conflict of interest.

References

1. Fúnez, G.E.; Carrasco, A.M.; Ordoñez-Ávila, J.L. Study Case: Electromechanical Design of a Magnetic Damper for Robotic Systems. *Int. J. Mech. Eng. Robot. Res.* **2022**. [CrossRef]
2. Zhang, X.; Trakarnchaiyo, C.; Zhang, H.; Khamesee, M.B. MagTable: A tabletop system for 6-DOF large range and completely contactless operation using magnetic levitation. *Mechatronics* **2021**, *77*, 102600. [CrossRef]
3. Zhou, R.; Yan, M.; Sun, F.; Jin, J.; Li, Q.; Xu, F.; Zhang, M.; Zhang, X.; Nakano, K. Experimental validations of a magnetic energy-harvesting suspension and its potential application for self-powered sensing. *Energy* **2022**, *239*, 122205. [CrossRef]
4. Aldawood, G.; Bardaweel, H. Self-Powered Self-Contained Wireless Vibration Synchronous Sensor for Fault Detection. *Sensors* **2022**, *22*, 2352. [CrossRef]
5. Widodo, P.J.; Budiana, E.P.; Ubaidillah, U.; Imaduddin, F.; Choi, S.B. Effect of Time and Frequency of Magnetic Field Application on MRF Pressure Performance. *Micromachines* **2022**, *13*, 222. [CrossRef]
6. Liu, J.; Li, W.; Jin, L.; Lin, G.; Sun, Y.; Zhang, Z. Analysis of linear eddy current brakes for maglev train using an equivalent circuit method. *IET Electr. Syst. Transp.* **2021**, *11*, 218–226. [CrossRef]
7. Bai, Y.W.; Lee, C.F. Magnetic Levitation Wireless Charging Platform with Adjustable Magnetic Levitation Height and Resonant Frequency for a Better Charging Efficiency. In Proceedings of the 2018 IEEE seventh Global Conference on Consumer Electronics (GCCE), Nara, Japan, 9–12 October 2018; pp. 784–785. [CrossRef]
8. Ebrahimi, N.; Bi, C.; Cappelleri, D.J.; Ciuti, G.; Conn, A.T.; Faivre, D.; Habibi, N.; Hošovský, A.; Iacovacci, V.; Khalil, I.S.; et al. Magnetic actuation methods in bio/soft robotics. *Adv. Funct. Mater.* **2021**, *31*, 2005137. [CrossRef]
9. Kazemzadeh Heris, P.; Khamesee, M.B. Design and Fabrication of a Magnetic Actuator for Torque and Force Control Estimated by the ANN/SA Algorithm. *Micromachines* **2022**, *13*, 327. [CrossRef]
10. Uvet, H.; Demircali, A.A.; Kahraman, Y.; Varol, R.; Kose, T.; Erkan, K. Micro-UFO (Untethered Floating Object): A Highly Accurate Microrobot Manipulation Technique. *Micromachines* **2018**, *9*, 126. [CrossRef]

11. Cheng, S.; Li, X.; Wang, Y.; Su, Y. Levitation Characteristics Analysis of a Diamagnetically Stabilized Levitation Structure. *Micromachines* **2021**, *12*, 982. [CrossRef]
12. Wang, Q.; Yang, L.; Zhang, L. Micromanipulation Using Reconfigurable Self-Assembled Magnetic Droplets With Needle Guidance. *IEEE Trans. Autom. Sci. Eng.* **2021**, *19*, 759–771. [CrossRef]
13. Wang, L.; Meng, Z.; Chen, Y.; Zheng, Y. Engineering magnetic micro/nanorobots for versatile biomedical applications. *Adv. Intell. Syst.* **2021**, *3*, 2000267. [CrossRef]
14. Golubev, I. Possibilities of 3D scanning of parts during repair of agricultural machinery. In *BIO Web of Conferences*; EDP Sciences: Les Ulis, France, 2021; Volume 37, p. 00003.
15. Mohanavel, V.; Ali, K.A.; Ranganathan, K.; Jeffrey, J.A.; Ravikumar, M.; Rajkumar, S. The roles and applications of additive manufacturing in the aerospace and automobile sector. *Mater. Today Proc.* **2021**, *47*, 405–409. [CrossRef]
16. Basgul, C.; Spece, H.; Sharma, N.; Thieringer, F.M.; Kurtz, S.M. Structure, properties, and bioactivity of 3D printed PAEKs for implant applications: A systematic review. *J. Biomed. Mater. Res. Part B Appl. Biomater.* **2021**, *109*, 1924–1941. [CrossRef]
17. Velásquez-García, L.F.; Kornbluth, Y. Biomedical Applications of Metal 3D Printing. *Annu. Rev. Biomed. Eng.* **2021**, *23*, 307–338. [CrossRef]
18. Tian, Y.; Chen, C.; Xu, X.; Wang, J.; Hou, X.; Li, K.; Lu, X.; Shi, H.; Lee, E.S.; Jiang, H.B. A review of 3D printing in dentistry: Technologies, affecting factors, and applications. *Scanning* **2021**, *2021*, 9950131. [CrossRef]
19. Li, J.; Pumera, M. 3D printing of functional microrobots. *Chem. Soc. Rev.* **2021**, *50*, 2794–2838. [CrossRef]
20. DebRoy, T.; Wei, H.; Zuback, J.; Mukherjee, T.; Elmer, J.; Milewski, J.; Beese, A.; Wilson-Heid, A.; De, A.; Zhang, W. Additive manufacturing of metallic components—Process, structure and properties. *Prog. Mater. Sci.* **2018**, *92*, 112–224. [CrossRef]
21. Alimardani, M.; Toyserkani, E.; Huissoon, J.P.; Paul, C.P. On the delamination and crack formation in a thin wall fabricated using laser solid freeform fabrication process: An experimental—Numerical investigation. *Opt. Lasers Eng.* **2009**, *47*, 1160–1168. [CrossRef]
22. Manoharan, M.; Kumaraguru, S. Path Planning for Direct Energy Deposition with Collaborative Robots: A Review. In Proceedings of the 2018 Conference on Information and Communication Technology (CICT), Jabalpur, India, 26–28 October 2018; pp. 1–6.
23. Palmero, E.M.; Bollero, A. 3D and 4D Printing of Functional and Smart Composite Materials. In *Encyclopedia of Materials: Composites*; Brabazon, D., Ed.; Elsevier: Oxford, UK, 2021; pp. 402–419. [CrossRef]
24. Huang, Yuze. Comprehensive Analytical Modeling of Laser Powder-Bed/Fed Additive Manufacturing Processes and an Associated Magnetic Focusing Module. Ph.D. Thesis, University of Waterloo, Waterloo, ON, Canada, 2019.
25. Huang, Y.; Khamesee, M.B.; Toyserkani, E. A comprehensive analytical model for laser powder-fed additive manufacturing. *Addit. Manuf.* **2016**, *12*, 90–99. [CrossRef]
26. Harkness, W.A.; Goldschmid, J.H. Free-Form Spatial 3-D Printing Using Part Levitation. U.S. Patent 9,908,288, 6 March 2018.
27. Bo, K.; Xiong, Y.; Yu, X.; Wu, J.; Wang, Y. Study on Transient Electromagnetic Characteristics Based on Typical Soft Magnetic Material for EMS Maglev Train. In Proceedings of the 2021 13th International Symposium on Linear Drives for Industry Applications (LDIA), Wuhan, China, 1–3 July 2021; pp. 1–4.
28. Huang, C.G.; Zhou, Y.H. Levitation properties of maglev systems using soft ferromagnets. *Supercond. Sci. Technol.* **2015**, *28*, 035005. [CrossRef]
29. Ghayoor, F.; Swanson, A. Modelling and analysis of electrodynamic suspension of an aluminium disc as a complex engineering problem. *Int. J. Electr. Eng. Educ.* **2018**, *55*, 91–108. [CrossRef]
30. Lee, S.M.; Lee, S.H.; Choi, H.S.; Park, I.H. Reduced modeling of eddy current-driven electromechanical system using conductor segmentation and circuit parameters extracted by FEA. *IEEE Trans. Magn.* **2005**, *41*, 1448–1451.
31. Nan, W.; Wen-Jun, X.; Bing-Bo, W. Physical characteristics of electromagnetic levitation processing. *Chin. Phys. B* **1999**, *8*, 503–513. [CrossRef]
32. Kumar, P.; Huang, Y.; Toyserkani, E.; Khamesee, M.B. Development of a Magnetic Levitation System for Additive Manufacturing: Simulation Analyses. *IEEE Trans. Magn.* **2020**, *56*, 8000407. [CrossRef]
33. Bird, J.O.; Chivers, P.J. 10—Electromagnetism and magnetic circuits. In *Newnes Engineering and Physical Science Pocket Book*; Bird, J.O., Chivers, P.J., Eds.; Newnes: London, UK, 1993; pp. 77–87. [CrossRef]
34. MSAM, U.O.W. Our Equipment & Facilities. Available online: https://msam.uwaterloo.ca/infrastructure/ (accessed on 1 May 2021).
35. Chemetals, S. Iron Rod, 99.6% Pure. Available online: https://www.surepure.com/Iron-Rod-99.6-Percent-pure-4.125-in.-diameter-x-12-inches-long-x-2-rods/p/5593 (accessed on 8 March 2022).
36. Carty, S.; Owen, I.; Steen, W.; Bastow, B.; Spencer, J. Catchment efficiency for novel nozzle designs used in laser cladding and alloying. In *Laser Processing: Surface Treatment and Film Deposition*; Springer: Berlin/Heidelberg, Germany, 1996; pp. 395–410.

Article

Design and Fabrication of a Magnetic Actuator for Torque and Force Control Estimated by the ANN/SA Algorithm

Pooriya Kazemzadeh Heris and Mir Behrad Khamesee *

Department of Mechanical and Mechatronics Engineering, University of Waterloo, Waterloo, ON N2L 3G1, Canada; pkazemza@uwaterloo.ca
* Correspondence: khamesee@uwaterloo.ca

Abstract: Magnetic manipulation has the potential to recast the medical field both from an operational and drug delivery point of view as it can provide wireless controlled navigation over surgical devices and drug containers inside a human body. The presented system in this research implements a unique eight-coil configuration, where each coil is designed based on the characterization of the working space, generated force on a milliscale robot, and Fabry factor. A cylindrical iron-core coil with inner and outer diameters and length of 20.5, 66, and 124 mm is the optimized coil. Traditionally, FEM results are adopted from simulation and implemented into the motion logic; however, simulated values are associated with errors; 17% in this study. Instead of regularizing FEM results, for the first time, artificial intelligence has been used to approximate the actual values for manipulation purposes. Regression models for Artificial Neural Network (ANN) and a hybrid method called Artificial Neural Network with Simulated Annealing (ANN/SA) have been created. ANN/SA has shown outstanding performance with an average R^2, and a root mean square error of 0.9871 and 0.0153, respectively. Implementation of the regression model into the manipulation logic has provided a motion with 13 μm of accuracy.

Keywords: electromagnetism; magnetic manipulator; magnetic actuator; deep learning; ANN; ANN/SA

Citation: Kazemzadeh Heris, P.; Khamesee, M.B. Design and Fabrication of a Magnetic Actuator for Torque and Force Control Estimated by the ANN/SA Algorithm. *Micromachines* **2022**, *13*, 327. https://doi.org/10.3390/mi13020327

Academic Editor: Aiqun Liu

Received: 27 January 2022
Accepted: 17 February 2022
Published: 19 Februray 2022

1. Introduction

Electromagnetism principles and magnetic actuators have been implemented in various applications such as metal forming, where a solenoid with 15 turns generates a magnetic force to push a light metal in order to form a cellphone case [1], real-time production of magnetic materials with the 3D printer in additive manufacturing is feasible by aligning the particles with the magnetic field [2]. In addition, magnetic levitation can be used to design a stable suspension for metal additive manufacturing, which results in substrate elimination [3]. For biological purposes, the electroless plating technique has been used to make magnetic microparts that can rotate at high speed using a magnetic actuator for functional lab-on-a-chip devices [4]. Microfluidics has also utilized magnetically actuated discs to control the flow, mixing, reaction, and separation of fluid [5]. Furthermore, the conversion of kinetic energy of human movement into electrical power is possible using a coil-spring system [6]. The density of dense non-magnetic materials, such as glass, is also measurable, by analyzing the strength of the levitation of a single iron-core coil (electromagnet) [7]. In another study, the variation of the eddy current produced by a coil is measured to detect the place and size of different defects [8]. In addition to the eddy current, electromagnetic impedance and magnetic field values in one direction are also utilized to develop various detection sensors, where one can detect a different contaminant in hydraulic oil [9] and measure the size and position of a contaminant during the casting process [10], respectively.

Magnetic actuators are also implemented in micro-electro-mechanical systems (MEMSs) such as micropumps [11], stable grasping [12], pick-and-place system [13], small force sensing [14], contactless delivery [15], and microsurgery [16]. Microsurgery can provide a minimally invasive surgery environment and offer benefits such as lower infection risk, fewer medical complications, and faster rehabilitation [17,18]. Bioengineering combined with the small-scale (millimeter and sub-millimeter) wireless robots manipulated via the magnetic field can potentially modernize the medical field [19] by replacing the current devices with safer alternatives that enable surgeons to access fragile organs such as the eye, heart, and brain to perform more precise and less invasive operations. It can also be utilized to navigate capsules inside the body for targeted delivery purposes and approach part of the body that was not safely accessible before [20].

Octomag is a magnetic actuator developed to enhance retinal surgery by eliminating irreversible damage to the eye through the control constraints that could even respond appropriately to unpredicted issues such as patient movement. This device contains eight electromagnets to perform a 5 degree of freedom (5 DOF), transition, and orientation, via magnetic force and torque control [21]. Octomag utilizes a linear relationship of fields generated by an individual coil to calculate the total magnetic field using the FEM approach. Then, calculated magnetic field values are calibrated with a few known values measured from the actual coil in order to form the control logic.

Yuan et al. have designed an eight metal-core coil actuator that adopted the same concept as the Octomag to navigate a capsule inside the stomach [22]. Minimag was also motivated by OctoMag, enabling 5 DOF inside a spherical working space surrounded by eight iron-core coils. This system also followed the same approach of calibrating the FEM simulations to develop its control logic [23].

Catheter positioning [24,25] is another main advantage of magnetic manipulators; the catheter guidance control and imaging (CGCI) systems use eight giant coils and have been implemented at an industrial size [26]. Moreover, Bigmag has used six moving coils to navigate a catheter [27].

Figure 1 indicates the developed system utilized in this study to manipulate a disk magnet with a diameter and thickness of 0.1 and 0.0625 inches, respectively. The manipulator contains eight coils in an open-asymmetric configuration that enables wider access to the working space than the mentioned systems and does not compromise torque-force controllability within the workspace [28]. This configuration also follows the square antiprism of Thomson's method [29] and can be modeled using the superposition theorem. This project assumes that the changes in a magnetic field are infinitely small; hence, Maxwell laws are applicable. Moreover, due to the high magnetic permeability of the pure iron core, the inserted core discharges almost instantly after getting magnetized; therefore, the slight non-linearity can be neglected, meaning that the superposition is still valid.

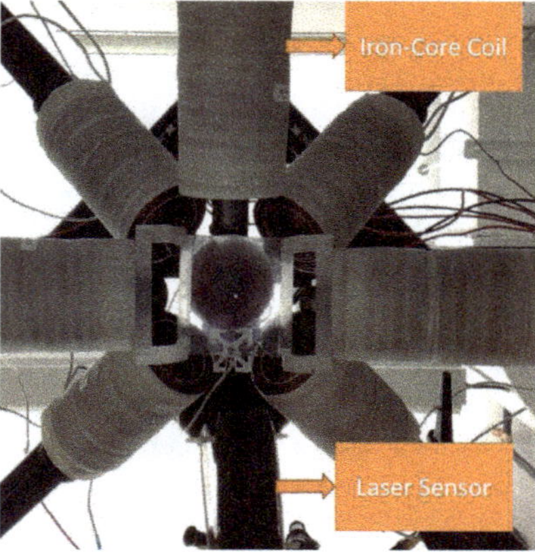

Figure 1. Magnetic manipulator consist of eight iron-core coils.

2. Method

2.1. Coil Design

As indicated in Figure 2, the radius of the spherical working space is correlated to the outer diameter of the coils. The surface area of coils is represented with planes, where the largest coil diameter is located at the intersection of the planes. The relationship between the largest outer radius of the coils, R_{out}, and spherical working space, R_{sphere}, in the open-asymmetric configuration, can be modeled as:

$$R_{out} = \tan\left(\frac{\pi}{6}\right) \times R_{sphere}. \tag{1}$$

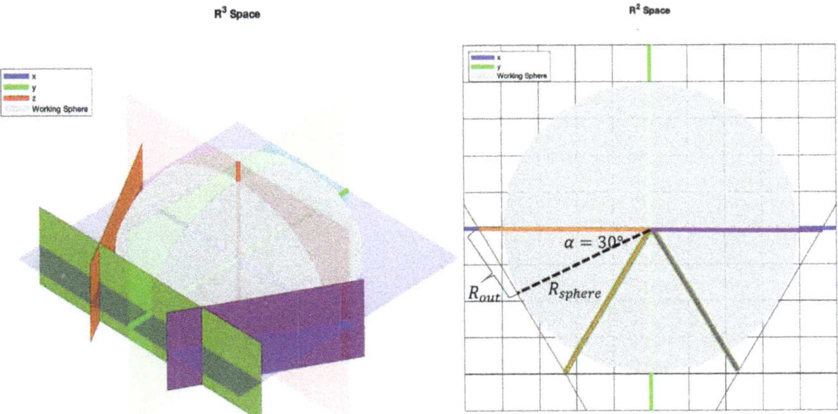

Figure 2. Working space and outer diameter of the coils.

By equating the spherical radius to 60 mm, the outer diameter of the coil should be less than 69.2820 mm. The selection of the outer diameter leads to using the Fabry factor, G, to measure the coils' inner radius and height.

$$G = \frac{1}{5}\sqrt{\frac{2\pi\beta}{\alpha^2 - 1}} \ln \frac{\alpha + \sqrt{\alpha^2 + \beta^2}}{1 + \sqrt{1 + \beta^2}} \tag{2}$$

$$\alpha = \frac{R_{out}}{R_{in}}$$

$$\beta = \frac{l}{2R_{in}}$$

where R_{in} and l are the inner radius and length of the coil, respectively. The optimum values for these factors are shown in Figure 3.

G factor parameterizes the coil's dimension, generating the least heat dissipation to produce a magnetic field. In this study, the G factor has been combined with the force generated on a 1-mm^3 magnet located at the center of the working space. Therefore, the coil's parameters that generate the most force value with minimum power consumption will be selected as the final coil.

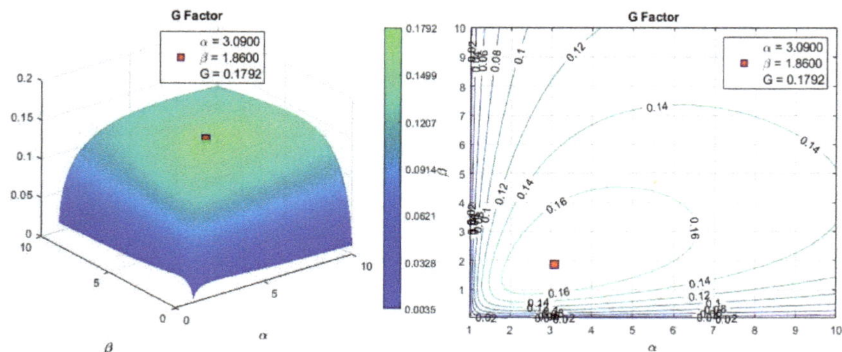

Figure 3. Ideal Fabry factor values for minimizing heat dissipation.

2.2. Magnetic Force and Torque

The generated magnetic force and torque of a coil can be calculated using Maxwell Law:

$$\vec{\nabla} \cdot \vec{B} = 0 \tag{3}$$

$$\vec{\nabla} \times \vec{B} = \mu_0 \vec{J} \tag{4}$$

where J, B, and μ_0 are the current density, magnetic field, and magnetic permeability of the free space. The force (F) and torque (τ) acting on an object with the dipole moment of $\vec{m}$ is equal to:

$$\vec{F} = (\vec{m} \cdot \vec{\nabla})\vec{B} \tag{5}$$

$$\vec{\tau} = \vec{m} \times \vec{B}. \tag{6}$$

Equations (5) and (6) can be represented in matrix form [30]:

$$\vec{F} = \begin{bmatrix} m_x & m_y & m_z & 0 & 0 \\ 0 & m_x & 0 & m_y & m_z \\ -m_z & 0 & m_x & -m_z & m_y \end{bmatrix} \begin{bmatrix} \frac{\delta B_x}{\delta x} \\ \frac{\delta B_x}{\delta y} \\ \frac{\delta B_x}{\delta z} \\ \frac{\delta B_y}{\delta y} \\ \frac{\delta B_y}{\delta z} \end{bmatrix} = h(\vec{m}) g(\vec{\nabla}\vec{B}^T) \tag{7}$$

$$\vec{\tau} = \begin{bmatrix} 0 & -m_z & m_y \\ m_z & 0 & -m_x \\ -m_y & m_x & 0 \end{bmatrix} \begin{bmatrix} B_x \\ B_y \\ B_z \end{bmatrix} = S(\vec{m})\vec{B} \tag{8}$$

and (7) and (8) can be integrated to form wrench, W:

$$\vec{W} = \begin{bmatrix} \vec{\tau} \\ \vec{F} \end{bmatrix} = \begin{bmatrix} S(\vec{m}) & O \\ O & h(\vec{m}) \end{bmatrix} \begin{bmatrix} \vec{B} \\ g(\vec{\nabla}\vec{B}^T) \end{bmatrix} \tag{9}$$

where O is the zero matrix with right dimensions, g is the gradient, and S is in skew functions. By applying superposition theorem to (9) the contribution of each coil carrying current I at point p in the working space can be calculated as:

$$\vec{W} = \begin{bmatrix} \vec{\tau} \\ \vec{F} \end{bmatrix}_{6\times1} = \begin{bmatrix} S(\vec{m}) & O \\ O & h(\vec{m}) \end{bmatrix}_{6\times8} \begin{bmatrix} B(\vec{p}) \\ G(\vec{p}) \end{bmatrix}_{8\times8} \times I_{8\times1} \tag{10}$$

if

$$\begin{bmatrix} S(\vec{m}) & O \\ O & h(\vec{m}) \end{bmatrix}_{6\times 8} \begin{bmatrix} B(\vec{p}) \\ G(p) \end{bmatrix}_{8\times 8} = A(m,p)_{6\times 8}$$

where A is called the actuation matrix then the amount of current flowing through each coil is:

$$I = A^{\dagger} \times W \tag{11}$$

where † is the pseudo inverse.

2.3. Deep Learning

As previously mentioned, FEM is used to predict the magnetic field at each point for experimental purposes. However, FEM models cannot capture the manufacturing error caused during the coil winding process. Coil winding is a delicate process and is consistently associated with unwanted gaps, reducing the number of designed turns in the actual coil. In addition, pure iron becomes deformed easily during core manufacturing and changes its dimensions after the annealing process; therefore, even a perfect model cannot represent the actual coil properties and model magnetic field values. In addition to the mentioned barriers, FEM itself is associated with an approximation error that is a part of this approach. Figure 4 indicates the insensibility of the design to permeability change of the iron core. Moreover, Figure 5 illustrates the difference between actual and calculated values of the magnetic field on the axial axis of a coil, with an average error of 17%.

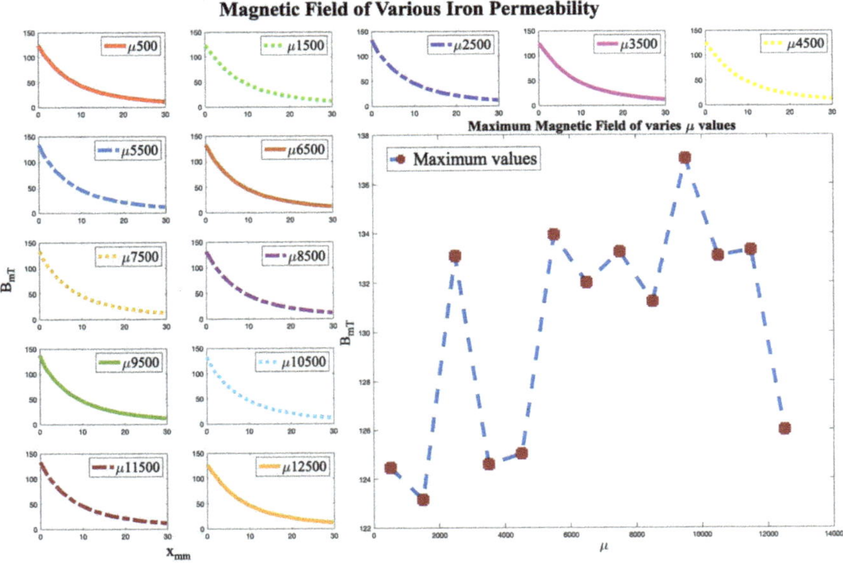

Figure 4. Simulated magnetic field for various magnetic permeability.

In order to enhance the modeling of the magnetic field and consider human error, deep learning has been used as an alternative to approximate the magnetic field values. Deep learning eliminates the need for collecting too many data points and can be developed with few values inside the region of the interest. Artificial Neural Network (ANN) and a hybrid model, Artificial Neural Network with Simulated Annealing (ANN/SA) algorithm, have been implemented in this study.

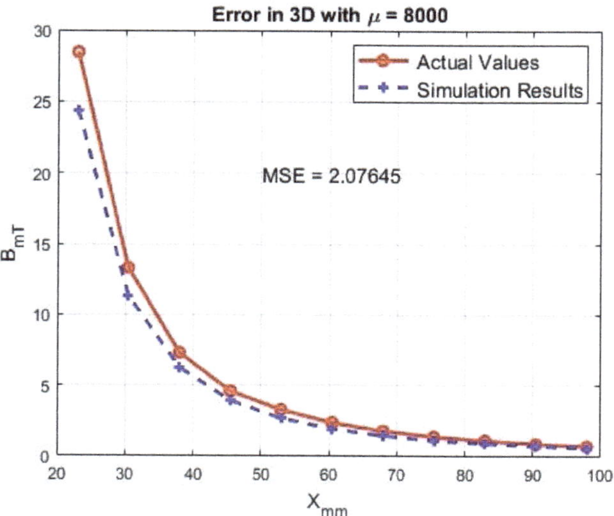

Figure 5. Actual and simulated magnetic field.

2.3.1. ANN

The artificial neural network is a deep learning approach, a subfield of machine learning, where the structure of the human brain inspires the foundation of the algorithm [31]. The ANN algorithm uses training data to recognize the patterns and predict the outputs for a new set of similar data [32]. The training data is fed to the network as an input layer; then, it moves along channels through hidden layers, which are the core processing units of the network, to the output layer. Figure 6 indicates the general structure of a network. Each channel is assigned to a number known as weight, which scales the neuron from the previous layer and feeds it to the neuron in the next layer. All the channels toward a particular neuron will be added together; then, another value known as the bias or interception point is added to this equation. Finally, an activation function will be applied to the neuron before it leaves toward the next layer through its channels. The weight adjustment process will be done through backpropagation, mainly using the gradient descent approach [33].

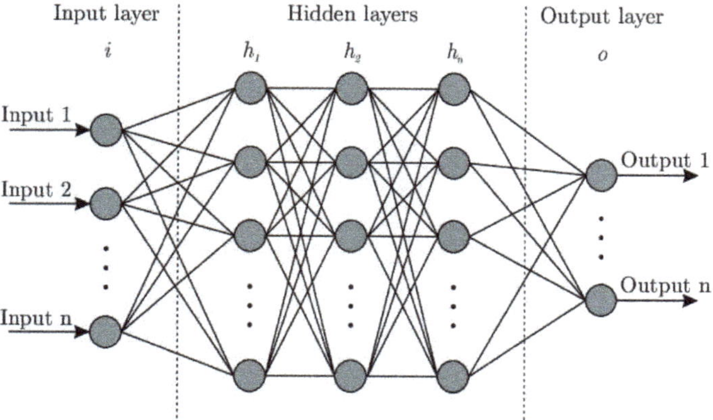

Figure 6. ANN structure.

For most regression problems, the outer layer is only a summation of the last layer's neuron values multiplied by their associated weights plus the interception point without any activation matrix. Figure 7 illustrates the scaling, bias (b), and activation function (f) over one neuron only; this process is known as forward propagation [34]. The weight associated with the channel that connects the jth neuron of the layer l to the ith element of the next layer will be shown as $\omega_{ji}^{(l)}$; the first number in the subscript is the neuron that the channel leaves from, and the second one is where it lands; the superscript also refers to the layer that the channel roots from, for example, the $\omega_{52}^{(3)}$, which connects the fifth neuron of the third layer to the second neuron of the fourth layer.

$$\begin{bmatrix} y_1^{(l+1)} \\ y_2^{(l+1)} \\ \ddots \\ y_n^{(l+1)} \end{bmatrix}_{n\times 1} = f\left(\begin{bmatrix} a_1^{(l+1)} \\ a_2^{(l+1)} \\ \ddots \\ a_n^{(l+1)} \end{bmatrix}_{n\times 1}\right) = \begin{bmatrix} \omega_{11}^{(l)} & \omega_{21}^{(l)} & \cdots & \omega_{k1}^{(l)} \\ \omega_{12}^{(l)} & \omega_{22}^{(l)} & \cdots & \omega_{k2}^{(l)} \\ \vdots & \vdots & \ddots & \vdots \\ \omega_{1n}^{(l)} & \omega_{2n}^{(l)} & \cdots & \omega_{kn}^{(l)} \end{bmatrix}_{n\times k} \begin{bmatrix} x_1^{(l)} \\ x_2^{(l)} \\ \ddots \\ x_k^{(l)} \end{bmatrix}_{k\times 1} + \begin{bmatrix} b_1^{(l)} \\ b_2^{(l)} \\ \ddots \\ b_n^{(l)} \end{bmatrix}_{n\times 1} \quad (12)$$

where n and k are the number of neurons at the $l+1$ and the l layers, respectively. The weight matrix can also be shown as ω^l.

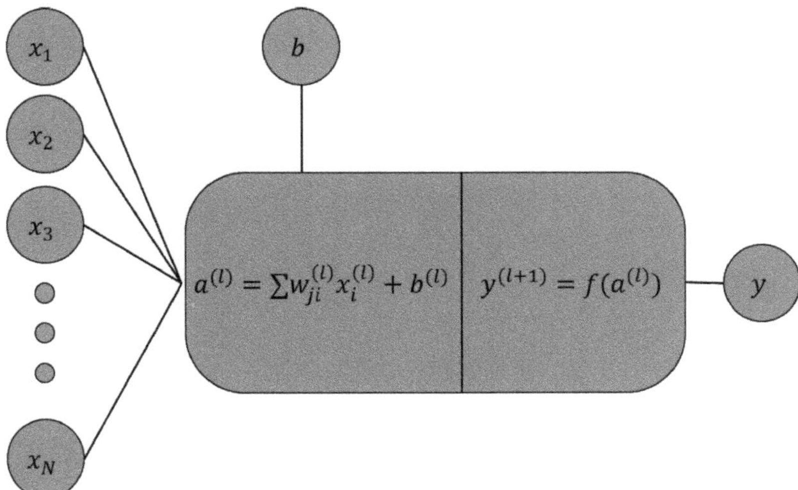

Figure 7. Neuron value calculation.

2.3.2. ANN/SA

The ANN algorithm's performance and convergence are heavily dependent on the initial weight values; therefore, it is beneficial to develop a method that can produce initial weights and biases that increase the probability of reaching the global minimum and increase the learning pace of the network.

ANN/SA optimizes the randomly generated initial weights and biases values; once the best initial values are selected, the network training process starts. The thermomechanical annealing process inspired the optimization of random values [35]. The metal is heated to a higher temperature and cooled down, resulting in a variation in the metal's atomic structure and material properties. The atomic structure and the temperature of the metal will be related together, and if the temperature drops slowly, it can be related to the energy change of the metal [36]. Figure 8 illustrates the explained process.

This method was used to optimize nonlinear functions involving multiple local minimums, and it uses the Metropolis algorithm to simulate the annealing process [37]. Since it is not a greedy process, the probability of reaching a local minimum is considerably low. During the training process, the SA will adjust the weights randomly, considering the algorithm's temperature change and evaluating the network's accuracy; once the accuracy has been calculated, the Metropolis algorithm decides whether the created solution is acceptable—the probability of acceptance P_a is:

$$P_a(\Delta E, y) = \begin{cases} \exp -\frac{k\Delta E}{y} & \Delta E > 0 \\ 1 & \Delta E \leq 0 \end{cases} \quad (13)$$

where ΔE is the error between the new solution and current solution, y is the current temperature, and k is the acceptance constant, based on the range of weights, biases, and inputs. As (13) indicates, the algorithm will frequently be accepting new results at high temperatures, however becomes more selective at the lower ones. The cooling schedule can be exponential, linear, and temperature cycling. In order to calculate the temperature at each state using the temperature cycling method, the following equations should be implemented [38].

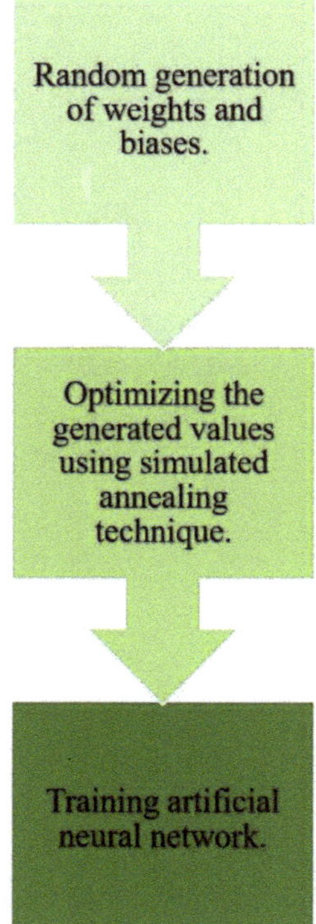

Figure 8. ANN/SA process.

This method first takes a series of $n \in [1, N]$ scenarios for temperature:

$$x[n+1] = \rho x[n] \quad (14)$$

where the total number of temperatures is N, the starting temperature is $x[1]$, the final temperature is $x[N]$, and cooling constant is ρ presented by:

$$\rho = e^{\log\left(\frac{(N-1)x[N]}{x[1]}\right)}. \quad (15)$$

If the number of cycles before each optimization is expressed as M and the temperature assigned to the simulated annealing indicated by $y[n]$, the schedule of the cooling process can be modeled as:

$$y[n] = \sum_{m=0}^{M-1} x[n - mN] \tag{16}$$

ultimately, the weight preparation of the layer l for the temperature can be formulated as:

$$\omega_{ij}^{l}[n+1] = \gamma(1-\lambda)\omega_{ij}^{l}[n] + \lambda u[n] - 0.5\lambda \tag{17}$$

where u is a random variable that is uniformly distributed and takes values between $[0, 1)$, γ is a value equal to 20 or 30, and λ is the perturbation ratio:

$$\lambda = \frac{y[n]}{x[1]}. \tag{18}$$

2.3.3. Validation

In order to evaluate the performance of each developed algorithm and select the best approximation, Root-Mean-Square-Error (RMSE) and R-Squared (R^2) are utilized.

$$RMSE = \sqrt{\frac{1}{N}\Sigma_{i=1}^{N}\left(y_i - \hat{y}_i\right)^2} \tag{19}$$

$$R^2 = 1 - \frac{\Sigma_{i=1}^{N}(y_i - \hat{y}_i)^2}{\Sigma_{i=1}^{N}(y_i - \bar{y}_i)^2} \tag{20}$$

where y_i, $\hat{y}_i$, $\bar{y}_i$, and N are the actual, predicted, mean of actual set values, and the number of samples, respectively. Once each model had been optimized, the final algorithm will be additionally evaluated using the values mentioned in Table 1 to check its predictability [39,40].

Table 1. Extra evaluation parameters.

Variable	Equation	Criteria
Mean Absolute Error (MAE)	$MAE = \frac{\Sigma_{i=1}^{N}\lvert(y_i - \hat{y}_i)\rvert}{N}$	As low as possible
Mean Absolute Percentage Error (MAPE)	$MAPE = \frac{100}{N}\Sigma_{i=1}^{N}\lvert\frac{y_i - \hat{y}_i)}{y_i}\rvert$	As low as possible
Slope regression line k	$k = \frac{\Sigma_{i=1}^{N}(y_i \times \hat{y}_i)}{y_i^2}$	$0.85 < k < 1.15$
Slope regression line k'	$k' = \frac{\Sigma_{i=1}^{N}(y_i \times \hat{y}_i)}{\hat{y}_i^{\,2}}$	$0.85 < k' < 1.15$
Squared correlation of actual vs predicted (Ro^2)	$(Ro^2) = 1 - \frac{\Sigma_{i=1}^{N}(\hat{y}_i - y^o{}_i)^2}{\Sigma_{i=1}^{N}(\hat{y}_i - \bar{\hat{y}}_i)^2}\ y_i{}^o = k \times \hat{y}_i$	Close to 1
Squared correlation of predicted vs actual $(Ro')^2$	$(Ro')^2 = 1 - \frac{\Sigma_{i=1}^{N}(y_i - \hat{y}_i{}^o)^2}{\Sigma_{i=1}^{N}(y_i - \bar{y}_i)^2}\ \hat{y}_i{}^o = k' \times y_i$	Close to 1
Predictability of model R_m	$R_m = R^2 \times (1 - \sqrt{\lvert R^2 - R_o^2\rvert})$	$R_m > 0.5$
Performance index m	$m = \frac{R^2 - Ro^2}{R^2}$	$\lvert m\rvert < 0.1$
Performance index n	$n = \frac{R^2 - Ro'^2}{R^2}$	$\lvert n\rvert < 0.1$

3. Results & Discussion

3.1. Designed Coil

The coil is wound using an AWG 14 wire with a diameter of 1.73228 mm with its isolation. The α value for the various number of vertical layers has been investigated considering the diameter of the wire and the geometric restriction of 67.69 mm of the coil's outer diameter. As indicated in Figure 9a, the optimum value of $\alpha = 3.3019$ occurs at 14 layers. Moreover, as Figure 9b illustrates, the calculation for optimum feasible alpha value starts when inner and outer radiuses are equal, meaning that the thickness is zero, then it continues by adding one layer at each step and calculates the alpha value accordingly. Figure 9 indicates that the optimum α value is associated with a 10.25-mm inner radius.

The coil's inner radius selection leads to β optimization. As indicated in Table 2, by varying the length of the coil and modeling each scenario on Ansys Maxwell 2020 the magnetic force had been calculated. According to the Figure 10, which is the illustration of the normalized force and Fabry factor indicated in Table 2, a length of 124 mm indicates a coil that provides almost as much force with less power consumption than the next scenario. Therefore, the optimum coil is a cylindrical iron-core coil with 3.3019, 6.0761, and 0.140 for α, β, and G values, respectively.

Table 2. Different coils with their generated force.

$Length_{mm}$	β	G	$Force_{mT}$
10.38	0.5063	0.129	0.013
20.76	1.0127	0.164	0.027
31.14	1.5190	0.177	0.080
41.52	2.0254	0.179	0.129
51.90	2.5317	0.176	0.207
62.28	3.0380	0.172	0.243
72.66	3.5444	0.166	0.284
83.04	4.0507	0.160	0.342
93.42	4.5571	0.155	0.373
103.80	5.0634	0.149	0.409
114.18	5.5698	0.144	0.454
124.56	6.0761	0.140	0.497
134.94	6.5824	0.135	0.536
145.32	7.0888	0.131	0.591
155.70	7.5951	0.128	0.625

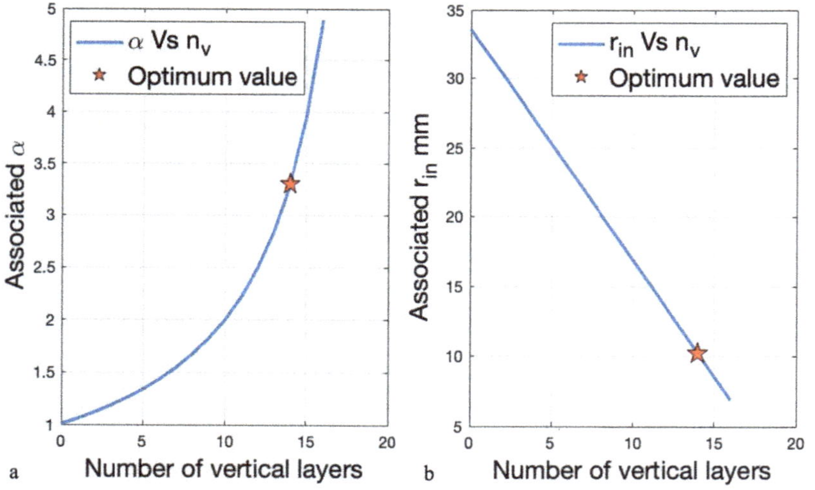

Figure 9. (**a**) Number of layers from the inner radius toward outer radius for various α. (**b**) Inner radius calculation based on the number of layers for different α values.

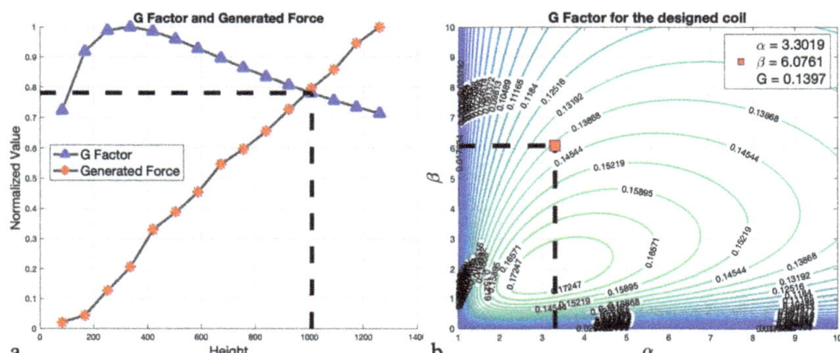

Figure 10. (**a**) Normalized values of different scenarios. (**b**) Optimum coil parameters.

3.2. Data Collection

As indicated in Figure 11, the magnetic field had been measured using a Gauss meter over a $7 \times 7 \times 6$ grid, where $y, z \in [-10 \text{ mm}, 10 \text{ mm}]$ and $x \in [52 \text{ mm}, 72 \text{ mm}]$. A total of 294 points for a coil running at one amp were collected for each magnetic field component. The input data is the location of the point of interest above the coil, and the target or label is the magnetic field components, B_x, B_y, and B_z.

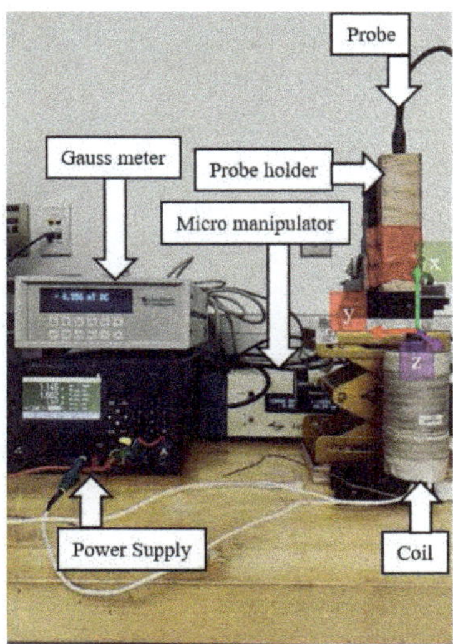

Figure 11. Magnetic field values collection setup.

In order to guarantee the convergence of the algorithms and reach the global minimum, the collected data have been normalized using the upper-lower approach between 0.05 and 0.95; Figure 12 shows the effect of normalization on targets.

3.3. Algorithm Development

Each target has its own algorithm developed separately for each method. However, to ensure that the algorithms receive the same samples, the data set associated with targets had been randomly shuffled and split into three categories: train, test, and validation sets. The train set consists 70 percent of the initial data, with 206 samples, whereas the test and validation sets are filled with only 15 percent, with 44 samples each. Each model is developed using the train set, where its hyperparameters are tuned using the test set.

Finally, the algorithm's performance is measured on validation as unseen data to ensure the accuracy of the model.

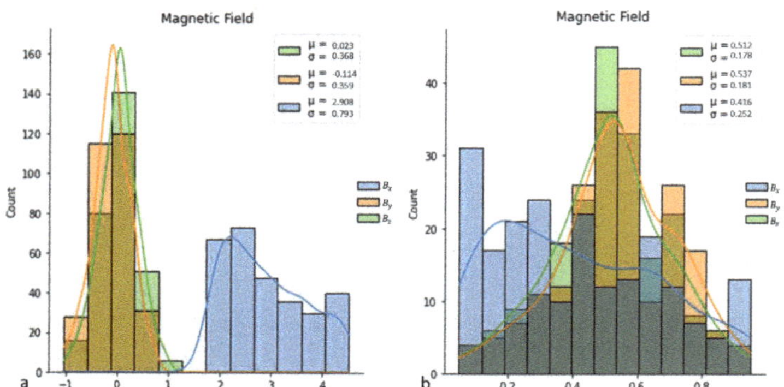

Figure 12. (**a**) Original data . (**b**) Normalized data.

3.3.1. ANN

The network has been developed using one hidden layer consisting of four to seven neurons. The Adam optimizer, stochastic gradient descent method, a learning rate of 0.001, and batch size of 64 was chosen. The hidden layer is formed by activating a relu function, and the output layer is the linear combination of the neurons and biases times to associated weights of the hidden layer. Figure 13 indicates the number of neuron selection process based on R^2 and $RMSE$ for each target. The highest R^2 and lowest $RMSE$ are associated with seven, four, and six neurons for B_x, B_y, and B_z, respectively. Figure 14 shows the implementation of the developed algorithm on the train, test, and validation sets. B_z reached $R^2 > 0.950$ on all the sets; whereas, B_x and B_y reached as far as $R^2 = 0.889$ and $R^2 = 0.876$.

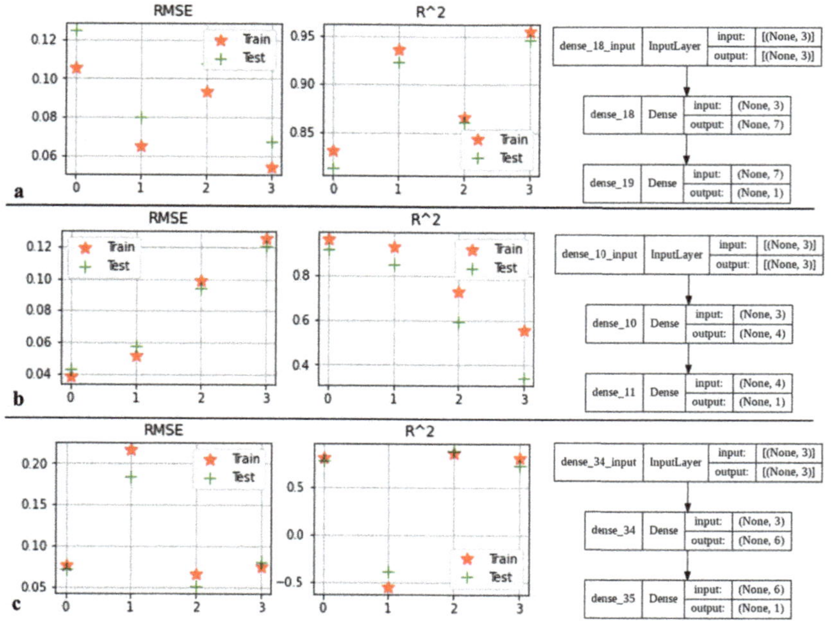

Figure 13. Different scenarios using ANN for (**a**) B_x, (**b**) B_y, and (**c**) B_z.

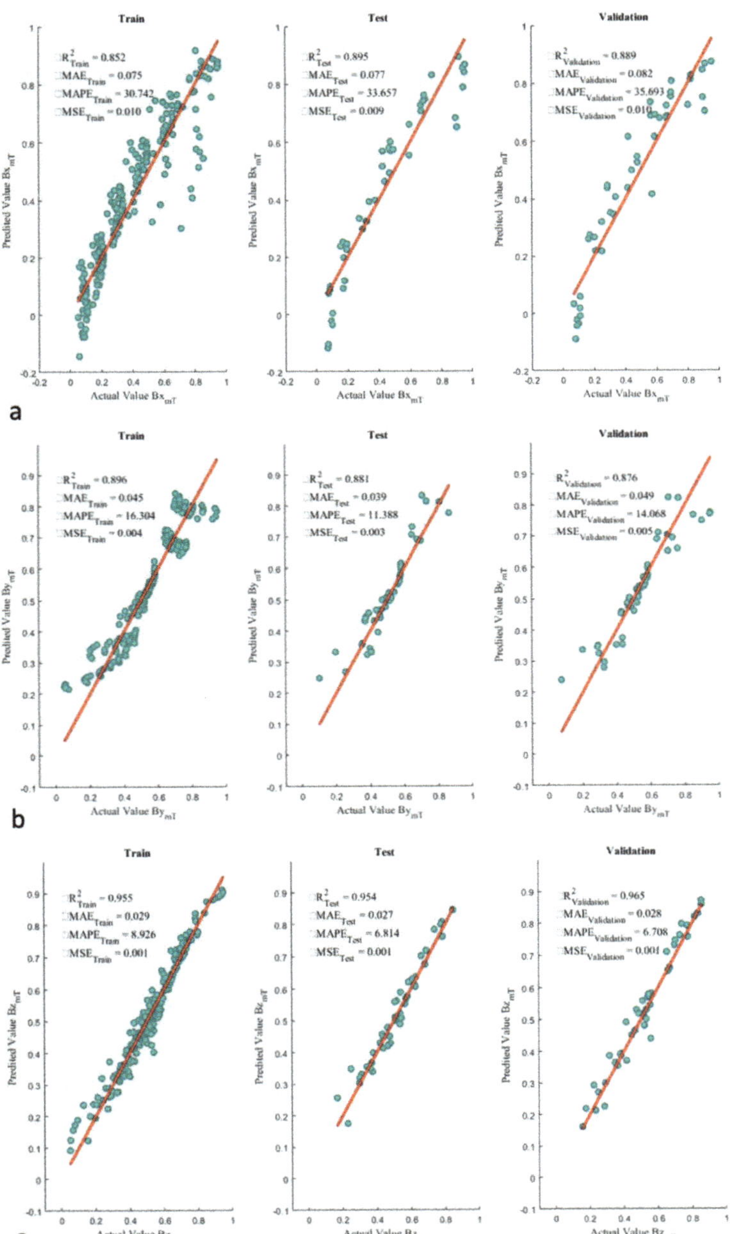

Figure 14. Scatter plot of optimized ANN algorithms for (**a**) B_x, (**b**) B_y, and (**c**) B_z.

3.3.2. ANN/SA

The network was created using similar data and structure as ANN. By using temperature cycling for the cooling cycle with initial and final temperatures of 15 and 0.015, the data set associated with each magnetic field component was trained and tested in NeuralLab v.3.1 software. Models indicated that the Levenberg–Marquardt algorithm had worked better in all cases than the gradient descent. According to Figure 15, B_x and B_y components have performed better with seven neurons than other scenarios. Moreover,

Figure 15 indicates that among four scenarios, a structure with seven neurons B_z is the optimal structure on the training data; however, six neurons are better than others on the test set. Since performance on the test set is more important than training and the six-neuron structure is the second best on the training set, the optimum structure will be six neurons.

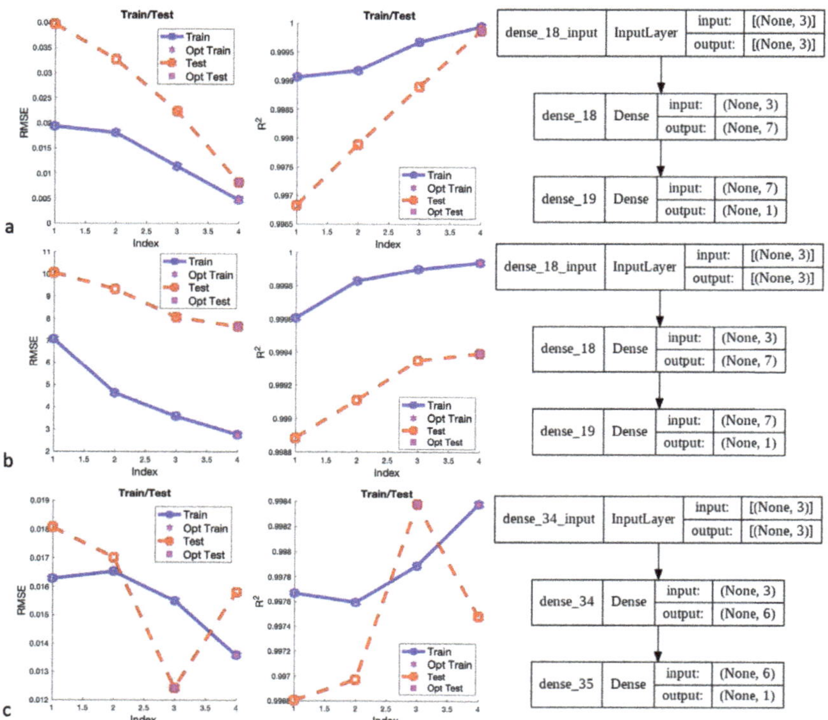

Figure 15. Different scenarios using ANN/SA for (**a**) B_x, (**b**) B_y, and (**c**) B_z.

Figure 16 illustrates the scatter plots with associated R^2 and error values for optimal structures. The performance of this approach shows a significant improvement compared to the ANN approach. The scatter plots closely follow the red line and has a minimal residual. However, the test performance of the Bx component is very similar to the ANN performance.

3.3.3. Extra Validation

According to Figures 14 and 16, the ANN/SA is a better approach to approximate the magnetic field than the other one. Therefore, the predictability of this algorithm on the validation set has been checked as the final performance measurement. As Table 3 indicates, all the variables are in the expected range, and the regression models can be used for manipulation purposes.

Figure 17 indicates the field values over the $7 \times 7 \times 6$ grid, where $y, z \in [-10\ \text{mm}, 10\ \text{mm}]$ and $x \in [52\ \text{mm}, 72\ \text{mm}]$. As can be seen, the actual values and approximated values are close to each other and the mean error percentage is 0.0633%.

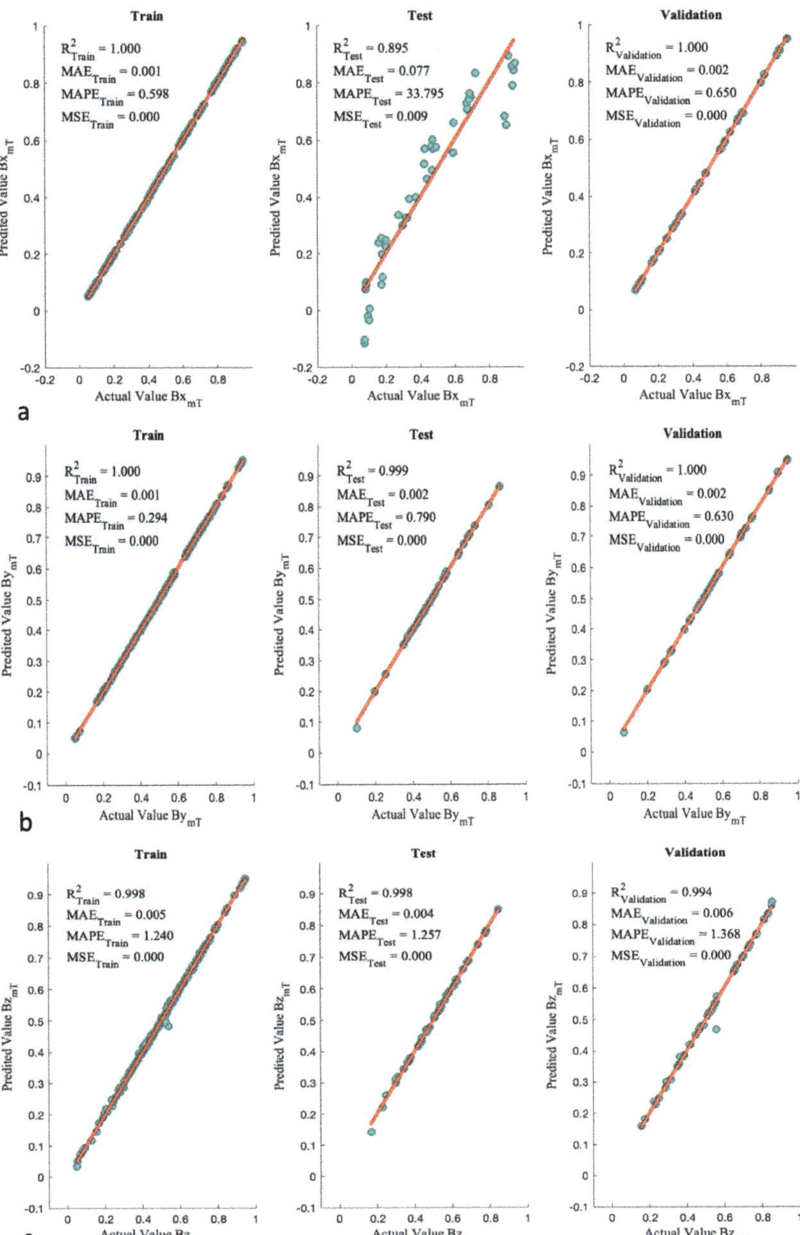

Figure 16. Scatter plot of optimized ANN/SA algorithms for (**a**) B_x, (**b**) B_y, and (**c**) B_z.

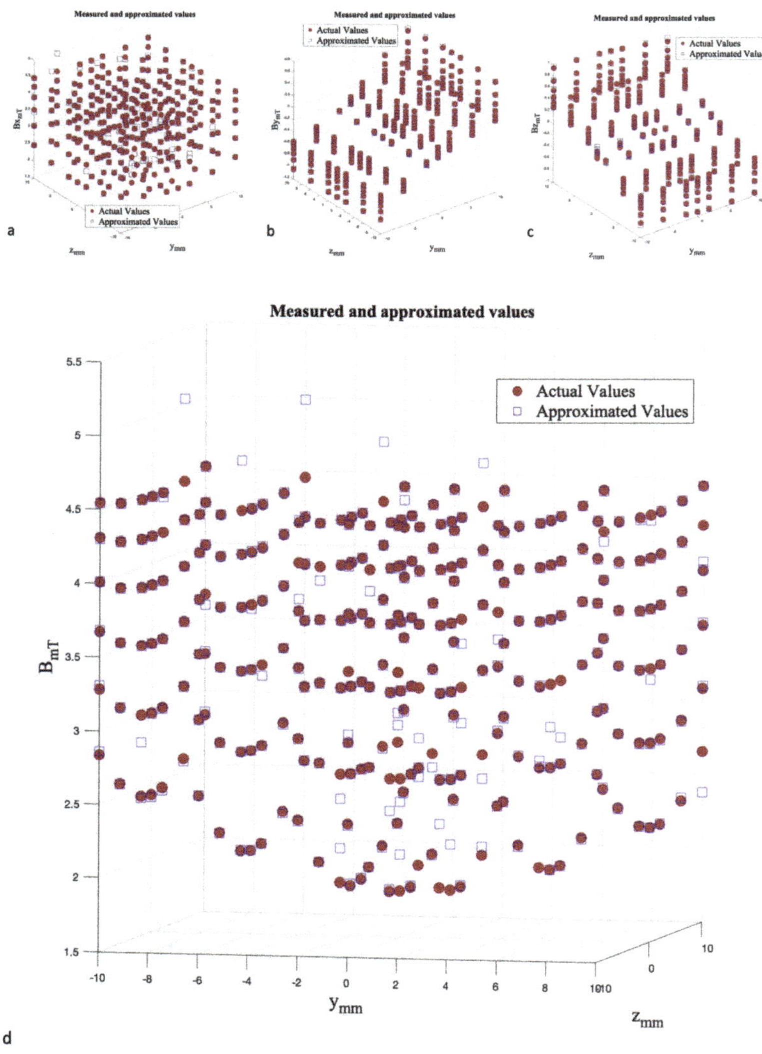

Figure 17. 3D Scatter plots of optimized ANN/SA algorithms for (**a**) B_x, (**b**) B_y, (**c**) B_z, and (**d**) B.

Table 3. Extra evaluations for ANN/SA.

Variable	B_x	B_y	B_z	Criteria
k	0.9995	0.9988	0.9985	$0.85 < k < 1.15$
k'	1.0005	1.0012	1.0007	$0.85 < k' < 1.15$
(Ro^2)	1.0000	1.0000	1.0000	Close to 1
$(Ro')^2$	1.0000	1.0000	1.0000	Close to 1
R_m	0.9912	0.9867	0.9161	$R_m > 0.5$
m	−0.0001	-0.1727×10^{-3}	−0.0062	$\|m\| < 0.1$
n	−0.0001	-0.1727×10^{-3}	−0.0062	$\|n\| < 0.1$

3.4. Motion

The developed algorithm has indicated reliable results and can be utilized for manipulation purposes. Each designed coil also can produce a maximum of 83.43 mH magnetic field at a maximum current of 10 amp, which is enough to manipulate at a milli/micro scale. Figure 18 is a representation of a magnet attached to a flexible road acting as the robot. This disk magnet has a dipole moment of 0.0084 Am^2 with a diameter and thickness of 0.1 and 0.0625 inches. The agent will be navigated only in one direction toward a laser sensor which provides real-time position feedback of the agent. Since there is no other information, such as the agent's orientation, provided by the laser sensor, the system can become unstable quickly. To overcome the rotational instabilities, the attached flexible part to the agent was designed to be heavier than its normal weight. Figure 19 indicates the fabricated system utilized to run the experiment.

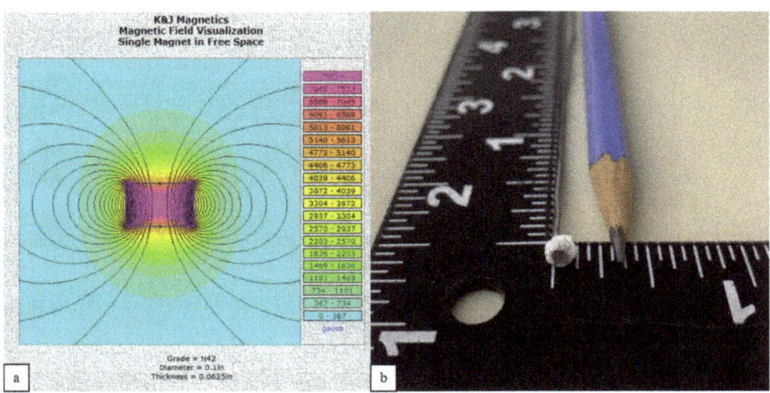

Figure 18. The agent's dimensions (**a**) and properties from the manufacturer [41] (**b**) with the attached flexible part in inch.

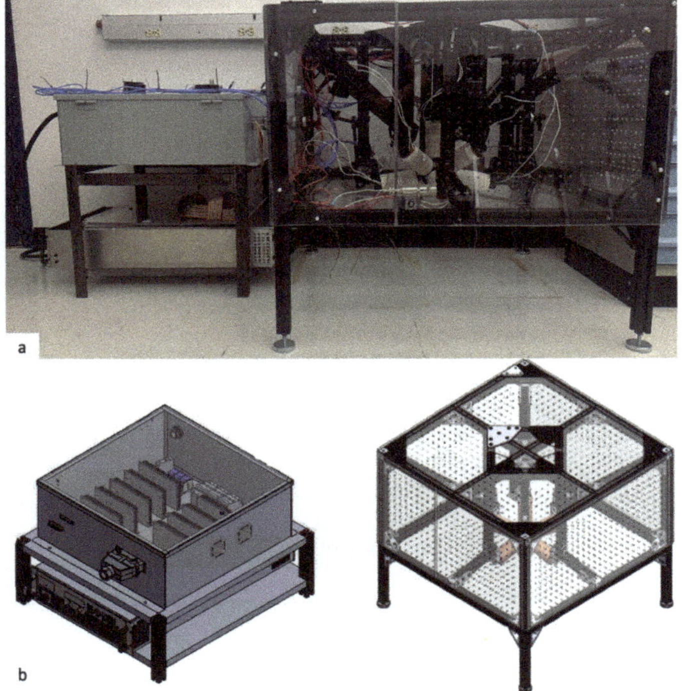

Figure 19. (**a**) Fabricated frame and (**b**) designed frame.

The provided position by the laser sensor will be used to calculate the amount of required current at each coil utilizing (11). The actuation logic, indicated in Figure 20, has resulted in the movement of the robot with an acceleration rate of 0.228 mm/s^2 toward a 0.5-mm point resulting in a movement accuracy of 13 μm. In this logic, the current state of the robot is collected from the laser sensor and compared to the desired point. The motion continues until the difference between the desired and current points is considerably small, meaning that the associated wrench cannot overcome the friction force. Figure 21 illustrates the movement of the robot with respect to time.

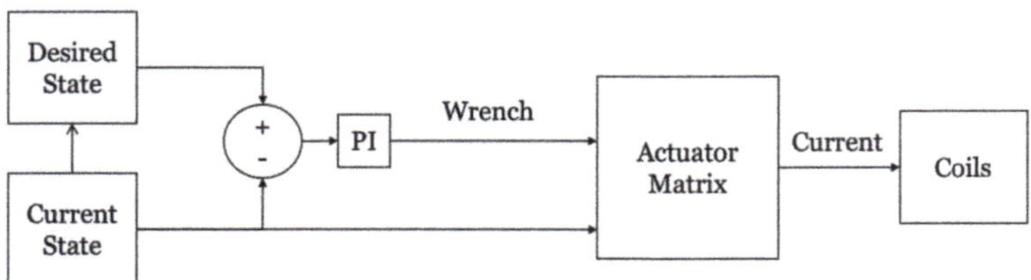

Figure 20. The actuation logic.

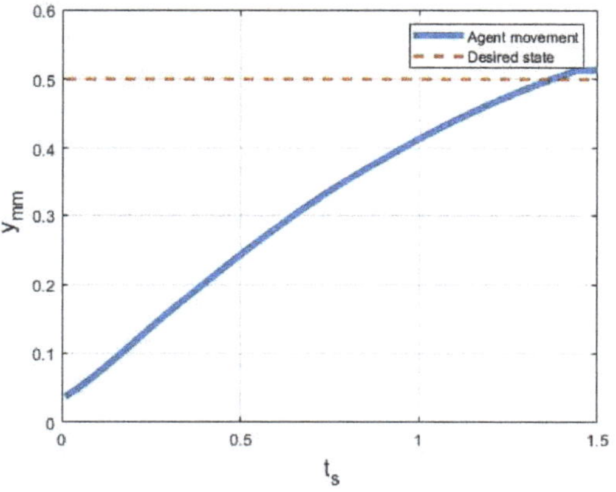

Figure 21. Robot movement toward 0.5 mm.

4. Conclusions

This work demonstrated a magnetic manipulator's design and implementation process with eight coils. The integration of the Fabry factor and force strength has led to an optimal coil with an inner radius, outer radius, and height of 20.5, 67.69, and 124.56 mm, respectively. Due to the error associated with FEM, 17% in this study, the magnetic field produced by this coil at one amp has been collected over a $7 \times 7 \times 6$ grid and approximated using deep learning techniques, ANN and ANN/SA. Although both models showed an acceptable performance, the ANN/SA showed more reliable performance values with an average of $RSME = 0.0153$ and $R^2 = 0.9871$, 76.61% less and 9.66% more than the ANN model, respectively. Utterly, the developed algorithm was used to approximate the magnetic field and navigate an agent from almost the center of the working space toward a half-millimeter away from it along a laser sensor direction. The manipulation had been implemented successfully with 13 μm of accuracy.

Author Contributions: P.K.H.: Conceptualization, formal analysis, data curation and algorithm development, writing original draft preparation. M.B.K.: Conceptualization, review and editing, supervision, project administration. All authors have read and agreed to the published version of the manuscript.

Funding: Natural Sciences and Engineering Research Council of Canada (NSERC).

Institutional Review Board Statement: Not applicable.

Informed Consent Statement: Not applicable.

Data Availability Statement: You can request data using the Maglev Laboratory at University of Waterloo.

Conflicts of Interest: The authors declare no conflict of interest.

Abbreviations

The following abbreviations are used in this manuscript:

FEM	Finite Element Method
AI	Artificial Intelligence
ANN	Artificial Neural Network
ANN/SA	Artificial Neural Network with Simulated Annealing
DOF	Degrees Of Freedom
CGCI	Catheter Guidance Control and Imaging

References

1. Kamal, M.; Shang, J.; Cheng, V.; Hatkevich, S.; Daehn, G. Agile manufacturing of a micro-embossed case by a two-step electromagnetic forming process. *J. Mater. Process. Technol.* **2007**, *190*, 41–50. [CrossRef]
2. Sarkar, A.; Somashekara, M.; Paranthaman, M.P.; Kramer, M.; Haase, C.; Nlebedim, I.C. Functionalizing magnet additive manufacturing with in-situ magnetic field source. *Addit. Manuf.* **2020**, *34*, 101289. [CrossRef]
3. Kumar, P.; Huang, Y.; Toyserkani, E.; Khamesee, M.B. Development of a Magnetic Levitation System for Additive Manufacturing: Simulation Analyses. *IEEE Trans. Magn.* **2020**, *56*, 1–7. [CrossRef]
4. Zandrini, T.; Taniguchi, S.; Maruo, S. Magnetically driven micromachines created by two-photon microfabrication and selective electroless magnetite plating for lab-on-a-chip applications. *Micromachines* **2017**, *8*, 35. [CrossRef]
5. Lin, Y.T.; Huang, C.S.; Tseng, S.C. How to Control the Microfluidic Flow and Separate the Magnetic and Non-Magnetic Particles in the Runner of a Disc. *Micromachines* **2021**, *12*, 1335. [CrossRef]
6. Li, X.; Meng, J.; Yang, C.; Zhang, H.; Zhang, L.; Song, R. A magnetically coupled electromagnetic energy harvester with low operating frequency for human body kinetic energy. *Micromachines* **2021**, *12*, 1300. [CrossRef]
7. Jia, Y.; Zhao, P.; Xie, J.; Zhang, X.; Zhou, H.; Fu, J. Single-electromagnet levitation for density measurement and defect detection. *Front. Mech. Eng.* **2021**, *16*, 186–195. [CrossRef]
8. Zhou, D.; Wang, J.; He, Y.; Chen, D.; Li, K. Influence of metallic shields on pulsed eddy current sensor for ferromagnetic materials defect detection. *Sens. Actuators A Phys.* **2016**, *248*, 162–172. [CrossRef]
9. Shi, H.; Huo, D.; Zhang, H.; Li, W.; Sun, Y.; Li, G.; Chen, H. An Impedance Sensor for Distinguishing Multi-Contaminants in Hydraulic Oil of Offshore Machinery. *Micromachines* **2021**, *12*, 1407. [CrossRef]
10. Nakai, T. Estimation of Position and Size of a Contaminant in Aluminum Casting Using a Thin-Film Magnetic Sensor. *Micromachines* **2022**, *13*, 127. [CrossRef]
11. Qi, C.; Han, D.; Shinshi, T. A MEMS-based electromagnetic membrane actuator utilizing bonded magnets with large displacement. *Sens. Actuators A Phys.* **2021**, *330*, 112834. [CrossRef]
12. Liu, D.; Liu, X.; Li, P.; Tang, X.; Kojima, M.; Huang, Q.; Arai, T. Magnetic Driven Two-Finger Micro-Hand with Soft Magnetic End-Effector for Force-Controlled Stable Manipulation in Microscale. *Micromachines* **2021**, *12*, 410. [CrossRef]
13. Shameli, E.; Craig, D.G.; Khamesee, M.B. Design and implementation of a magnetically suspended microrobotic pick-and-place system. *J. Appl. Phys.* **2006**, *99*, 08P509. [CrossRef]
14. Ştefănescu, D.M. Application of Electromagnetic and Optical Methods in Small Force Sensing. In *Handbook of Force Transducers*; Springer: Berlin/Heidelberg, Germany, 2020; pp. 41–50.
15. Hong, D.K.; Lee, K.C.; Woo, B.C.; Koo, D.H. Optimum design of electromagnet in magnetic levitation system for contactless delivery application using response surface methodology. In Proceedings of the 2008 18th International Conference on Electrical Machines, Vilamoura, Portugal, 6–9 September 2008; pp. 1–6.
16. Jeon, S.; Hoshiar, A.K.; Kim, K.; Lee, S.; Kim, E.; Lee, S.; Kim, J.Y.; Nelson, B.J.; Cha, H.J.; Yi, B.J.; et al. A magnetically controlled soft microrobot steering a guidewire in a three-dimensional phantom vascular network. *Soft Robot.* **2019**, *6*, 54–68. [CrossRef] [PubMed]
17. Nelson, B.J.; Kaliakatsos, I.K.; Abbott, J.J. Microrobots for minimally invasive medicine. *Annu. Rev. Biomed. Eng.* **2010**, *12*, 55–85. [CrossRef] [PubMed]
18. Zhang, J. Evolving from laboratory toys towards life-savers: Small-scale magnetic robotic systems with medical imaging modalities. *Micromachines* **2021**, *12*, 1310. [CrossRef]
19. Giltinan, J.; Sitti, M. Simultaneous six-degree-of-freedom control of a single-body magnetic microrobot. *IEEE Robot. Autom. Lett.* **2019**, *4*, 508–514. [CrossRef]

20. Sitti, M.; Ceylan, H.; Hu, W.; Giltinan, J.; Turan, M.; Yim, S.; Diller, E. Biomedical applications of untethered mobile milli/microrobots. *Proc. IEEE* **2015**, *103*, 205–224. [CrossRef]
21. Kummer, M.P.; Abbott, J.J.; Kratochvil, B.E.; Borer, R.; Sengul, A.; Nelson, B.J. OctoMag: An electromagnetic system for 5-DOF wireless micromanipulation. *IEEE Trans. Robot.* **2010**, *26*, 1006–1017. [CrossRef]
22. Yuan, S.; Wan, Y.; Mao, Y.; Song, S.; Meng, M.Q.H. Design of a novel electromagnetic actuation system for actuating magnetic capsule robot. In Proceedings of the 2019 IEEE International Conference on Robotics and Biomimetics (ROBIO), Dali, China, 6–8 December 2019; pp. 1513–1519.
23. Cole, G.A.; Harrington, K.; Su, H.; Camilo, A.; Pilitsis, J.G.; Fischer, G.S. Closed-loop actuated surgical system utilizing real-time in-situ MRI guidance. In *Experimental Robotics*; Springer: Berlin/Heidelberg, Germany, 2014; pp. 785–798.
24. Le, V.N.; Nguyen, N.H.; Alameh, K.; Weerasooriya, R.; Pratten, P. Accurate modeling and positioning of a magnetically controlled catheter tip. *Med. Phys.* **2016**, *43*, 650–663. [CrossRef]
25. Edelmann, J.; Petruska, A.J.; Nelson, B.J. Estimation-based control of a magnetic endoscope without device localization. *J. Med. Robot. Res.* **2018**, *3*, 1850002. [CrossRef]
26. Nguyen, B.L.; Merino, J.L.; Gang, E.S. Remote navigation for ablation procedures—A new step forward in the treatment of cardiac arrhythmias. *Eur. Cardiol.* **2010**, *6*, 50–56. [CrossRef]
27. Sikorski, J.; Dawson, I.; Denasi, A.; Hekman, E.E.; Misra, S. Introducing BigMag—A novel system for 3D magnetic actuation of flexible surgical manipulators. In Proceedings of the 2017 IEEE International Conference on Robotics and Automation (ICRA), Singapore, 29 May–3 June 2017; pp. 3594–3599.
28. Pourkand, A.; Abbott, J.J. A critical analysis of eight-electromagnet manipulation systems: The role of electromagnet configuration on strength, isotropy, and access. *IEEE Robot. Autom. Lett.* **2018**, *3*, 2957–2962. [CrossRef]
29. Thomson, J.J. XXIV. On the structure of the atom: An investigation of the stability and periods of oscillation of a number of corpuscles arranged at equal intervals around the circumference of a circle; with application of the results to the theory of atomic structure. *Lond. Edinb. Dublin Philos. Mag. J. Sci.* **1904**, *7*, 237–265. [CrossRef]
30. Petruska, A.J.; Nelson, B.J. Minimum bounds on the number of electromagnets required for remote magnetic manipulation. *IEEE Trans. Robot.* **2015**, *31*, 714–722. [CrossRef]
31. Yu, Q.; Tang, H.; Tan, K.C.; Yu, H. A brain-inspired spiking neural network model with temporal encoding and learning. *Neurocomputing* **2014**, *138*, 3–13. [CrossRef]
32. Put, R.; Perrin, C.; Questier, F.; Coomans, D.; Massart, D.; Vander Heyden, Y. Classification and regression tree analysis for molecular descriptor selection and retention prediction in chromatographic quantitative structure–retention relationship studies. *J. Chromatogr. A* **2003**, *988*, 261–276. [CrossRef]
33. Dongare, A.; Kharde, R.; Kachare, A.D. Introduction to artificial neural network. *Int. J. Eng. Innov. Technol.* **2012**, *2*, 189–194.
34. Alaloul, W.S.; Qureshi, A.H. Data processing using artificial neural network. In *Dynamic Data Assimilation-Beating the Uncertainties*; IntechOpen: London, UK, 2020.
35. Abramson, D.; Krishnamoorthy, M.; Dang, H. Simulated annealing cooling schedules for the school timetabling problem. *Asia Pac. J. Oper. Res.* **1999**, *16*, 1–22.
36. Alavi, A.H.; Ameri, M.; Gandomi, A.H.; Mirzahosseini, M.R. Formulation of flow number of asphalt mixes using a hybrid computational method. *Constr. Build. Mater.* **2011**, *25*, 1338–1355. [CrossRef]
37. Gandomi, A.H.; Alavi, A.H.; Shadmehri, D.M.; Sahab, M. An empirical model for shear capacity of RC deep beams using genetic-simulated annealing. *Arch. Civ. Mech. Eng.* **2013**, *13*, 354–369. [CrossRef]
38. Alavi, A.H.; Gandomi, A.H. Prediction of principal ground-motion parameters using a hybrid method coupling artificial neural networks and simulated annealing. *Comput. Struct.* **2011**, *89*, 2176–2194. [CrossRef]
39. Golbraikh, A.; Tropsha, A. Predictive QSAR modeling based on diversity sampling of experimental datasets for the training and test set selection. *Mol. Divers.* **2000**, *5*, 231–243. [CrossRef] [PubMed]
40. Golbraikh, A.; Tropsha, A. Beware of q2! *J. Mol. Graph. Model.* **2002**, *20*, 269–276. [CrossRef]
41. Magnetics, K. Single Magnet in Free Space. Available online: https://www.kjmagnetics.com/magfield.asp?D=0.1&T=0.0625&L=&W=&OD=&ID=&calcType=disc&GRADE=42&surf_field=5154&rsurfC=&rsurfR= (accessed on 10 December 2021).

MDPI
St. Alban-Anlage 66
4052 Basel
Switzerland
Tel. +41 61 683 77 34
Fax +41 61 302 89 18
www.mdpi.com

Micromachines Editorial Office
E-mail: micromachines@mdpi.com
www.mdpi.com/journal/micromachines

www.ingramcontent.com/pod-product-compliance
Lightning Source LLC
LaVergne TN
LVHW070108170726
843515LV00007B/1639

* 9 7 8 3 0 3 6 5 7 0 6 9 3 *